D1175664

ENVIRONMENTAL
MICROBIOLOGY

Wiley Series in
ECOLOGICAL AND APPLIED MICROBIOLOGY

ENVIRONMENTAL
MICROBIOLOGY

Edited by RALPH MITCHELL

Division of Applied Sciences, Harvard University, Cambridge, Massachusetts

WILEY-LISS

A JOHN WILEY & SONS, INC., PUBLICATION
New York • Chichester • Brisbane • Toronto • Singapore

Address all Inquiries to the Publisher
Wiley-Liss, Inc., 605 Third Avenue, New York, NY 10158-0012

Second Printing, January 1993
Third Printing, June 1993

Library of Congress Cataloging-in-Publication Data
Environmental microbiology / edited by Ralph Mitchell.
    p. cm. — (Wiley series in ecological and applied microbiology)
  Includes bibliographical references and index.
  ISBN 0-471-50647-8 (casebound)
  ISBN 0-471-59587-X (paperback)
   1. Microbial ecology. I. Mitchell, Ralph, 1934–   . II. Series
QR100.N48 1992
576'.15—dc20
                                        91-24761
                                            CIP

Life is a germ,
and a germ is life.
Louis Pasteur

CONTENTS

CONTRIBUTORS

JAMES E. BAUER, Department of Oceanography, Florida State University, Tallahassee, FL 32306 [191]

GABRIEL BITTON, Department of Environmental Engineering Sciences, University of Florida, Gainesville, FL 32611 [103]

FRED C. BOOGERD, Department of Microbiology and Enzymology, Kluyver Laboratory of Biotechnology, Faculty of Chemical Engineering and Material Sciences, Delft University of Technology, 2628 BC Delft, The Netherlands [375]

PIETER BOS, Department of Microbiology and Enzymology, Kluyver Laboratory of Biotechnology, Faculty of Chemical Engineering and Material Sciences, Delft University of Technology, 2628 BC Delft, The Netherlands [375]

E.J. BOUWER, Department of Geography and Environmental Engineering, The Johns Hopkins University, Baltimore, MD 21218 [287]

MICHAEL BOYLE, Division of Applied Sciences, Harvard University, Cambridge, MA 02138 [319]

S. CAIRNCROSS, London School of Hygiene and Tropical Medicine, University of London, London WC1E 7HT, England [157]

DOUGLAS G. CAPONE, Chesapeake Biological Laboratory, University of Maryland, Solomons, MD 20688 [191]

ILAN CHET, Faculty of Agriculture, The Hebrew University of Jerusalem, 76100 Rehovot, Israel [335]

HUGH W. DUCKLOW, Horn Point Laboratory, University of Maryland, Cambridge, MD 21613 [1]

M.J.R. FASHAM, Deacon Laboratory, Institute of Oceanographic Sciences, Wormley, Godalming, Surrey GU8 5UB, England [1]

The numbers in brackets are the opening page numbers of the contributors' articles.

MELVIN S. FINSTEIN, Department of Environmental Sciences, Cook College, Rutgers University, New Brunswick, NJ 08903 [355]

TIM FORD, Division of Applied Sciences, Harvard University, Cambridge, MA 02138 [83]

CYNTHIA C. GILMOUR, Benedict Estuarine Research Laboratory, The Academy of Natural Sciences, Benedict, MD 20612 [33]

RONALD W. HARVEY, United States Geological Survey, Menlo Park, CA 94025 [103]

J. GIJS KUENEN, Department of Microbiology and Enzymology, Kluyver Laboratory of Biotechnology, Faculty of Chemical Engineering and Material Sciences, Delft University of Technology, 2628 BC Delft, The Netherlands [375]

GORDON A. McFETERS, Department of Microbiology, Montana State University, Bozeman MT 59717 [125]

RALPH MITCHELL, Division of Applied Sciences, Harvard University, Cambridge, MA 02138 [83]

DAVID D. MYROLD, Department of Crop and Soil Science, Oregon State University, Corvallis, OR 97731; present address: Department of Plant Physiology, University of Umea, S-901 87 Umea, Sweden [59]

G.E. NASON, Alberta Environment, Lethbridge, Alberta, Canada, T1J HC7 [59]

BETTY H. OLSON, Program in Social Ecology, University of California, Irvine, Irvine, CA 92717 [239]

BRUCE E. RITTMANN, Department of Civil Engineering, University of Illinois at Urbana-Champaign, Urbana, IL 61801 [265]

AJAIB SINGH, Bureau of Laboratories, City of Milwaukee Health Department, Milwaukee, WI 53202 [125]

ALEX SIVAN, The Institute for Agriculture and Applied Biology, The Institutes for Applied Research, Ben-Gurion University of the Negev, Beer-Sheva 84110, Israel [335]

YU-LI TSAI, Program in Social Ecology, University of California, Irvine, Irvine, CA 92717 [239]

PREFACE

It is almost twenty years since I edited the first volume of Water Pollution Microbiology. My objective in that volume was to provide a text for advanced courses as a means of applying modern microbiological concepts to the new and rapidly developing field of water pollution control. The second volume of Water Pollution Microbiology concentrated on new developments in microbial ecology. It emphasized the developing synthesis between microbiology and environmental engineering, as a means of understanding processes leading to contamination of natural waters, and prevention of pollution.

It is my intention in the current volume to build on this solid foundation. The field of environmental microbiology has matured during the past decade, encompassing, in addition to water pollution, contamination of the soil, the atmosphere, and the stratosphere. Since the publication of the two volumes of Water Pollution Microbiology a revolution has occurred in biology. Molecular genetics has provided new techniques for the detection of microorganisms, for the degradation of hazardous chemicals, and as a means of safely controlling agricultural pests. In this volume I have invited contributors to speculate on the increasingly important role that molecular genetics is playing in the detection and control of environmental contaminants.

Throughout the book I have attempted to emphasize new concepts. It is my hope that only by bringing new and sometimes controversial theories to the attention of students, researchers, and even practicing engineers, can we hope to further our understanding of the complex microbial processes underlying environmental deterioration, its control, and ultimately its prevention.

The editor is grateful to Tim Ford for the cover photo of the thermophilic bacterium *Thermus*, and to Springer-Verlag, New York, for permission to quote from R. Dubos, *Pasteur and Modern Science*.

RALPH MITCHELL
Cambridge, Massachusetts, 1991

1

BACTERIA IN THE GREENHOUSE: MODELING THE ROLE OF OCEANIC PLANKTON IN THE GLOBAL CARBON CYCLE

HUGH W. DUCKLOW

Horn Point Laboratory, Cambridge, Maryland 21613

M.J.R. FASHAM

Institute of Oceanographic Sciences, Deacon Laboratory, Wormley, Godalming, Surrey GU8 5UB, England

1. ANTHROPOGENIC CARBON DIOXIDE POLLUTION AND THE OCEAN

Invisible, odorless, and nontoxic, anthropogenic carbon dioxide (CO_2) released into the atmosphere as the waste from energy production is nonetheless at least potentially one of the gravest pollutants challenging ecosystems and society. This is due to its behavior as a ''greenhouse'' gas, now well-documented (1,2), whereby it absorbs long-wave infrared radiation emitted from the earth's surface, causing a rise in the equilibrium temperature of the atmosphere. Under preanthropogenic circumstances, the atmospheric burden of CO_2 was beneficial. The ''greenhouse effect'' driven by atmospheric concentrations of 200–300 ppm permits the existence and flourishing of life on the planet by maintaining the

Environmental Microbiology, pages 1–31, © 1992 Wiley-Liss, Inc.

mean global temperature within the range 5°–15°C compared with −50°C in the absence of CO_2 or ~200°C in the absence of life (3). But since about 1850 industrial emissions have caused a rise in the atmospheric concentration from ~280 ppm to ~350 ppm today. This may already have caused a warming of ~0.5°C, and projections suggest that a further rise of 5°C is possible by the middle of the twenty-first century (1,4). This would be a greater, and faster, temperature increase than any which has occurred in the last 200,000 years and would cause large changes in sea level, coastal development, agricultural practice, vegetation patterns, weather, and global economy.

The sheer size of the problem, and the cost of efforts to alleviate it, or even to coexist in a warmer world would be enormous. Today about 25×10^9 tons (25 gigatons [gT]) of CO_2 are released to the atmosphere annually. No other commodity in the world is handled on this scale. Annual world food production and steel production are 1–2 orders of magnitude smaller, and the value of the raw materials that CO_2-producing industries process annually is about $2–3 trillion, or twice the U.S. federal budget (5). To plan effectively to deal with the greenhouse problem, we need a fundamental understanding of the biogeochemical and physical machinery that cycles carbon in the global system, and we need models of the carbon cycle to project the effects of increasing CO_2. In this chapter, we discuss one element of such a model. We describe our efforts to simulate the cycling of carbon and nitrogen in the upper ocean, concentrating on the model's treatment of marine bacterioplankton, and what it tells us about their role in the biogeochemical cycling of carbon between the ocean and atmosphere.

Our focus is on the upper ocean because oceanic uptake appears to regulate the level of CO_2 in the atmosphere. Photosynthetic consumption lowers the partial pressure of CO_2 (pCO_2) dissolved in seawater, permitting the passage of atmospheric CO_2 across the air–water interface by simple diffusion along the concentration gradient. The ocean is the largest reservoir of freely exchangeable carbon on the planet (carbonate sediments being excluded because of their long turnover times), amounting to ~35,000 gT, or about 50 times the size of the atmospheric reservoir (Table 1.1) (6). Clearly, even a small redistribution of the inorganic carbon in the ocean can amount to a large perturbation in the atmospheric concentration. Analysis of oceanic box models has shown that changes in ocean circulation and carbon cycling could lower the atmospheric level to 165 ppm, or raise it to 425 ppm, i.e., from ~40% lower to 50% higher than the preindustrial value of 280 ppm (7). The striking covariance among ice core records of atmospheric CO_2 in trapped air bubbles and sediment core records of $\sigma^{13}C$ in foraminifera shells signifies the connection between CO_2, temperature, and ocean productivity over the last 160,000 years (8,9). Changes in the biogeochemical and physical processes governing the partitioning of carbon within the ocean–atmosphere system are the driving force behind climate change on time scales of centuries to millennia (10).

TABLE 1.1. The Global Carbon Cycle and
Atmospheric Carbon Budget

Reservoirs	Gigatons (gT) of carbon
Atmosphere	700
Marine	
Total inorganic	35,000
Particulate organic	30
Plankton	3
Dissolved organic	1,000[a]
Sediments	60,000,000
Terrestrial	
Land biota	600
Soil humus	3,000
Fossil fuels	5,000

Fluxes to atmosphere	gT of carbon per year
Sources	
Fossil fuel CO_2	5
Deforestation (net)	0.4–2.6
Observed atmospheric increase	3
Sinks	
Oceanic uptake	1.6–2.5
Terrestial (required to balance sources	
minus sinks against observed increase	0–3

[a]The DOC value is uncertain to a factor greater than 2 as a result of recent developments in the chemical analysis of DOC (cf. 17,66).

Data are from refs. 6, 11, 13, 86.

What we do not know are the magnitude of the oceanic uptake and the precise operation of the physical and biological mechanisms regulating it. Until quite recently, it was generally thought that the ocean took up about 30%–50% of the fossil CO_2 released annually, with another 60% showing up in the atmosphere and 0%–20% unaccounted for (6,11). Such analyses suggested that large uncertainties are present in the global carbon budget. For example, if the budgets are to balance, the numbers leave no room for terrestrial sources of CO_2, such as deforestation (Table 1.1). Now new analyses of the atmospheric and oceanic CO_2 distributions using general circulation models suggest that observed north–south gradients in atmospheric CO_2 can only be simulated if the major CO_2 sinks are in the northern hemisphere. Most extant data do not support an oceanic sink in the northern hemisphere of sufficient strength to account for the observed budget, and on the basis of these data a major terrestrial sink is postulated (11).

However, the undersampling of oceanic CO_2, the uncertainties in the present data, and our incomplete understanding of ocean biogeochemistry are such that the question is not resolved. In the next section we review the biological and ecological processes thought to contribute to the oceanic removal of atmospheric CO_2.

2. PLANKTON DYNAMICS AND OCEANIC CO₂ REMOVAL

An important key to understanding the oceanic uptake of CO_2 is a mechanistic knowledge of the annual cycles of plankton and nutrients in the oceanic euphotic zone (6,7,12–14). During the growing season, which is set by the combined effects of the cycles of solar irradiance and vertical mixing that limit phytoplankton photosynthesis (15), the utilization of dissolved CO_2 during plant growth drives the ocean surface pCO_2 below the atmospheric level (*i.e.,* the air–sea pCO_2 becomes negative), permitting CO_2 to diffuse into the ocean. The resumption of vertical mixing toward the end of the growth season restores the surface pCO_2 to equilibrium with the atmosphere and mixes atmospheric CO_2 into the sea (16). Most of the CO_2 fixed into plant biomass is quickly respired back into solution, either by the phytoplankton themselves or by animals and bacteria after consumption, resulting in no net removal and no lowering of the pCO_2. But some fraction of the fixed carbon is transported deeper into the sea by sinking of particles, vertical mixing, or diffusion, and this loss of carbon from the surface is responsible for net CO_2 removal. If the biogenic carbon is transported far enough into the oceanic interior, it will be removed from contact with the atmosphere for decades to millennia. Some of the fixed CO_2 is also lost to a long-lived pool of dissolved organic carbon, the origins, fates, and magnitudes of which are quite poorly understood, though it is fair to say that bacterial oxidation is probably the major mechanism for converting the dissolved organic carbon (DOC) back to CO_2 (17). Particles are consumed by other animals and bacteria as they settle through the water column, and subsequent respiration releases CO_2 back into the water. The net result of this removal of CO_2 at the surface and addition at depth is the generation of a vertical CO_2 gradient with $\sim$2,000 μEq at the surface and 2,300 μEq below 1,000 m (18). Because biological processes thus move CO_2 against the concentration gradient, the process is called the *biological pump* (19). It is one of three different "pumps" that together maintain the CO_2 depth gradient (20,21).

The biological pump, then, is really the combination of the cycles of growth, mixing, consumption, and decomposition as they are driven by seasonal changes in physical forcing of the surface ocean by winds and heat transport. The cogs of the pump machinery are planktonic organisms (22). The detailed construction and operation of the pump, or its strength in different regions of the world ocean, are not well known. But limits to its magnitude, and the temporal and spatial scales of its operation, appear to be set by upwelling and vertical mixing. Eppley and Peterson (23), in a seminal paper, suggested that the loss rate of biogenic materials from the surface by vertical transport had to be balanced over some large time and space scales (usually assumed to be around 1 year for an oceanic gyre or basin) by the uptake of nitrate (NO_3^-) supplied from below the euphotic zone. This is because nitrogen is the principal limiting nutrient for plant production; the losses of fixed nitrogen cannot long exceed the rate of supply without the system running down.

In a sense, then, all we appear to need to assess the strength of the biological pump are measurements of the vertical transport of nitrate in different ocean regions. However, under closer scrutiny, this expectation falls short for at least three reasons:

1. primary production is not entirely supported by the "new" nitrogen supplied from exogenous sources
2. loss rates are not perfectly coupled to the nitrate supply
3. the primary producers are not capable of using all the nitrate supplied in some oceanic areas

The mix of nitrogenous nutrients supporting the plant growth, their utilization, the loss rates, and the composition of the export from the euphotic zone are all functions of the ecological, size, and taxonomic structures of the plankton community (24).

Primary production in the sea is supported by two kinds of nitrogen species: new and regenerated forms supplied from outside and inside the foodweb, respectively (25,26). The new, or allochthonous, nitrogen is mostly nitrate from deep water, though nitrogen fixation may supply a fraction of the total input in some areas. The regenerated forms, principally ammonium (NH_4^+) and urea, are produced by the excretory and respiration processes of bacteria, protozoans, zooplankton, and other grazers as a result of metabolism of plant biomass and other organic matter. The flow of these regenerated forms of nitrogen back from consumers to the primary producers forms the primary recycling loop of the plankton system (27). The amount of the total primary production supported by regenerated *versus* new forms of nitrogen varies regionally and seasonally as well as at smaller scales. The fraction of the primary production supported by new nutrients has been called the f ratio (23), and it varies from about 0.1 to 0.8. In the long term, f also approximates the fraction of the total production that is exported from the surface, carrying fixed CO_2 down with it. But on shorter time and space scales, the new production and export fluxes can diverge widely, and these seasonal asymmetries may influence regional differences in CO_2 removal.

TABLE 1.2. Estimates of Sea to Air CO_2 Flux (gT of carbon per year)[a]

Ocean	Location	ΔpCO_2	Flux
North Atlantic	>50°N; 90°W to 20°E	−37	−0.23
Atlantic gyre	15°N to 50°N; 90°W to 20°E	−15	−0.30
North Pacific	>15°N; 110°E to 90°W	2	−0.06[b]
Equatorial	15°S to 15°N; 180°W to 180°E	33	1.62
Southern gyres	50°S to 15°S; 180°W to 180°E	−17	−2.39
Antarctic	>50°S	−17	−0.20
Global		1	−1.6

[a]The fluxes are computed from the pCO_2 values, wind speeds, and basin areas. Negative fluxes are into the ocean.

[b]The apparent discrepancy between the positive ΔpCO_2 and negative flux is due to a seasonal assymetry in wind speed driving gas exchange rates during late winter when ΔpCO_2 is −11.

Data are from ref. 11.

In some oceanic areas like the North Atlantic and in many temperate and boreal shelf regions, the seasonal plankton cycle is characterized by conspicuous spring blooms of phytoplankton, summertime depletion of the NO_3^- pool, and periodic episodes of sedimentation (28,29). In other regions, including the sub-arctic North Pacific (30) and much of the open Southern Ocean, phytoplankton stocks vary little, and NO_3^- remains well above limiting concentrations throughout the year. The reasons for these large-scale differences are debatable (31,32), but appear to be due in part to differences in the couplings or coadaptation between phytoplankton producers and their zooplankton consumers (24,30). Interestingly, bacterial dynamics seem to differ between the two areas as well: In the Atlantic, bacterial biomass is usually 10%–20% of the plant biomass, with turnover rates on the order of 1–2 days (33), whereas in the North Pacific bacterial biomass equals or exceeds the plant biomass, but turns over only once in >10 days (34,35). These differences in community structure that produce large-scale differences in the plankton and nutrient cycles of these ocean basins also may influence the different roles the two oceans play in CO_2 removal. The North Atlantic, though smaller in area than the North Pacific, develops a more intense pCO_2 gradient and removes about four times more CO_2 on an annual basis (Table 1.2).

3. MODELING NUTRIENT CYCLES AND PLANKTON PROCESSES

To understand the role plankton dynamics plays in governing the rates of new production, export from the upper ocean, and CO_2 removal, we have begun to construct models of euphotic zone plankton systems and to study their behavior under various choices of community structure and biological parameter values. Our purpose is twofold: 1) to provide a tool for examining the effects of community structure and biological variability on primary production and plankton cycles and 2) to develop a basic ecosystem model that can be used in concert with supercomputer-based general circulation models (GCMs) for the purpose of constructing basin or global-scale simulations of plankton systems. Our basic model is described by Fasham et al. (36) and below. The preliminary results with a GCM including the ecosystem model are discussed by Sarmiento et al. (14). In this section, we discuss some previous models and then present ours.

3.1. Previous Plankton Models

We believe that models of plankton dynamics must distinguish explicitly between new and regenerated production if they are to advance our understanding of the role of upper ocean biogeochemistry as it governs the atmosphere–ocean CO_2 balance. In addition, we believe that simplicity is desirable, or even necessary, because only rather simple plankton models can be inserted into GCMs given current hardware limitations on speed and memory capabilities. Therefore we sought an irreducible minimum number of model compartments that are required to analyze or reproduce the essential features of ocean biogeochemistry (27,37). In the following paragraphs, we present not an exhaustive review of plankton models, but instead a selection of previous models that we have found useful and stimulating. The reader may consult the references for a deeper exploration of the field. Some of the early history is discussed by Mills (38).

The first modern plankton models were developed by John Steele (39,40) in the late 1950s, drawing on early theoretical and modeling work by Gordon Riley (41). Besides being generally influential, Steele's work emphasized the importance of being able to vary individual factors independently to investigate model (and real-world) behavior and identified the stabilizing properties of thresholds in grazing equations. Most current models trace their key formulations back to Steele. His results echoed one of the classic paradoxes of oceanography: It was difficult to reconcile observed fishery yields in the North Sea with observed or modeled levels of primary production, given accepted ideas about ecological transfer efficiencies between trophic levels. There did not seem to be sufficient food produced to support the fisheries, even when all the annual primary production was eaten.

Steele's models included nutrient regeneration, but did not distinguish between new and regenerated nutrients, and they did not include bacterial or dissolved organic matter compartments. A better treatment of recycling was required. By the mid-1970s it was also becoming apparent that the microbial components of ecosystems controlled a majority of the fluxes of nutrients and organic matter (42). This suggested another problem: If there was not enough primary production for the fish even when zooplankton ate all the phytoplankton, how could high metabolism by bacteria and protozoans be supported too? And it was becoming clear that in many, perhaps most, places zooplankton seldom ate all the phytoplankton. There were substantial losses of ungrazed phytoplankton to mortality and sinking (43).

Williams (44,45) reconciled these views with a conceptual model that included bacteria, bacteriovores, and dissolved organic matter (DOM). In recognizing that all organisms released or leaked DOM, and by invoking the exquisite bacterial adaptations for DOM uptake, he accounted for substantial bacterial production and also showed how to increase the recycling efficiency of the foodweb by drawing on protozoan metabolism of bacterial biomass to support regenerated production. Notably, this extra production could be inserted into the same traditional foodweb paradigm used by Steele, without robbing the fish. These ideas were later formalized by Fasham (46), using linear flow analysis. Pace *et al.* (47) included most of the same concepts and observations in a numerical simulation model with the first realistic description of bacterioplankton in a plankton foodweb model.

Subsequent contributions built on these beginnings to explore the effects of foodweb structure and dynamics on recycling efficiency and nutrient cycles. Evans and Parslow (31) and Frost (32) investigated the factors responsible for persistence of high NO_3^- levels in the North Pacific, with the former authors demonstrating a simple formulation for reproducing repeating seasonal cycles. In general it appears that when microzooplankton grazers can be maintained at population levels sufficiently high to exert grazing pressure on phytoplankton, or when grazers are capable of rapid growth responses to increased food supplies, phytoplankton blooms are suppressed and nutrient utilization rates are kept too low for NO_3^- depletion to occur during the growth season. Ulanowicz (48) and Fasham (46) stressed the importance of recovering information on intercompartmental flows or fluxes for deeper understanding of model and real foodweb dynamics. Vezina and Platt (37) and Ducklow et al. (27) used inverse modeling and diagnostic flow analysis, respectively, to identify the importance of recycling nutrients through microbial and zooplankton populations and to characterize the variation of the f ratio.

As a result of this work, it became clear that the following needed to be included in a new plankton simulation model:

   1. Separate pools for NO_3^-, NH_4^+, and dissolved organic nitrogen (DON)

   2. A bacterial component that recycles NH_4^+, but also competes with phytoplankton for it

   3. A treatment of phytoplankton mortality and phytodetrital recycling

4. An easily managed facility for varying community composition and structure

In the next section we describe a new model that incorporates these features. A more detailed description is given by Fasham et al. (36).

3.2. A Model of Plankton and Nutrient Dynamics in the Oceanic Mixed Layer

3.2.1. The Model Structure

Compartmental ecosystem models of the mixed layer have been widely used in marine ecology, beginning with the pioneering work of Steele (39). They make the basic assumption that the mixed layer can be considered biologically homogeneous, which is equivalent to assuming that the physical mixing rate is fast compared with the growth rates of the organisms. This assumption is probably robust in most cases but may break down in areas, such as the northeast Atlantic, that have very deep mixed layers in the winter. We use an annual cycle model in which ecosystem seasonality is driven by seasonal changes in incident photosynthetically active radiation (PAR) and mixed layer depth (31). Our initial model is based on nitrogen because it is the limiting nutrient for primary production and because we wanted to partition primary production, new production (fueled by nitrate), and regenerated production (fueled mainly by ammonium), as discussed above. The compartments, flows between compartments, and equations governing those flows are defined at run-time and do not require program changes. This was achieved by building on the general ecosystem equations developed by Wiegert (49). The software was designed so that intercompartmental flows, averaged over any specified time period, could be easily calculated. Thus the model yields estimates of mean annual fluxes such as primary production, which are still very difficult to obtain by observation. The model is written in FORTRAN and runs on an IBM or compatible PC equipped with a DOS and a math coprocessor. The model software is available from the authors on receipt of a 3.5 inch IBM diskette.

In our original version, we considered the simplest model capable of capturing the essence of nitrogen cycling in the mixed layer (36). It had seven compartments: phytoplankton, zooplankton, bacteria, nitrate, ammonium, labile DON, and detritus. We model only the labile fraction of DON that is rapidly utilized (with a time scale of hours to days) by bacteria, not the large, long-lived pool of DON recently discussed by Williams and Druffel (50), Sugimura and Suzuki (87) and Suzuki *et al.* (51), and Toggweiler (17). The "ammonium" pool implicitly includes other forms of regenerated nitrogen, principally urea. In this simplified model, we have not specified a separate behavior for urea, though

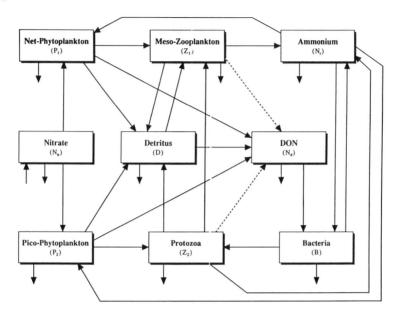

Fig. 1.1. Diagrammatic representation of a nitrogen-based model of mixed layer plankton and nitrogen cycling showing the compartments and the modeled nitrogen flows among compartments and between compartments and the deep ocean (arrows exiting boxes). The dashed lines represent optional DON release from the grazer compartments.

there is support for such a formulation. Bacteria cannot use it as a growth substrate (52,53). A more realistic model would have a separate urea compartment, with utilization by phytoplankton only. This would be analogous to the DON, which is taken up only by bacteria. The detritus compartment was defined as comprising fecal materials and dead phytoplankton and zooplankton. Detritus can be recycled within the mixed layer by two mechanisms, reingestion by zooplankton (54–56) or breakdown into DON and subsequent uptake by bacteria. In reality the latter process may be mediated by bacteria (57,58) attached to the surface of the detritus, and some models (e.g., 47) have included a separate compartment for attached bacteria. The sensitivity of the model to detrital breakdown and sinking rates is considered below.

The original version had single, "portmanteau," or aggregated compartments for phytoplankton and zooplankton. However, to model changes in community structure and their effects of biogeochemical fluxes more realistically, we elaborated the model to a nine-compartment version including two populations of phytoplankton and two of zooplankton (Fig. 1.1). These are distinguished primarily by the feeding behavior of the two zooplankton populations and also by choices of the parameter values governing the model processes (see Table 1.4, below). The net phytoplankton (P_1) have a sinking rate of 3 m per day and are eaten by the mesozooplankton (Z_1), which also eat protozoans (Z_2) and detritus (D). The picophytoplankton (P_2) have no sinking rate until they die and aggregate in detrital flocs. They are eaten, along with the bacteria (B), by the protozoans. Detritus is derived from the death of the phytoplankton and protozoans and defecation by mesozooplankton. Mesozooplankton mortality produces ammonium and an instantaneously exported fraction, as a way of parameterizing losses to unmodeled, higher trophic levels. In the main version discussed here, DON is derived from photosynthetic release from the phytoplankton and by decay or "mortality" of the detritus. This latter process is a way of parameterizing decomposition by attached bacteria and protozoans. Metabolism of detritus also produces ammonium following its reingestion by mesozooplankton. Release of DON via "sloppy feeding" or excretion by grazers is optional. The details of parameter selection, nutrient uptake feeding kinetics, and photosynthetic physiology are discussed at length by Fasham et al. (36). For the development of the two size–class model, the parameters α, V_p, K_1–K_4, g, μ_{ij}, and V were adjusted to produce initial runs in which the net phytoplankton ("diatoms") bloomed in the spring and were followed by growth of the picophytoplankton. Since our purpose here is to investigate the effects of changing community structure, *e.g.*, the dominance of the biomass, by "large" or "small" organisms rather than to generate variations in community structure with realistic external forcings, we did not attempt to match measured parameter values for the different populations beyond our earlier work. However, we stress that all parameter values, as well as the feeding, excretion, and mortality destinations (*i.e.*, the model topology), can easily be modified. The system of equations governing the compartmental dynamics for Figure 1.1 are shown in Table 1.3.

Following Evans and Parslow (31), we have not attempted to model the mixed layer dynamics explicitly, but instead assume that data are available to define the seasonal change in mixed layer depth, M, as a function of time, t (days). Mathematically, this can be written as

$$\frac{dM}{dt} = h(t). \tag{1}$$

Evans and Parslow (31) pointed out that the effect of the deepening or shallowing of the mixed layer on the concentration of a state variable will depend on whether that variable describes a nonmotile biotic, or abiotic, entity such as phytoplankton or detritus, or instead a motile biotic entity such as migrating mesozooplankton. In the latter case it might be assumed that the zooplankton are able actively to maintain themselves within the mixed layer when its depth changes, and thus the volumetric concentration of zooplankton will decrease and

TABLE 1.3. The Equations Governing Dynamics of the Nine-Compartment, "Two Size–Class" Plankton Foodweb Model[a]

Net phytoplankton (P_1):

$$\frac{dP_1}{dt} = (1 - \gamma_{11})\, \sigma_1(t,M,N_n,N_r)P_1 \; - \; G_{11} \; - \; \mu_{11}P_1 \; - \; \frac{[m + h^+(t)]P_1}{M}$$

Picophytoplankton (P_2):

$$\frac{dP_2}{dt} = (1 - \gamma_{12})\, \sigma_2(t,M,N_n,N_r)P_2 \; - \; G_{22} \; - \; \mu_{12}P_2 \; - \; \frac{[m + h^+(t)]P_2}{M}$$

Mesozooplankton (Z_1):

$$\frac{dZ_1}{dt} = \beta_{11}G_{11} + \beta_{41}G_{41} + \beta_{61}G_{61} - \mu_{21}Z_1 - \mu_{51}Z_1 - h(t)\frac{Z_1}{M}$$

Protozoans (Z_2):

$$\frac{dZ_2}{dt} = \beta_{22}G_{22} + \beta_{52}G_{52} - G_{41} - \mu_{22}Z_2 - \mu_{52}Z_2 - \frac{[m + h^+(t)]}{M}$$

Bacteria (B):

$$\frac{GB}{dt} = U_1 + U_2 - G_{52} - \mu_3 B - \frac{[m + h^+(t)]B}{M}$$

Detritus (D)

$$\frac{dD}{dt} = (1 - \beta_{11})G_{11} + (1 - \beta_{22})G_{22} + (1 - \beta_{41})G_{41} + (1 - \beta_{52})G_{52}$$

$$- \beta_{61}G_{61} - \mu_4 D + \mu_{11}P_1 + \mu_{12}P_2 - \mu_{52}Z_2 - \frac{[m + h^+(t) + V]D}{M}$$

Nitrate (N_n)

$$\frac{dN_n}{dt} = -\sum_{i=1}^{2} [J_i(t,M)Q_{1i}(N_n,N_r)P_i] \; + \; \frac{[m + h^+(t)]}{M}(N_0 - N_n)$$

Ammonium (N_r)

$$\frac{dN_r}{dt} = -\sum_{i=1}^{2} J_i(t,M)Q_{2i}(N_r)P_i \; - \; U_2 + \mu_{22}Z_2 + [\epsilon\mu_{21}$$

$$+ (1 - \Omega)\mu_{51}]Z_1 - \frac{[m + h^+(t)]}{M}N_r$$

DON (N_d)

$$\frac{dN_d}{dt} = \sum_{i=1}^{2}[\gamma_r J_i(t,M)Q_i(N_n,N_r)P_i] \; + \; \mu_4 D + (1 - \epsilon)\mu_{21}Z_1$$

$$+ (1 - \epsilon)\mu_{22} - U_1 - \frac{[m + h^+(t)]}{M}N_d$$

[a]See Table 1.4. for definitions of symbols; see also ref. 36.

increase when the mixed layer depth increases and decreases, respectively. However, in the case of nonmotile entities a deepening of the mixed layer will dilute the volumetric concentration, whereas, when the mixed layer shallows, material is left behind, or detrained, but the volumetric concentration in the mixed layer will remain unchanged. Evans and Parslow (31) dealt with this asymmetry by defining the variable $h^+(t) = \max[h(t),0]$ and by using $h^+(t)$, rather than $h(t)$, in equations representing nonmotile entities. In our model, all compartments except the mesozooplankton are nonmotile. We defined $h(t)$ based on the observed cycle of mixed layer depth off Bermuda.

3.2.2. The Bacteria Equation

Because we explore the bacterial dynamics in some detail below, the bacterial equation is discussed more fully here. Bacteria are assumed to take up ammonium and DON and also to excrete ammonium. However, DON and ammonium are not equally interchangeable forms of nitrogen for bacteria. Bacteria obtain their carbon from DON and take up ammonium mainly to obtain sufficient nitrogen to synthesize cell protein. Therefore, in a balanced growth situation, the ammonium uptake depends on the carbon/nitrogen ratio of the DON and bacteria and on the bacterial gross growth efficiency (GGE) for nitrogen and carbon (27,53,59). To quantify this concept, we define the bacterial uptake of ammonium and DON as e and d, respectively. If the carbon/nitrogen ratios of bacteria and DON are defined as R_b and R_d, respectively, and the bacterial conversion efficiencies for carbon and nitrogen are g_c and g_n, respectively, then the bacterial production, h, in nitrogen units, will be given by

$$h = g_n(e + d). \tag{2}$$

Similarly, the bacterial production, H, in carbon units, will be given by

$$H = R_b h = g_c R_d d. \tag{3}$$

Dividing equation 2 by 3 and rearranging, we obtain

$$\frac{e}{d} = \frac{g_c R_d}{g_n R_b} - 1 \tag{4}$$

In a balanced growth situation, the ratio of bacterial ammonium uptake to DON uptake should be constant in order to ensure that bacterial biomass of the required carbon/nitrogen ratio is produced from DON with a given carbon/nitrogen ratio. In this way DON can be made a proxy for DOC in the model. This concept can be incorporated into a Michaelis-Menten model of bacterial uptake (G.T. Evans, personal communication) by first defining a total bacterial nitrogenous substrate, S, as

$$S = min\ (N_r, \eta N_d),\tag{5}$$

where $\eta = \dfrac{g_c R_d}{g_n R_b} - 1$. Then the bacterial DON uptake (U_1) and ammonium uptake (U_2) can be written as

$$U_1 = \frac{V_b B N_d}{K_4 + S + N_d}\tag{6}$$

$$U_2 = \frac{V_b B S}{K_4 + S + N_d},\tag{7}$$

where V_b is the maximum bacterial uptake rate and K_4 is the half-saturation coefficient for uptake. This formulation ensures that the uptake of ammonium will be η times the DON uptake, as required by the balanced growth model, as long as there is sufficient ammonium present to meet this demand. However, if the there is insufficient ammonium, the uptake rate of both DON and ammonium will be reduced accordingly.

The full equation for bacteria can now be written as

$$\frac{dB}{dt} = U_1 + U_2 - G_{52} - \mu_3 B - \frac{[m + h^+(t)]B}{M},\tag{8}$$

where μ_3 is the bacterial-specific excretion rate. Note that we have not attempted to parameterize bacterial mortality separately, but have included it within bacterial excretion. This simple formulation does not allow shifts from one growth state to another (60), although the specific growth rate,

$$\frac{1}{B}\frac{dB}{dt},\tag{9}$$

does vary depending on changes in the concentrations of bacteria, DON, and NH_4^+. Because the excretion is biomass dependent and is independent of the uptake of substrates, the conversion efficiency, defined as

$$(U_1 + U_2 - \mu_3 B)/(U_1 + U_2),\tag{10}$$

will also vary. For example, when the standing stock is high and the concentration of DON is low so that uptake is low, the conversion efficiency could be low, or even negative, as it might be during starvation. The use of mass-specific, or first-order, mortality coefficients is common (*cf.* 61), but models may be very sensitive to the functional form of these closure terms (Steele and Henderson, unpublished data).

3.2.3. Bacterial Parameters

A maximum bacterial uptake rate, V_b, of 2.0 d^{-1} was chosen as representative of oceanic bacteria (62). Estimates of the half-saturation parameter K_4 were obtained from Carlucci *et al.* (63) and Fuhrman *et al.* (64). Reliable estimates of the specific excretion rate μ_3 are not available for bacteria. However, the magnitude of excretion will affect the conversion efficiency of the bacteria, and thus an excretion rate of 0.05 d^{-1} was chosen, which gave reasonable values throughout the seasonal cycle. The parameter η specifies the ratio of bacterial uptake of ammonium to DON, and its magnitude depends on the bacterial conversion efficiencies for carbon and nitrogen and the carbon/nitrogen ratios for bacteria and DON. We have assumed that the two efficiencies are equal and that the carbon/nitrogen ratios of bacteria and DON are 5 (59) and 8 (51), respectively, yielding a value for η of 0.6. R_b and R_d may be lower (4) and higher (10), respectively, yielding $\eta > 1$ (53).

These models have been explored in some detail using a mixed layer cycle, nitrate concentrations, and irradiance characteristic of Station S near Bermuda in the Sargasso Sea (36). Most runs with various parameter settings yield a spring bloom in which primary production and phytoplankton biomass rise quickly around days 50–100. Following the cessation of nitrate entrainment after the mixed layer stabilizes at its summer levels, the phytoplankton decline because of the combined effects of detrainment, nutrient depletion, and grazing (36) (*cf.* Fig. 1.6). The subsequent behavior of the plankton cycle through the summer and fall is critically dependent on the timing, amplitude, and rate of decline of the bloom.

4. COMMUNITY STRUCTURE, BACTERIAL DYNAMICS, AND DETRITUS RECYCLING

The structure of plankton communities *i.e.,* the relative proportions of different taxonomic or functional groups, or of different-sized members of those groups, and the trophic exchanges or linkages connecting them, is determined by a complex interplay of physical, ecological, and genetic factors that we understand only poorly. We are interested in two aspects of this problem: 1) the differential responses of different groups (particularly size classes) or organisms to physical forcings and 2) the consequences of the resulting community structures for biogeochemical cycling in the ocean. A well-known example concerns the changes in community composition that characterize the seasonal succession of plankton groups during phytoplankton blooms. In late winter, large-celled phytoplankton (diatoms) predominate and grow faster than their large-bodied predators (copepods) (28). Community structure is relatively simple, and the components are not tightly coupled. Losses from the system are high (65). Following the collapse and export of the phytoplankton bloom, a complex mixed community dominated by microbial plankton including smaller celled algae, protozoan grazers, copepods, and other carnivores and bacteria replaces the earlier diatom-based assemblage. Partly because of the multiplicity of trophic exchanges and partly because of the reduction in sizes, the plankton system is more retentive. Nitrogen is more extensively recycled, and losses are a smaller proportion of the total production, *i.e.,* the f ratio decreases. The changes in cycling and retentiveness caused by successional changes in plankton community structure may have great significance for ocean biogeochemistry.

4.1. Phytoplankton Mortality and Community Structure

To explore the relationships between community structure and nutrient cycling, we have performed an extensive series of model sensitivity analyses on the basic two size–class model depicted in Figure 1.1 and defined in Tables 1.3 and 1.4. The intrinsic rates of natural mortality of phytoplankton are almost completely unknown, yet potentially exert a powerful control over plankton dynamics (43). Though the true rates for healthy cells in nature are unknown, common sense argues that they must be less than at most about 0.1 per day. Earlier we varied the phytoplankton-specific mortality rate in a one-size class model to adjust the annual primary production to observed values (36). Here we vary the mortality coefficients of the two phytoplankton size classes to yield a series of model runs with different patterns of dominance by large and small phytoplankton and their grazers, then examine the behavior of the detritus and bacteria compartments in runs with different size structures. Each point in Figures 1.2. to 1.5. and Figure 1.8. represent the annual average fluxes for a single model run, with different parameter values.

Figure 1.2.A,B shows the annual total primary production as a function of the two mortality coefficients. As expected, there are levels of natural mortality beyond which each of the phytoplankton compartments cannot sustain itself over the annual cycle. Both compartments cease to be sustainable above mortality rates between ~0.4 and 0.6 per day, and the annual production is very low above rates of 0.2 per day unless one of the rates is quite low (Fig. 1.2.A). In general, production declines as mortality rises because losses of phytoplankton to the sinking detrital pool increase, even though netplankton mortality increases retention because living cells sink faster than the detritus. In the region below mortality rates of ~0.1 per day, the relationship between annual production and the mortality rates is complex. Figure 1.2.B shows two areas of interest. When the picoplankton mortality is held constant at 0–0.03 and the netplankton mortality is increased, the total productivity first declines as expected, then increases, then declines, then increases, and then declines again (Fig. 1.2.A,B).

TABLE 1.4. Model Parameters

Parameter	Symbol	Value
PAR/total irradiance	—	0.41
Cloudiness	—	4 Oktas
Light attenuation caused by water	k_W	0.04 m^{-1}
Cross-thermocline mixing rate	m	0.1 m d^{-1}
Netphytoplankton maximum growth rate	V_{p1}	1.48 d^{-1}
Picophytoplankton maximum growth rate	V_{p2}	2.56 d^{-1}
Initial slope of P–I curve (P_1)	α_1	0.035 (W m^{-2})$^{-1}$ d^{-1}
Initial slope of P–I curve (P_2)	α_2	0.023 (W m^{-2})$^{-1}$ d^{-1}
Half-sat. for P_1 NO$_3^-$ uptake	K_1	0.5 mmol m^{-3}
Half-sat. for P_1 NH$_4^+$ uptake	K_2	0.5 mmol m^{-3}
Half-sat. for P_2 NO$_3^-$ uptake	K_3	0.25 mmol m^{-3}
Half-sat. for P_2 NH$_4^+$ uptake	K_4	0.5 mmol m^{-3}
P_1 specific mortality rate	μ_{11}	0.025 or 0.1 d^{-1}
P_2 specific mortality rate	μ_{12}	0.1 or 0.025 d^{-1}
Light atten. by phytoplankton	k_c	0.03 m^2 (mmol N)$^{-1}$
P_1, P_2 exudation fraction	γ_1	5%
NH$_4^+$ inhibition parameter (Phy)	Ψ	1.5 (mmol N)$^{-1}$
Mesozoo. maximum growth rate	g_1	1.4 d^{-1}
Protozoo. maximum growth rate	g_2	4 d^{-1}
Z_1 assim. eff. feeding on P_1	β_{11}	75%
Z_1 assim. eff. feeding on Z_2	β_{41}	75%
Z_1 assim. eff. feeding on D	β_{61}	75%
Z_2 assim. eff. feeding on P_2	β_{22}	75%
Z_2 assim. eff. feeding on B	β_{52}	75%
Mesozoop. spec. excretion rate	μ_{21}	0.1 d^{-1}
Mesozoop. spec. mortality rate	μ_{51}	0.05 d^{-1}
Protozoo. spec. excretion rate	μ_{21}	0.2 d^{-1}
Protozoo. spec. mortality rate	μ_{51}	0.05 d^{-1}
Z_1, Z_2 half-sat. for ingestion	K_{3i}[a]	1.0 mmol m^{-3}
Exported fraction of Z_1 mortality	Ω	33%
Ammonium fraction of Z_1 excretion	ϵ	100%
Bacterial maximum growth rate	V_b	2.0 d^{-1}
Bacterial specific excretion rate	μ_3	0.05 d^{-1}
Bacterial Half-sat. rate for uptake	K_4	0.5 mmol N m^{-3}
NH$_4^+$/DON uptake ratio	η	0.6
Detrital breakdown rate	μ_4	0 to 0.05 d^{-1}
Detrital sinking rate	V	1 to 10 m d^{-1}

[a]Where i = 1,2. . . .

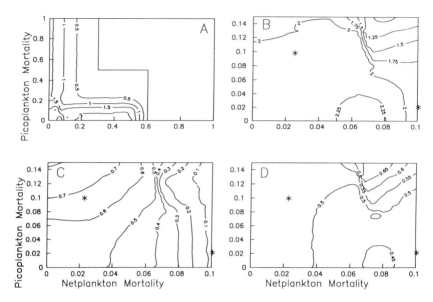

Fig. 1.2. Contour plots showing sensitivity to changes in **A** and **B:** phytoplankton mortality rates (d^{-1}) for annual average daily primary production (mmol N m^{-2} d^{-1}); **C:** fraction of annual production contributed by the netplankton; and **D:** annual f ratio. B is a rescaled version of A. Both phytoplankton compartments are extinct in blank region in A. Asterisks indicate model runs based on two combinations of mortality coefficients (0.025/0.1 and 0.1/0.025) chosen for further analysis (Figure 1.3).

This nonlinear behavior is caused by the interactions between outcomes of competition for nutrients between the two phytoplankton groups and nutrient regeneration from the detrital pool, which is fed by phytoplankton mortality. There is an abrupt gradient between low and high annual productivity values centered near the mortality rates of (0.08, 0.10), where a change of just 0.001 produces about a 30%–50% change in annual production. This suggests that the model behaves fundamentally differently in its high and low productivity regions. Interestingly, the average daily mixed layer production near Bermuda, the region on which this model is based, is about 1.5–2 mmol N m^{-2} d^{-1}, right in the gradient region. To the extent that this model simulates the plankton productivity regime of the Sargasso Sea, we might say that this area is poised between two rather different states of ecosystem organization.

To examine the effects of the mortality-induced changes in community structure further, we picked two different runs from the mortality phase plane shown in Figure 1.2.B for additional sensitivity studies. The complementary points (0.025, 0.1) and (0.1, 0.025), which are depicted by dots or asterisks in Figure 1.2., represent model runs in which the values of annual production (Fig. 1.2.B) and f ratio (Fig. 1.2.D) are quite similar. However, the community structure, or the routes through the annual cycle over which the productivity is integrated, are very different (Fig. 1.2.C). In the "netplankton run" (mortality values 0.025, 0.1), about 70% of the total production is by the larger netplankton; in the other run, the "picoplankton run," only 10% is by the netplankton. Because the two phytoplankton classes have different grazers, and because the netplankton can be lost directly from the model domain by sinking, without entering any other compartment, the other compartments behave quite differently in the two scenarios. The effects of variations in parameters governing detritus turnover were explored for these two runs.

4.2. Detritus Turnover and Bacterial Dynamics

In our earlier work we noted that model behavior was also moderately sensitive to two other poorly evaluated parameters, the detrital sinking and breakdown rates. Detritus, or dead organic nitrogen, acts in the plankton system as a kind of capacitor, or slowly cycling reservoir for potential regenerated production. Energy and nutrients are stored in the detritus and become available for further utilization depending on the rates at which the material is exported from the system, decomposed by microbial action, or reingested by detritovores and coprophages (56). In our model, the detrital compartment is composed of the remains of dead phytoplankton, protozoans, and zooplankton and of their fecal products. Thus the compartment undoubtedly contains particles with different sinking rates and susceptibility to breakdown or reingestion. The parameter values governing all these fates are merely mean values that may or may not be representative of the particle spectra existing at any given point in the annual cycle of a given run. An alternative, but much more complex, approach would be to assign independent compartments for each detrital type (66). Here we examine the sensitivity of the model's bacterial compartment to changes in detrital sinking and breakdown rates. These two parameters also control the action of the biological pump in this model. To the extent that sinking is a major export route, the detrital parameters govern the rate at which detrital particles enter the deep sea and how deep into it their carbon is injected before decomposition and respiration back to CO_2. These processes in turn determine how much of the primary production is removed from contact with the atmosphere and for how long.

The sensitivity of total production and several bacterial processes for the netplankton-dominant run are shown in Figure 1.3; the picoplankton-dominant results are shown in Figure 1.4. The patterns differ quantitatively, but not qualitatively, between the two basic runs. In general, ecosystem productivity is greater at most points in the sinking–breakdown phase plane in the picoplankton-dominated run, than it is at the corresponding points in the netplankton-domi-

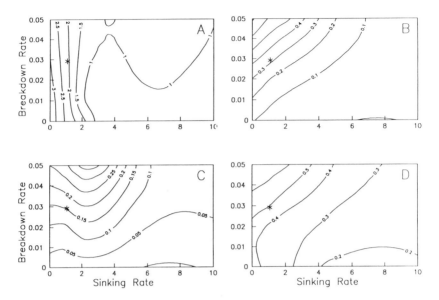

Fig. 1.3. Contour plots showing sensitivities to changes in detrital breakdown rate (d^{-1}) and sinking rate (m^{-1}) of **A:** annual average daily primary production (mmol N m^{-2} d^{-1}); **B:** bacterial production (mmol N m^{-2} d^{-1}); **C:** bacteria as a fraction of primary production; and **D:** bacterial conversion efficiency (equation 10). These runs are based on the mortality pair (0.025, 0.1) favoring netplankton shown on the left side of each plot in Figure 1.2. Runs represented by asterisks here are shown in Figures 1.6 and 1.7.

nated run. The annual primary productivity decreases more slowly as detrital sinking rate increases more in the picoplankton-dominated run (Fig. 1.4A) than in the netplankton-dominated run (Fig. 1.3A). Bacterial production declines as the sinking rate increases, and increases as the detrital breakdown rate increases (Figs. 1.3B, 1.4B). This is because, in this model, most of the DON for bacterial metabolism is derived from detrital breakdown. Bacterial production is greater in all the picoplankton-dominated runs than in their netplankton-dominated counterparts (compare the values marked by the asterisk), but is more sensitive to changes in the parameters in the picoplankton-dominated runs. Annual bacterial production ranges from about 5% to 35% of annual primary production in both sets of runs (Figs. 1.3C, 1.4C), though "high" percentages are about twice as common in the picoplankton-dominated runs (compare the area above the 0.25 lines in Figs. 1.3C, 1.4C). Similarly, the bacterial conversion efficiency is higher in the picoplankton-dominated runs, ranging up to annual average values in excess of 0.7 (Fig. 1.4D) compared with a more restricted range in the runs dominated by netplankton. In general, the model runs derived from picoplank-

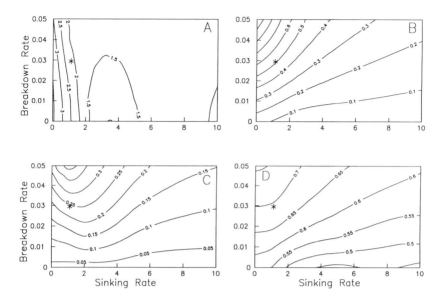

Fig. 1.4. As in Figure 1.3, except that these runs are based on the mortality pair (0.1, 0.025) favoring picoplankton shown on the right side of each plot in Figure 1.2. Runs represented by asterisks here are shown in Figures 1.6 and 1.7.

ton-dominated foodwebs have a more productive, efficient, and responsive bacterial component than their counterpart runs derived from netplankton-dominated foodwebs. This is because the picoplankton sustain the bacteriovorous protozoans, decreasing the bacterial biomass and turnover times.

The modern view is that bacteria in the sea both regenerate, or supply ammonium for subsequent uptake, and compete with phytoplankton for ammonium (45,52,67). Although bacteria are generally more productive in both absolute and relative terms in the picoplankton-dominated runs, they are much more important in annual nitrogen dynamics in the netplankton-dominated runs. Figure 1.5 shows the bacterial proportion of the total annual ammonium uptake (by phytoplankton and bacteria; Fig. 1.5A,B) and the bacterial contribution to total ammonium regeneration (by bacteria and zooplankton; Fig. 1.5C,D). In both sets of runs, most of the annual ammonium uptake is usually by the phytoplankton, but the bacterial share rises to over 50% in some runs of the netplankton-dominated model. In contrast, bacteria are clearly the dominant remineralizers in the netplankton-dominated runs over much of the parameter domain explored, supplying upwards of 60%–80% of the total ammonium flux (Fig. 1.5C) compared with a strikingly uniform 10% in the picoplankton-dominated runs (Fig. 1.5D). The bacteria act much more as net remineralizers in the netplankton-

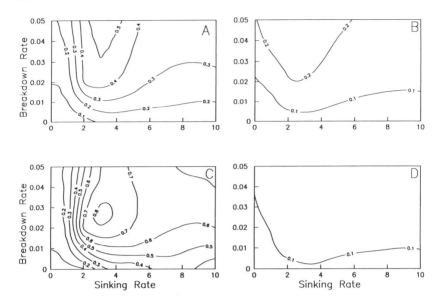

Fig. 1.5. Contour plots showing sensitivities to changes in detrital breakdown rate (d^{-1}) and sinking rate $(m \ d^{-1})$ of **A,B:** fraction of total annual ammonium uptake contributed by bacteria, and **C,D:** fraction of total annual ammonium regeneration contributed by bacteria. Runs in A and C are based on the mortality pair (0.025, 0.1) favoring netplankton; runs in B and D are based on the mortality pair (0.1, 0.025) favoring picoplankton.

dominated foodwebs, supplying ammonium in excess of their own uptake. In the picoplankton-dominated runs, the bacteria are net ammonium utilizers, taking it up in excess of their regenerative activity, but in fact compete poorly with the phytoplankton in the annual ammonium budget. This is because bacteriovores keep B and thus $\mu_3 B$ lower in picoplankton-dominated runs.

4.3. Annual Plankton Cycles and Community Structure

To illustrate the differences in the seasonal dynamics that follow from the parameter variations discussed here, we present the annual cycles for two individual model runs taken from the sensitivity studies shown in Figures 1.3 to 1.5. A parameter set with a sinking rate of 1 m d^{-1} and a detrital breakdown rate of 0.03 d^{-1} was chosen from the netplankton-dominated and picoplankton-dominated series. These particular runs are identified by the asterisks in Figures 1.3 and 1.4. Each of the two runs has a average annual primary productivity of about 2 mmol N m^{-2} d^{-1} (Figs. 1.3A, 1.4A), but the bacterial production is about 60% higher in the picoplankton-dominated run (Fig. 1.4B) and the bacteria are

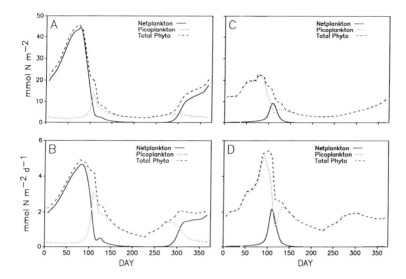

Fig. 1.6. Model output for annual cycles of **A,C:** phytoplankton stocks and **B,D:** primary production. A and B are the netplankton-dominated runs derived from mortality pair (0.025, 0.1); C and D are the picoplankton-dominated runs derived from the pair (0.1, 0.025). All these runs have a detrital breakdown rate of 0.03 d^{-1} and a sinking rate of 1 m d^{-1}) as shown in Figures 1.3 and 1.4.

about twice as efficient on an annual basis (Fig. 1.4D). The annual cycles of the phytoplankton are shown in Figure 1.6, and the cycles of bacteria and protozoans in Figure 1.7.

Most, but not all, of the runs we have investigated are characterized by a marked spring bloom, a period of rapid accumulation of phytoplankton biomass, and high primary production driven by high nitrate concentrations and increasing light availability in the relative absence of grazing pressure. Figure 1.6A shows a classic spring bloom dominated by netplankton, during which time the high primary production (Fig. 1.6) is well in excess of losses because of mortality, sinking, grazing, and detrainment. The mesozooplankton grazers start the year at very low population levels, and their low growth rates (Table 1.4) prevent them from closer coupling to their netplankton prey. This in turn allows the explosive growth we call a *bloom*. In this run with high picoplankton mortality, the picoplankton have a small bloom after the crash of the netplankton around day 100, and then they in turn constitute most of the phytoplankton biomass and production during the summer. Although their mortality is lower, the netplankton stocks fall to low levels in summer because of the combined losses from sinking of healthy cells and removal by grazers.

In the contrasting, low picoplankton mortality run, the roles and dominant periods for the two phytoplankton groups are somewhat reversed. The spring

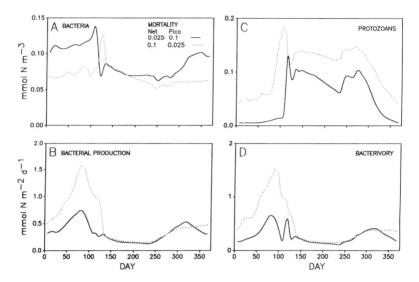

Fig. 1.7. Model output for annual cycles of **A:** average mixed layer bacterial biomass, **B:** bacterial production, **C:** protozoan biomass, and **D:** protozoan bacterivory (ingestion rates on bacteria) for the netplankton-dominated runs (solid lines) and the picoplankton-dominated runs (dashed lines). All these runs have a detrital breakdown rate of 0.03 d^{-1} and a sinking rate of 1 m d^{-1}), as shown in Figures 1.3 and 1.4. Note that the biomass figures are m^{-3} not standing stocks m^{-2} as in Figure 1.6.

bloom is smaller and is dominated by the picoplankton (Fig. 1.6). The levels of primary production (Fig. 1.6D) are about the same as in the netplankton case, though the peak is attained somewhat later. Biomass is kept low throughout the year because the rapid growth rate (Table 1.4), of the protozoans allows them to maintain control over their picoplankton prey. The annual cycle and annual total of primary production are very similar in the two cases (especially after day 150), which indicates that the picoplankton-dominant case, with its intensive grazing pressure, has a greater rate of organic matter turnover. The netplankton have just a brief period of high biomass in this run, when they outcompete the picoplankton for light and nutrients briefly after day 100.

The consequences of these structural and dynamic differences in the two models for bacterial cycles are shown in Figure 1.7. The average mixed layer concentration of bacterial biomass is quite similar in the two cases, especially in summer. However, in autumn, winter, and spring, the bacterial biomass is about 50% higher in the netplankton case (Fig. 1.7A), in spite of much higher bacterial production in the picoplankton-dominant system (Fig. 1.7B). Production is higher in the picoplankton-dominated run as a result of the combined, and interacting, effects of picoplankton dominance on the fluxes through the detrital,

DON, and ammonium pools (data not shown). In comparison to the netplankton-dominated run, the fluxes into the DON pool from the breakdown of detritus, the bacterial NH_4^+, and DON uptake are greater and the bacterial excretion considerably lower than in the netplankton-dominated case. Protozoan biomass is higher in this run (Fig. 1.7C) because of the greater supply of their main food, picophytoplankton. Detritus is more abundant because the lack of netplankton over most of the year keeps mesozooplankton detritovory low, and bacterial excretion is low because protozoan bacterivory keeps bacterial biomass low (Fig. 1.7D).

5. SYNTHESIS: COMMUNITY STRUCTURE, BACTERIA, AND CARBON REMOVAL

Thus we see that the nutrient and grazing dynamics associated with particular foodweb structures result in qualitative and quantitative differences in the overall behavior of the bacterial component. But what is the role of bacterial processes in the net fixation and removal of carbon from the atmosphere into the deep sea? In our model bacteria have an explicitly modeled role of converting reduced forms of dissolved nitrogen into biomass or excreting ammonium. The bacterial biomass itself is largely ingested by protozoans and subsequently remineralized as well. In general, the bacteria act like a ''sink'' in the foodwebs, returning nitrogen and carbon to solution rather than passing it on up to higher order consumers (68–71) or exporting it. In the models described here, the bacteria obtain their DON principally from the breakdown of detritus and from healthy phytoplankton via photosynthetic release. It is easy to add two other routes of DON supply, by having the grazers excrete some of their waste nitrogen as DON (Fig. 1.1., dashed lines). This results in marginally higher bacterial production (36) and lower sensitivity to changes in detrital recycling. The model behavior we specified is derived from the ''free bacterial paradigm'' in which the great majority of bacterial biomass in the ocean is seen as free-living rather than as attached to particles and sustained by the flux of DOM rather than by exoenzymatic decomposition of particulate substrates (44,72–73,88). The changes in bacterial cycles and stocks wrought by changes in community structure are modulated through shifts in the fluxes on dissolved matter.

The presence and activity of attached bacteria is implicit in our treatment of detritus breakdown. Rather than modeling explicitly the dynamics of bacterial attachment, growth, and metabolism on detrital particles, we assigned a mass-specific breakdown rate to the detritus, which results in the release of DON. The observations of Pett (75) on decomposing algae and the studies by Jacobsen and Azam (58), who found than bacteria colonizing ^{14}C-labeled fecal pellets released more dissolved ^{14}C into the water than they assimilated, support this formula-

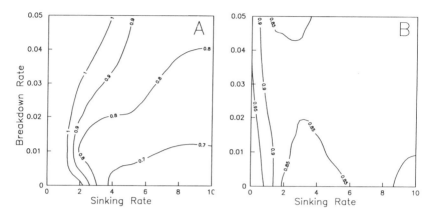

Fig. 1.8. Rates of new production (NO$_3^-$ assimilation by phytoplankton) as a function of detrital breakdown and sinking rates for the netplankton-dominated (**A**) and picoplankton-dominated (**B**) runs, respectively. These runs are derived from those marked by asterisks in Figure 1.2. and are part of the data set as Figures 1.3 and 1.4.

tion. Bacteria are associated with particulate matter in the sea (74,76–78) and are responsible for its formation (79) and its breakdown (57). Although it is generally accepted that bacteria participate in particle breakdown by the liberation of proteases and other exoenzymes at the cell surface (80–82), the mechanisms are still unclear. For example, it seems hard to reconcile the generally low incidence of attachment with independently estimated rates of particle turnover (83,84). Cho and Azam (85) showed that, in the midwater column of the North Pacific, the production of free-living bacteria equalled the flux of sedimenting particulate matter, suggesting a coupling between the vertical flux and the bacteria mediated by unexplained particle decomposition and DOC release, exactly the process we modeled. Whether this dissolution of the rapidly sinking particles is by a large population of attached bacteria whose presence or activity has thus far escaped notice or by remote free-living bacteria that accomplish the feat by still obscure means is not known.

A better knowledge of these relationships is crucial to our understanding of the ocean's role in regulating the atmospheric carbon balance for at least three reasons. First, the capability of bacteria to take up and mineralize dissolved organic matter must be the major control on the size of the DOC and DON pools in the sea. These are among the largest pools of fixed carbon on the planet and may represent a major route governing the sequestering of carbon in the deep sea (17). Second, the breakdown rates of detrital particles in the surface layer, in conjunction with particle size and sinking rates, limits the amount of particulate mass that can be exported. Third, the dissolution of sinking particles within or below the surface layer, whether by attached or—somehow—by free-living bacteria, determines the rate and depth to which the particles penetrate the oceanic interior (66). Finally, as we argued above, a primary control on the partitioning of carbon between atmosphere and ocean is oceanic new production, which drives the pCO_2 negative and thus pumps atmospheric CO_2 into the sea. The processes we have discussed, changes in plankton community structure, detrital sinking rate, and breakdown, all influence the annual rates of new production (Fig. 1.8). Variations in community structure influence the sensitivity of the new production system to changes in detrital sinking and breakdown rates. We require new observations and experimental approaches to define phytoplankton mortality and detrital breakdown in order to model these processes more realistically and to gain a better capability to predict the future course of atmospheric CO_2 accumulation.

ACKNOWLEDGMENTS

The research reported in this chapter was supported in part by NSF grant OCE 8914229 to H.W.D. and by U.S. DOE subcontract 19X-SC167V to M.J.R.F. Rick Slater prepared Figure 1.1.

REFERENCES

1. Bolin B, Doos BR, Jager J, Warrick RA: The Greenhouse Effect Climatic Change and Ecosystems. SCOPE Report 29, New York: Wiley, 1986.

2. Schneider SH: The greenhouse effect: Science and policy. Science 243:771–781, 1989.

3. Lovelock J: The Ages of Gaia. New York: Norton, 1988, 252 pp.

4. Dickinson RE, Cicerone RJ: Future global warming from atmospheric trace gases. Nature 319:109–115, 1986.

5. Maddox J: The greenhouse question (cont'd). Nature 345:473, 1990.

6. Moore B, Bolin B: The oceans, carbon dioxide and global change. Oceanus 29 4:9–15, 1986.

7. Sarmiento JL, Toggweiler JR, Najjar R: Ocean carbon cycle dynamics and atmospheric pCO_2. Philos Trans R Soc Lond [A] 325:3–21, 1988.

8. Barnola JM, Raynaud D, Korotkevich YS, Lorius C: Vostok ice core provides 160,000 year record of atmospheric CO_2. Nature 329:408–414, 1987.

9. Mix AC: Pleistocene paleoproductivity: Evidence from organic carbon and foraminiferal species. In Berger WH, Smetacek VS, Wefer G (Eds): Productivity of the Ocean: Present and Past. Chichester: Wiley, 1990, pp 313–340.

10. Berger WH, Smetacek VS, Wefer G: Ocean productivity and paleoproductivity—An overview. In Berger WH, Smetacek VS, Wefer G (eds). Productivity of the Ocean: Present and Past. Chichester: Wiley, 1990, pp 1–34.

11. Tans PP, Fung IY, Takahashi T: Observational constraints on the global atmospheric CO_2 budget. Science 247:1431–1438, 1990.

12. National Academy of Sciences: Global Ocean Flux Study. Proceedings of a Workshop. Washington, DC: National Academy Press, 1984.

13. McCarthy JJ, Brewer PG, Feldman G: Global ocean flux. Oceanus 29(4):16–26, 1986.

14. Sarmiento JL, Fasham MJR, Slater R, Toggweiler JR, Ducklow HW: The role of biology in the chemistry of CO_2 on the ocean. In Farrell M (ed): Chemistry of the Greenhouse Effect. New York: Lewis Publ. (in press).

15. Wolf KU, Woods JD: Langrangian simulation of primary production in the physical environment—the deep chlorophyll maximum and nutricline. In Rothschild BJ (ed): Toward a Theory of Biological–Physical Interactions in the World Ocean. Dordrecht: Reidel, D, 1988, pp 51–70.

16. Brewer PG: What controls the variability of carbon dioxide in the surface ocean? A plea for complete information. In Burton JD, Brewer PG, Chesselet R (eds): Dynamic Processes in the Chemistry of the Upper Ocean. New York, Plenum, 1986.

17. Toggweiler JR: Is the downward dissolved organic matter (DOM) flux important in carbon transport? In Berger WH, Smetacek VS, Wefer G (eds): Productivity of the Ocean: Present and Past. Chichester: Wiley, 1989.

18. Broecker WS, Peng TH: Tracers in the Sea. Palisades: Eldigio Press, 1982, 690 pp.

19. Berger WH, Vincent E: Deep-sea carbonates: Reading the carbon-isotope signal. Geol. Rundschau 75:249–269, 1986.

20. Broecker WS, Peng T-H: The oceanic salt pump: Does it contribute to the glacial-interglacial difference in atmospheric CO_2 content. Global Biogeochem Cycles 1:251–259, 1987.

21. Volk T, Liu Z: Controls of CO_2 sources and sinks in the scale surface ocean: Temperature and nutrient. Global Biogeochem Cycles 2:73–90, 1988.

22. Longhurst AR, Harrison WG: The biological pump: Profiles of plankton production and consumption in the upper ocean. Progr Oceanogr 22:47–123, 1989.

23. Eppley RW, Peterson BJ: Particulate organic matter flux and planktonic new production in the deep ocean. Nature 282:677–680, 1979.

24. Peinert R, von Bodungen B, Smetacek VS: Food web structure and loss rate. In Berger WH, Smetacek VS, Wefer G (eds): Productivity of the Ocean: Present and Past. Chichester: Wiley, 1990, pp 35–48.

25. Dugdale RC, Goering JJ: Uptake of new and regenerated forms of nitrogen in primary productivity. Limnol Oceanogr 12:196–206, 1967.

26. Goering JJ: The role of nitrogen in eutrophic processes. In Mitchell R (ed): Water Pollution Microbiology. New York: Wiley, 1972, pp 43–68.

27. Ducklow HW, Fasham MJR, Vezina AF: Flow analysis of open sea plankton networks. In Wulff F, Field JG, Mann KH (eds): Network Analysis in Marine Ecology. Coastal and Estuarine Studies, Vol 32. New York: Springer-Verlag, 1989, pp 159–205.

28. Smetacek V: Role of sinking in diatom life history cycles: Ecological, evolutionary, and geological significance. Mar Biol 84:239–251, 1985.

29. Wefer G: Particle flux in the ocean: Effects of episodic production. In Berger WH, Smetacek VS, Wefer G (eds): Productivity of the Ocean: Present and Past. Chichester: Wiley, pp 139–153, 1990.

30. Parsons TR, Lalli CM: Comparative oceanic ecology of the plankton communities of the Subarctic Atlantic and Pacific Oceans. Oceanogr Mar Biol Annu Rev 26:317–359, 1988.

31. Evans GT, Parslow JS: A model of annual plankton cycles. Biol Oceanogr 3:327–347, 1985.

32. Frost BW: Grazing control of phytoplankton stock in the open Subartic Pacific Ocean: A model assessing the role of mesozooplankton particularly the large calanoid copepods Neocalanus. Mar Ecol Progr Ser 39:49–68, 1987.

33. Ducklow HW: Bacterial Biomass in warm Gulf Stream ring 82-B: mesoscale distributions, temporal changes and production. Deep Sea Res 33:1789–1812, 1986.

34. Cho BC, Azam F: Major role of bacteria in biogeochemical fluxes in the ocean's interior. Nature 332:441–443, 1988.

35. Kirchman DL, Keil RG, Simon M, Welschmeyer NA: Biomass production of heterotrophic bacterio-plankton in the oceanic Subartic Pacific: SUPER 1987 and 1988. Mar Ecol Progr Ser (in submission), 1991.

36. Fashan MJR, Ducklow HW, McKelvie SM: A nitrogen-based model of plankton dynamics in the oceanic mixed layer. J Mar Res 48:591–639, 1990.

37. Vezina AF, Platt T: Foodweb dynamics in the ocean. Part I. Best-estimates of flow networks using inverse methods. Mar Ecol Progr Ser 42, 269–287, 1987.

38. Mills EL: Biological Oceanography, An Early History, 1870–1960. Ithaca: Cornell University Press, 1989, 378 pp.

39. Steele JH: Plant production in the northern North Sea, Scottish Home Department. Mar Res Rep No. 7. Edinburgh: HMSO, 1958.

40. Steele JH: The Structure of Marine Ecosystems. Cambridge: Harvard, 1974.

41. Riley GA, Stommel H, Bumpus DF: Quantitative ecology of the plankton of the western North Atlantic. Bull Bingham Oceanogr Coll 12(3):1–169, 1949.

42. Pomeroy LR: The ocean's food web. BioScience 24:499–504, 1974.

43. Walsh JJ: Death in the sea: Enigmatic phytoplankton losses. Prog Oceanogr 12:1–86, 1983.

44. Williams PJLB: Incorporation of microheterotrophic processes into the classical paradigm of the planktonic food web. Kieler Meeresforsch 5:1–28, 1981.

45. Williams PJLB: Bacterial production in the marine food chain: The emperor's new suit of clothes? Flows of energy and materials in marine ecosystems. In Fasham M (ed): Flows of Energy and Materials in Marine Ecosystems. Theory and Practice. New York: Plenum, 1984.

46. Fasham MJR: Flow analysis of materials in the marine euphotic zone. In Ulanowicz RE, Platt T (eds): Ecosystem Theory for Biological Oceanography. Can Bull Fish Aquat Sci 213:139–162, 1985.

47. Pace ML, Glasser JE, Pomeroy LR: A simulation analysis of continental shelf food webs. Mar Biol 82:47–63, 1984.

48. Ulanowicz RE: Community measures of food networks and their possible applications. In Fasham M (ed): Flows of Energy and Materials in Marine Ecosystems. Theory and Practice. New York: Plenum, 1984, pp 23–48.

49. Wiegert RG: Population models: experimental tools for the analysis of ecosystems. In Horn DJ, Mitchell R, Stairs GR (eds): Proceedings of Colloquium on Analysis of Ecosystems. Columbus: Ohio State University Press, 1979, pp 239–275.

50. Williams PM, Druffel ERM: Dissolved organic matter in the ocean: comments on a controversy. Oceanography 1:14–17, 1988.

51. Suzuki Y, Sugimura T, Itoh T: A catalytic oxidation method for the determination of total nitrogen dissolved in seawater. Mar Chem 16:83–97, 1985.

52. Wheeler PA, Kirchman DL: Utilization of inorganic and organic nitrogen by bacteria in marine systems. Limnol Oceanogr 21:998–1009, 1986.

53. Goldman JC, Caron DA, Dennett MR: Regulation of gross growth efficiently and ammonium regeneration in bacteria by substrate C:N ratio. Limnol Oceanogr 32:1239–1252, 1987.

54. Paffenhofer G-A, Knowles SC: Ecological implications of fecal pellet size, production and consumption by copepods. J Mar Res 37:35–49, 1979.

55. Poulet SA: Factors controlling utilization of non-algal diets by particle-grazing copepods: A review. Oceanol Acta 6:221–234, 1983.

56. Lampitt RS, Noji T, von Bodungen B: What happens to zooplankton fecal pellets? Implications for material flux. Marine Biol 104:15–23, 1990.

57. Newell RC, Linley EAS: Significance of microheterotrophs in the decomposition of phytoplankton: estimates of carbon and nitrogen flow based on the biomass of plankton communities. Mar Ecol Progr Ser 16:105–119, 1984.

58. Jacobsen TR, Axam F: Role of bacteria in copepod fecal pellet decomposition: Colonization, growth rates and mineralization. Bull Mar Sci 35:495–502, 1984.

59. Fenchel T, Blackburn TH: Bacteria and Mineral Cycling. New York: Academic, 1979.

60. Chin-Leo G, Kirchman DL: Unbalanced growth in natural assemblages of marine bacterioplankton. Mar Ecol Progr Ser 63:1–8, 1990.

61. Billen G: Heterotrophic utilization and regeneration of nitrogen. In Hobbie JE, Williams PJLB. Heterotrophic Activity in the Sea. New York: Plenum, pp 313–355, 1984.

62. Ducklow HW, Hill SM: The growth of heterotrophic bacteria in the surface waters of warm core rings. Limnol Oceanogr 30(2):239–259, 1985a.

63. Carlucci AF, Craven DB, Heinrichs SM: Diel production and microheterotrophic utilization of dissolved free amino acids in waters off Southern California. Appl Environ Microbiol 48(1): 165–170, 1985.

64. Fuhrman JA: Close coupling between release and uptake of dissolved free amino acids in seawater studied by an isotope dilution approach. Mar Ecol Progr Ser 37:45–52, 1987.

65. Peinert R, von Bodungen B, Smetacek VS: Food web structure and loss rate. In Berger WH, Smetacek VS, Wefer G (eds): Productivity of the Ocean: Present and Past. Chichester: Wiley, 1989, pp 35–48.

66. Andersen V, Nival P: A pelagic ecosystem model simulating production and sedimentation of biogenic particles: Role of salps and copepods. Mar Ecol Progr Ser 44:37–50, 1988.

67. Tupas L, I Koike: Amino acid and ammonium utilization by heterotrophic marine bacteria grown in enriched seawater. Limnol Oceanogr 35:1145–1155, 1990.

68. Ducklow HW, Purdie DA, Williams PJLB, Davies JM: Bacterioplankton: A sink for carbon in a coastal plankton community. Science 232:865–867, 1986.

69. Pomeroy LR, Wiebe WJ: Energetics of microbial food webs. Hydrobiologia 159:7–18, 1988.

70. Ducklow HW: The passage of carbon through microbial foodwebs: results from flow network models. Mar Microbiol Foodwebs 5:1–16, 1991.

71. Hagstrom A, Axam F, Andersson A, Wikner J, Rassoulzadegan F: Microbial loop in an oligotrophic pelagic marine ecosystem: possible roles of cyanobacteria and nanoflagellates in the organic fluxes. Mar Ecol Progr Ser 49:171–178, 1988.

72. Azam F, Hodson RE: Size distribution and activity of marine microheterotrophs. Limnol Oceanogr 22:492–501, 1977.

73. Azam F, Fenchel T, Field JG, Gray JS, Meyer-Reil LA, Thingstad F: Mar Ecol Progr Ser 10:257, 1983.

74. Ducklow H, Kirchman D: Bacterial dynamics and distribution during a spring diatom bloom in the Hudson River. J Plankton Res 5:333–355, 1983.

75. Pett RJ: Kinetics of microbial mineralization of organic carbon from detrital Skeletonema costatum cells. Mar Ecol Progr Ser 52:123–128, 1989.

76. Wiebe WJ, Pomeroy LR: Microorganisms and their association with aggregates and detritus in the sea: A microscopic study. Mem Ist Ital Idrobiol 29(Suppl):325–352, 1972.

77. Kirchman DL, Ducklow HW: Trophic dynamics of particle-bound bacteria in pelagic ecosystems: A review. In Pullin RSV, Moriarty DJW (eds): Detrital Systems in Aquaculture. Manila: ICLARM, 1988, pp 54–82.

78. Iriberri J, Unanue M, Barcina I, Egea L: Seasonal variation in population density and heterotrophic activity of attached and free-living bacteria in coastal waters. Appl Environ Microbiol 53:2308–2314, 1987.

79. Biddanda B: Microbial synthesis of macroparticulate matter. Mar Ecol Progr Ser 20:241–251, 1985.

80. Hoppe H-G: Significance of exoenzymatic activities in the ecology of brackish-water: Measurements by means of methyl-umbelliferyl substrates. Mar Ecol Progr Ser 11:299–308, 1983.

81. Vives-Rigo J, Billen G, Fontigny A, Somville M: Free and attached proteolytic activity in water environments. Mar Ecol Progr Ser 21:245–249, 1985.

82. Azam F, Cho BY: Bacterial utilization of organic matter in the sea. In Fletcher M, Gray TRG, Jones JG (eds): Ecology of Microbial Communities. Cambridge, Cambridge Univ. Press. 1987, pp 262–281.

83. Ducklow HW, Hill S, Gardner WD: Bacterial growth and the decomposition of particulate organic carbon collected in sediment traps. Continental Shelf Res 4(4):445–464, 1985.

84. Karl DM, Knauer GA, Martin JH: Downward flux of particulate organic matter in the ocean: A particle decomposition paradox. Nature 332:438–441, 1988.

85. Cho BY, Azam F: Biogeochemical significance of bacterial biomass in the ocean's euphotic zone. Mar Ecol Progr Ser 63:253–259, 1990.

86. Mopper K, Degens ET: Organic carbon in the ocean: Nature and cycling. In Bolin B, Degend ET, Kempe S, Ketner P (eds): The Global Carbon Cycle. Chichester: Wiley, 1979, pp 293–316.

87. Sugimura Y, Suzuki Y: A high-temperature catalytic oxidation method for the determination of non-violatile dissolved organic carbon in seawater by direct injection of a liquid sample. Mar Chem 24:105—131, 1988.

88. Ducklow HW: Production and fate of bacteria in the oceans. BioScience 33(8):494–499, 1983b.

EFFECTS OF ACID DEPOSITION ON MICROBIAL PROCESSES IN NATURAL WATERS

CYNTHIA C. GILMOUR

Benedict Estuarine Research Laboratory, The Academy of Natural Sciences, Benedict, Maryland 20612

1. INTRODUCTION

The acidification of lakes and streams in Europe and North America has been related to atmospheric deposition of strong acids (1,2). The areas most affected by this phenomenon are the most heavily industrialized regions of the world and those land masses downwind of these regions. Sulfuric acid is the major strong acid component of deposition in most of these regions, with nitric acid second. Nitric acid constitutes a particularly large fraction of acid deposition in western North America and Europe (3). However, the relative contribution of nitric acid in the United States is increasing because of controls on sulfur emissions and continued increases in NO_x emissions from internal combustion engines (4). (Summaries of emission and deposition patterns can be found in refs. 5–8.) Acidification of natural waters can also arise from nonatmospheric sources, including acid mine drainage (9), naturally occurring bogs and peatlands (10), and weathering of acid-producing minerals, particularly naturally occurring pyrite (11).

Environmental Microbiology, pages 33–57, © *1992 Wiley-Liss, Inc.*

The influence of acid deposition on a lake or stream depends on both the amount of strong acid deposited and the ability of the ecosystem to buffer the incoming acids. Natural waters with low buffering capacities are most at risk of acidification and its ensuing effects. These effects include changes in biological species composition and densities and changes in the biogeochemical cycling of a number of elements. In addition to changes in the cycling of the ions introduced, acid deposition can result in the mobilization and bioaccumulation of some metals and a decrease in dissolved organic carbon (DOC) with a concomitant build-up of organic matter at the sediment–water interface (12,13).

Some of these changes are exemplified by the specific responses of experimentally acidified Lake 223 in the Experimental Lakes Area in Ontario, Canada (13,14). This small lake was brought from pH 6.6 to 5.4 over a period of 4 years, using direct sulfuric acid additions. The lake began to lose invertebrate and fish species at an average lake water pH just below 6, particularly a benthic crustacean and minnow. Like many acid-sensitive species, these organisms are most at risk during reproduction. At pH 5.6, mats of filamentous algae (genus *Mougeotea*) invaded surfaces in the littoral zone. Free-living algal species composition also changed, with decreases in *Chrysophyta* and increases in *Chlorphyta*. Transparency in the lake increased because of precipitation of suspended organic carbon. Because of this increase in light penetration and faster warming of the lake, algal biomass was actually increased. Of course, sulfate and H^+ concentrations increased, but Mn, Na, Zn and Al concentrations rose as well. Bacterial sulfate reduction in sediments, a process that generates alkalinity through the reduction of sulfate to sulfide and permanent storage of reduced S in sediments, increased, partially mitigating the acidification of the lake. This process was reflected in depletion of sulfate (Fig. 2.1) and a lack of decrease of pH in the hypolimnion. As discussed below, a number of bacterial processes counteract acidification through alkalinity generation. In Lake 223, sulfate reduction in both hypolimnetic and epilimnetic sediments slowed acidification by about one-third relative to the rate calculated from the natural alkalinity of the lake plus acid additions (14).

The rates of basic microbial processes such as primary production, organic matter decomposition, and nutrient regeneration do not appear to be seriously affected until the pH of a water body is well below that which would affect higher organisms (15). The effect of acid deposition and acidification on bacterial communities is more often a change in species composition or a change in the rate of use of various electron acceptors among sediment bacteria rather than a reduction in biomass or overall metabolic rate. For example, lake acidification with sulfuric acid can increase the rate of bacterial sulfate reduction in sediments without changing the overall rate of organic matter decomposition within sediments (15). Fish reproduction can be reduced at pH values just above 6 (7),

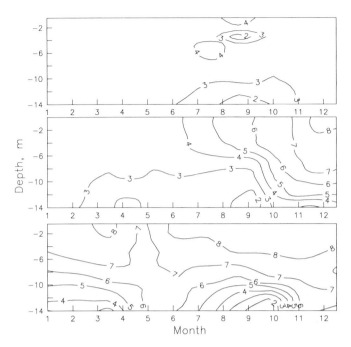

Fig. 2.1. Sulfate distribution in experimentally acidified Lake 223, 1974–1977. The upper panel is an isopleth map for a preacidification season, 1974. Acidification with sulfuric acid began in 1976. The lower two panels are 1976 and 1977, respectively. Data are given in mg SO_4^{2-} L^{-1}. (Adapted from Schindler et al. [14].)

whereas microbial organic matter decomposition is generally not significantly reduced unless the pH is below 5.0 (15–17). When considering the effects of pH, it is important to remember that a lake at pH 5.0 is 10 times more acidic than one at pH 6.0.

The most important in-lake effects of acid deposition on bacteria may be changes in the bacterially mediated cycling of nitrate and sulfate. Microbial uptake and reduction of sulfate, nitrate, and other electron acceptors including Fe and Mn are alkalinity-generating reactions that can mitigate the effects of acid deposition on pH change. The sum of all bases that can be titrated with strong acid is referred to as *acid-neutralizing capacity* (ANC). Sulfur and nitrogen cycles in acid-impacted waters are discussed in more detail below. Other effects of acidification on bacterial processes that are reviewed include nitrate stimulation of primary production in systems where it is the limiting nutrient; alteration in bacterial processes that play a role in the biogeochemistry of metals; and decreased bacterial activity, including decreased rates of organic matter decomposition and secondary production at more extreme pH. The effects of acid deposition on microbial processes within natural waters have been studied at many levels, from cellular to ecosystem. Because of complex interactions between ecosystem components, and because changes that result from acid deposition frequently appear at the population level, many experimental studies have been conducted using microcosm, experimental lake or stream, and field approaches. Results from studies using these approaches are emphasized in this chapter.

2. MICROBIAL PROCESSES AFFECTING ALKALINITY

Whole-lake acidification experiments and mass-balance studies of atmospherically acidified waters have shown that the amount of acid needed to deplete alkalinity and to depress pH is greater than expected based on existing alkalinity (14,18–20). In many of the experimental lakes studied (small lakes with relatively small watersheds), this phenomenon is mainly a result of biological alkalinity generation within the lakes (21,22). Alkalinity generation in watersheds (*e.g.*, weathering, ion exchange, and biological reactions) becomes more important as the size of the watershed relative to the lake increases. Biological processes may consume or produce alkalinity within lakes, although the net effect in acidified waters is usually alkalinity generation. The acid–base chemistry of some of these reactions is shown in Table 2.1.

Because carbonate acts as the primary buffer for most natural waters, reactions that produce or consume CO_2 may alter the pH temporarily, until equilibrium with atmospheric CO_2 is reestablished. Organic carbon oxidation by heterotrophic bacteria using oxygen as the electron acceptor and CO_2 fixation by phototrophs are examples of reactions that alter CO_2 balance (23). Respiration and photosynthesis can produce enough excess CO_2 or remove enough CO_2, respectively, to alter water pH temporarily by as much as 2 to 3 pH units from neutral. Increased pH is frequently seen in surface waters during summer algal blooms, whereas depressed pH may occur because of heterotrophic metabolism under ice cover when equilibrium with the atmosphere is blocked.

Reactions in which nitrate or sulfate are reduced, through either assimilatory or dissimilatory processes, consume H^+. Assimilation is the dominant process in the oxic water column, where these ions act as nutrients and are assimilated during carbon fixation (24). In dissimilatory reduction, sulfate and nitrate act as electron acceptors during the oxidation of organic matter by bacteria. Dissimilatory sulfate reduction and denitrification are anoxic processes and occur mainly in sediments. Mineralization of organic nitrogen to ammonium also produces alkalinity. Conversely, oxidation of reduced sulfur or ammonia generates acidity

**TABLE 2.1. In-Lake and In-Stream Microbial Processes That
Modify the Acid–Base Chemistry of Surface Waters**[a]

Process	Product	Ratio of ΔANC (Eq)/ Δreactant (mole)
Nitrogen cycle		
Ammonium assimilation	Acidity	$-14/16$[b]
Nitrate assimilation	Alkalinity	$+18/16$[b]
Denitrification[c]	Alkalinity	$+92.3/94.4$[b]
Nitrification	Acidity	$-2/1$
Ammonification	Alkalinity	$+1/1$
Sulfur cycle		
Dissimilatory sulfate reduction	Alkalinity	$+120/53$
Sulfate assimilation		
Ester sulfate formation	Alkalinity	$+1/1$
Reduction to organic sulfides	Alkalinity	$+1/1$
HS-oxidation	Acidity	$-1/1$
Iron cycle		
Fe(II) oxidation to Fe(OH)$_3$	Acidity	$-1/1$
Fe(III) oxidation	Alkalinity	$+57/53$

[a]Adapted from Turner et al. (8).
[b]The ratio deviates from 1 because of the presence of HPO_4^{2-} as either a product or
reactant.
[c]With oxidation of organic nitrogen.

so that permanent alkalinity generation is ony achieved when sulfide is fixed in
sediments or nitrate is reduced to N_2, as during denitrification.

Other than atmospheric disposition, a major source of nitrate in lakes is the
oxidation of NH_4^+ through microbial nitrification. This autotrophic process appears to be inhibited at pHs somewhat higher than those of most other microbial
processes. During whole-lake acidification studies, nitrification ceased at an
average lake pH of 5.4 to 5.7 (25), resulting in accumulation of NH_4^+ and lack
of NO_3^- build-up under ice during winter. Although nitrification is an acidity-
generating process, its blockage also results in decreased substrate for denitrification, which generates permanent alkalinity through the reduction of NO_3^- to N_2
gas. The role of the N and S cycles in lake acidification are discussed in more
detail below.

On an areal basis, these within-lake processes can generate more alkalinity
than watershed processes (19,26,27). The relative importance of each depends
on the size of a lake relative to its watershed, weathering rates in the watershed,

and the hydraulic residence time of the lake (21,28). Because most in-lake alkalinity generation in acidified lakes occurs in sediments (see below), its relative importance in overall alkalinity generation increases when water and sediments are in contact for longer times and in shallow lakes where the ratio of sediment surface area to water volume is high. In-lake ANC generation is particularly important in seepage lakes (29). In-lake (stream or river) processes are of less importance in lakes with large or calcareous watersheds, or short hydraulic residence times, and in streams and rivers.

2.1. Water Column Processes Affecting Alkalinity Generation

The biological processes most likely to influence acid–base chemistry in the water column are nitrate, sulfate, and ammonium assimilation and ammonium and sulfide oxidation. Sulfate and nitrate assimilation generate alkalinity, whereas the other processes generate acidity (Table 2.1). The assimilatory processes are often limited by the availability of other nutrients. Phosphate generally limits algal growth in freshwaters, whereas S and N are in excess, so that additions of sulfate and nitrate from acid deposition do not result in increased algal production (30–32). Sulfate assimilation is only important to overall alkalinity generation in very low sulfate lakes or in eutrophic lakes where sulfate becomes limiting (29,33). The same is true for nitrate and ammonium assimilation in most freshwaters. For both sulfate and nitrate, the fraction of added ion that results in water column assimilation and hence water column alkalinity generation diminishes as their concentration increases (22,31,34). Microbial oxidation of NH_4^+ (nitrification) or sulfide is limited by the rate of production or input of these ions to the lake. Dissolved sulfide produced by sediment sulfate reduction may diffuse out of sediments and be reoxidized either chemically by O_2 or microbially in the water column. The relative rates of sulfate reduction and sulfide oxidation determine the overall amount of alkalinity derived from microbial sulfur cycling; these important processes are discussed in more detail below.

In contrast to the situation in many freshwaters, N is often the limiting nutrient in marine and estuarine ecosystems (35). In these systems, atmospheric N deposition can stimulate primary production—a potential problem in many coastal waters that are already eutrophied. Addition of rainfall to estuarine waters from coastal North Carolina stimulated primary production in some seasons (36,37) (Fig. 2.2). However, the importance of atmospheric N deposition to estuarine eutrophication is speculative at this time. It has been estimated that 20%–30% of the annual N load to Chesapeake Bay is atmospheric (38), although it is very difficult to assess what fraction of N deposited on land reaches the estuary. Nevertheless, direct deposition of N onto the surface of the bay alone is equivalent to approximately 8% of its annual N loading. Nitrate loadings to U.S. east coast estuaries, the Gulf of Mexico, and the Great Lakes (which are likely

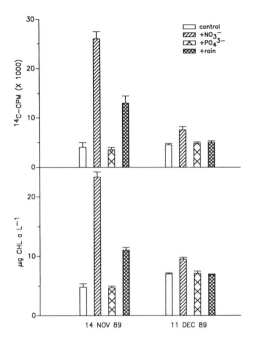

Fig. 2.2. Bioassay experiments with nutrient and precipitation additions to Neuse River (North Carolina) estuary water. **Top:** Phytoplankton $^{14}CO_2$ assimilation in counts per minute averaged over a 4 day experiment. **Bottom:** Chlorophyll a concentrations averaged over a 4 day experiment. Control, no nutrient addition; $+NO_3^-$, 14.3 μM NO_3^- added; $+PO_4^{3-}$; 3.2 μM PO_4^{3-} added; +rain, addition of rainwater collected just prior to incubations. Numbers above bars indicate percent rain added to treatments. (Adapted from Paerl et al. [37].)

associated with atmospheric nitrate deposition) have increased in recent years, whereas P loadings have shown no clear trend (39).

In lakes impacted by acid deposition, microbial alkalinity generation in the water column is quantitatively of less importance than in sediments (18,32). While the assimilatory alkalinity-generating reactions in the water column are limited by concentrations of other nutrients, dissimilatory reductions of sulfate and nitrate in sediments are functions of the concentrations of these ions and their flux rate into sediments.

2.2. Sediment Microbial Processes That Alter Acid–Base Chemistry

Early diagenesis of organic matter in sediments takes place via oxic and anoxic bacterial respiration, which consumes substantial amounts of various electron acceptors. Organic matter is also degraded via microbial fermentations, which may either produce or consume acidity through the production and consumption of various organic acids (40). Oxygen is the most energetically favorable electron acceptor, and it is consumed quickly near the sediment water interface. NO_3^-, MnO_2, $FeO(OH)$, SO_4^{2-}, and CO_2 are sequentially less favorable and are depleted in the above order with depth in sediments (41). Microbial reduction of MnO_2, $FeO(OH)$, NO_3^- and SO_4^{2-} all generate alkalinity (Table 2.1); respiration of organic matter with oxygen as the electron acceptor and methanogenesis neither generate nor consume alkalinity. Reduction of MnO_2, $FeO(OH)$, or SO_4^{2-} consumes 2 mol of H^+ for 1 mol of electron acceptor reduced, whereas denitrification consumes 1 mol H^+ per 1 mol nitrate (42).

These processes are competitive. An increase in the availability of a more favorable electron acceptor decreases the rate of utilization of less energetic acceptors. As a result, deposition of sulfate and nitrate in natural waters can result in increased sediment sulfate-reduction and denitrification rates at the expense of methanogenesis (22,41,43,44). This shift away from organic carbon degradation through methanogenesis results in increased alkalinity production within sediments, mitigating the rate of lake acidification. Alkalinity production in sediments is often manifested by increased pH in the static hypolimnion relative to the epilimnion (*e.g.*, 14,18) (see Fig. 2.1).

The rates of net ANC generation through sulfate reduction and denitrification generally increase with the concentrations of these ions in the overlying water. Their availability to the sediment microflora depends on their rate of flux across the sediment water interface, which is a function of the concentration gradient, molecular diffusion, bioturbation, and advection associated with groundwater flow (45). However, once alkalinity is generated through these reactions, it may be lost again through reoxidation of the reduced end product. Denitrification produces inert N_2 gas, which equilibrates with atmospheric N_2, and therefore results in permanent alkalinity generation. Sulfate reduction, however, results in geochemically active sulfides, which may diffuse into surface sediments and overlying water and be reoxidized and which often undergo seasonal oxidation within sediments.

2.2.1. Sulfate Reduction

Although algal assimilation may be the major sulfate-removal process in lakes that do not receive substantial sulfate deposition, dissimilatory reduction in sediments is primarily responsible for in-lake sulfate removal in sulfuric acid–acidified lakes. Researchers have found this to be the case in both experimentally and atmospherically acidified lakes (18,28,31,32). In both types of lakes, sediment sulfate reduction increases substantially with increasing lake water sulfate concentrations (13,31,29,46,47), while algal sulfate assimilation becomes saturated and remains fairly constant (31).

Sulfate-reduction rates in freshwater sediments are typically limited by sulfate concentration rather than by organic substrates (48–50). This is in contrast to estuarine and marine sediments, where sulfate-reduction rates may be limited by the availability of organic matter in sediments (41). Most studies have shown that sulfate-reduction rates in freshwater sediments are first order in water-column sulfate concentration (21,30,47,51), suggesting that flux rate into sediments limit sulfates reduction. Freshwater sulfate concentrations typically range from 20 to 200 μM, increasing with sulfate deposition or eutrophication. There is a lower limit sulfate concentration beneath which sulfate-reducing bacteria cannot compete for organic substrates. Various laboratory and field studies have estimated this concentration to be < 5 to 30 μM (41,51).

The end product of sulfate reduction is sulfide, which is chemically reactive, forming solid and dissolved complexes within sediments. Some of these products can be reoxidized upon diffusion out of sediments into oxic waters or if sediments become reoxidized during winter. The amount of permanent alkalinity generated by sulfate reduction is therefore affected by the extent of these reoxidation processes, since they generate acidity, neutralizing the alkalinity generated by the reductive process. The sulfidic end products formed through microbial sulfate reduction appear to vary among ecosystems and with sediment type. The two most important classes of end products other than dissolved HS^- are iron-sulfur and organic S compounds. Formation of both stabilizes reduced sulfur within sediments.

Total solid S concentrations in lake sediments often increase toward the surface (in sediments deposited since 1900), mimicking increases in atmospherically deposited sulfate (*e.g.,* Adirondack lakes [52]; Sudbury, Ontario lakes [53]; Little Rock Lake, WI [54]; Quebec lakes [55]; and New England lakes [56]). Much of this increase is often due to inorganic sulfides (54–57). Total and inorganic S concentrations are generally higher in hypolimnetic sediments, which remain reduced for more of the year, than in epilimnetic sediments (54–56). However, increased primary production through eutrophication can also lead to increases in sediment S concentrations because of increased deposition of S-containing organic matter (56).

Either organic or inorganic sulfides may predominate in lake sediments. Organic sulfides arise from deposition of organic matter to the sediment surface, incorporation of sulfate into ester sulfates within sediments (58), and diagenesis of inorganic sulfides within sediments (59). David and Mitchell (59) found that ester sulfates are the dominant constituent of seston S, followed by C-bonded S. Ester sulfates deposited on sediments undergo fairly rapid remineralization (59), whereas C-bonded organic sulfides appear to be fairly persistent within sediments (60). Assimilation of sulfate into ester sulfates and their remineralization do not appear to affect alkalinity; assimilation of sulfate into C-bonded organics is an alkalinity-generating reaction, but alkalinity is lost upon remineralization.

Although dissimilatory sulfate reduction does not produce organic sulfides directly, they may be formed indirectly by the attack of HS^- on organic molecules. This has been implied from an increase in C-bonded organic sulfur compounds with depth in sediments (57) and from stable S isotope profiles (53). The extent of organic-S formation from the end products of sulfate reduction varies among and within lakes, however. In long-term incubations of sediment cores with $^{35}SO_4^{2-}$ from an ELA lake, Rudd et al. (60) found that organic S was often the major end product of sulfate reduction. Landers and Mitchell (58) obtained the same result in sediment of an Adirondack lake, with the major end product being ester sulfates. In a number of lakes, however, little of the sulfide formed through dissimilatory sulfate reduction is incorporated into organic sulfur (56,60).

Solid inorganic sulfur forms are dominated by pyrite (FeS_2) and "acid-volatile sulfides," (AVS) which consist primarily of iron monosulfides. Elemental sulfur makes up a smaller fraction of the solid reduced inorganic sulfur pool. Pyrite is generally the dominant form of reduced inorganic S (56,60). Iron availability can limit the formation of Fe–S compounds in some lakes, particularly in lakes with high organic carbon content in sediments or low rates of new Fe supply (29,55,56), although Fe is in excess in many lake sediments. The conventional paradigm for sulfur diagenesis, that HS^- is immobilized initially as AVS, followed by pyrite formation through diagenesis, has changed substantially in the past few years. It is now recognized that AVS, pyrite, and C-bonded organic S compounds all may be important immediate products of dissimilatory sulfate reduction.

Organic sulfides seem to be more stable in sediments than are inorganic forms (60). Among the latter, iron monosulfides are more readily oxidized than is pyrite (FeS_2). Factors that influence the relative amount of inorganic and organic sulfide formation include lake water pH, trophic status, sulfate loading, and the amount of iron available to form complexes. Low pH, and high sulfate and Fe concentrations seem to favor inorganic sulfide formation (29,60). Organic S may be a more important end product of sulfate reduction in systems like bogs, where inorganic S concentrations are relatively low compared with organic matter (61,62). In marine sediments, in which dissimilatory sulfate reduction is quantitatively more important than in freshwaters, and carbon/sulfur ratios are much lower than in most freshwater sediments, organic S compounds contribute minimally to the total reduced S (41,52). Although there is a relationship between S storage in sediments and sulfate concentration in overlying water among sets of lakes with constrained characteristics (28,29), there appears to be no clear relationship among lakes with a wider range of trophic status, or size (56).

2.2.2. Denitrification and Other Processes Producing Reduced N

In a manner similar to sulfate reduction in lakes acidified by sulfuric acid, denitrification is primarily responsible for nitrate removal in nitric acid acidified lakes (31). Denitrification also becomes an increasingly important contribution to total ANC production as nitrate concentrations increase. Rudd et al. (31) estimate that denitrification becomes significant at NO_3^- concentrations above about 1 μM. At this level, nitrate loading generally exceeds that utilized by algae in oligotrophic waters, and excess nitrate becomes available for denitrification. Although denitrification itself is relatively insensitive to acidity (63), denitrification may be slowed in acidified lakes through blockage of another part of the N cycle, nitrification, which generates NO_3^- from NH_4^+ (25).

NH_4^+ is regenerated from sediments through remineralization of organic matter as well as from dissimilatory nitrate reduction. Both processes generate ANC, although it may be temporary because of oxidation of ammonium in the water column. Also called *ammonification,* dissimilatory nitrate reduction by some bacteria produces NH_4^+ (64), an end product that may be reoxidized. Since ammonium oxidation produces acidity, nitrate reduction to NH_4^+ results in less net alkalinity generation than denitrification. However, denitrification to N_2 appears to be quantitatively much more important in sediments than nitrate reduction to NH_4^+ (31,65). Ammonium remineralization and release from sediments may contribute significantly to ANC generation in systems where algal uptake of N and subsequent sedimentation is important. An example is Dart's Lake in the Adirondacks (USA), where denitrification contributed 45% and ammonium release from sediments contributed 32% of ANC generation in the hypolimnion (19).

2.2.3. Relative Efficiency of Sulfuric and Nitric Acid Acidification

The efficiency of either sulfate reduction or denitrification in alkalinity generation from atmospherically deposited acids depends on the concentration of the anion, on the hydraulic residence time of the lake, and on its mean depth (21,28). In shallow lakes and in those with long residence times, there is more opportunity for nitrate and sulfate reduction to remove these ions before flushing. For example, sulfate reduction is much more important in controlling lake water alkalinity in lakes with residence times of years than in lakes with shorter water residence times (28,56).

Although sulfate reduction consumes $2H^+$ per 1 mol whereas denitrification consumes only 1, denitrification often results in more permanent alkalinity generation on a molar basis than sulfate reduction, since the end product of denitrification is generally not reoxidized. Nitric acid is consumed more rapidly with depth in lake sediments than sulfate (60) (Fig. 2.3). Denitrification is energetically more favorable than sulfate reduction and should out-compete sulfate

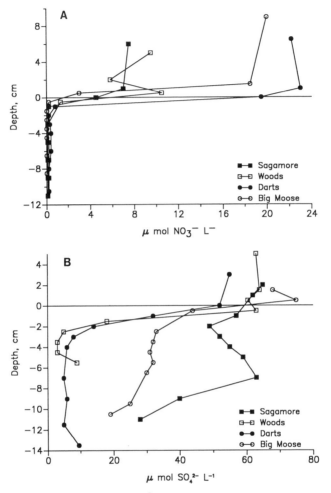

Fig. 2.3. Concentrations of NO_3^- (**A**) and SO_4^{2-} (**B**) in the pore water of epilimnetic sediments of four atmospherically acidified (lake water pH 4.7–5.5), softwater Adirondack lakes. (Reprinted from Rudd et al. [60], with permission of the publisher.)

reducers for organic substrate until nitrate is depleted (41). The relative importance of the two processes in a lake depends mainly on the relative concentrations of the two anions, but also on trophic status, and the efficiency of S storage in sediments. Sulfate is the major strong acid in rain in the eastern United States, Canada, and Europe, although the percentage of nitrate is increasing (6) as sulfate emission controls have been imposed. Since sulfate deposition exceeds

TABLE 2.2. **Midsummer Nitrate Concentrations, Percentages of Nitrate Removed Within the Lake, Water Residence Times (t_w), and Mass Transfer Coefficients (S_N) for Lakes Located in the Experimental Lakes Area (302 North, 302 South, 223, and 227), in Northern Wisconsin (Crystal), in Southern Ontario (Plastic and Harp), and in Southern Norway (Langtjern)**[a]

Lake	Years	Midsummer NO_3^- (μmol L^{-1})	% NO_3^- removed	t_w (year)	S_N mass (year^{-1})
302 South (pre-acid)	1981	0.13	120	8.3	440
302 North (pre-acid)	1981	0.14	130	5.8	220
227	1971–1982	0.45	100	4.1	210
Crystal	1984	ND	99	25	42
223	1976–1983	<0.4	98	8.7	25
Plastic	1984–1986	<0.15	81	3.0	11
239	1981–1983	0.2	57	2.5	6.8
Harp	1984–1986	<0.15–1.2	57	2.5	6.8
Langtjern	1972–1978	2	36	0.2	6.8
302 South (acid added)	1982–1985	0.8–1	89–93	8.3	6.9
302 North (acid added)	1982–1985	20–40	69	5.8	5
Dart's	1982–1984	20	7	0.6	0.89

[a]Lakes are listed in order of decreasing values of S_N. Adapted from Kelly et al. (71).

nitrate, and watersheds (at last in the northeastern United States) retain NO_3^- better than SO_4^{2-} (66,67), this suggests that sulfate reduction rather than denitrification is the more important contributor to in-lake ANC generation in most lakes. Many acid-impacted lakes have insignificant nitrate concentrations, as the amount of loading is less than that required by algae. In these lakes, sulfate reduction is the dominant ANC generation process (8).

However, in lakes with significant nitrate concentrations, denitrification may be the dominant ANC-generating process. Nitrate retention efficiency in watersheds decreases with loading and length of exposure (68). Once saturated, nitrate runoff to lakes becomes important, and nitrate concentrations may be sufficient for denitrification to become important in ANC generation. In the northeastern United States, for example, export of nitrate from watersheds is important and contributes to elevated nitrate concentrations in surface waters (66). Nitrate levels are elevated sufficiently to stimulate sediment denitrification in many Adirondack lakes (19,69,70). In a comparative study of a number of naturally and experimentally acidified lakes with significant nitrate concentrations, Rudd et al. (63) found that H^+ consumption through denitrification exceeded consumption by sulfate reduction by factors of 1 to 6.

Using roughly equimolar additions of nitric and sulfuric acids to two separate experimental lakes, Rudd et al. (31) found that the acidification efficiency of

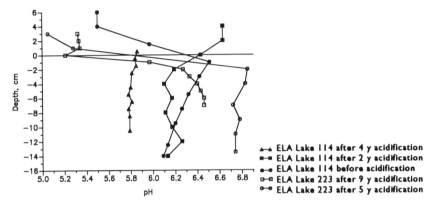

Fig. 2.4. Pore water pH profiles in epilimnetic sediments of experimentally acidified Lakes 223 and 114. (Reprinted from Rudd et al. [63], with permission of the publisher.)

nitrate was about 50% of that of sulfate. Sulfuric acid was probably a more efficient acidifier because 1) not all sulfate reduction results in permanent alkalinity generation because of reoxidation of some sulfides and 2) a larger fraction of added nitrate (70%) was removed from the nitric acid–acidified lake than sulfate was removed from the sulfuric acid–acidified lake (57%). However, neither sulfate reduction nor denitrification is 100% efficient in removing sulfate or nitrate from lakes (28,71). Once the supply of these anions exceeds algal needs, water column concentrations increase, and ANC is depleted, despite removal and ANC generation by sediment bacterial processes. This relationship is illustrated by the decreasing fractions of NO_3^- removal in a set of similar lakes of increasing NO_3^- concentration (Table 2.2).

2.2.4. Effect of Acidity on Sulfate Reduction and Denitrification

Neither of these alkalinity-generating processes appears to be significantly inhibited by the pH of most acidified waters. Both sulfate reduction and denitrification take place at relatively rapid rates even in lakes with a water column pH as low as 4.5 (63). Because of the alkalinity-generating processes within sediments discussed above, the pH of sediment pore waters is often near neutral, even in low pH lakes, as shown in Figure 2.4. (47,72). However, pH values near the sediment–water interface may be lower, and pore waters may become acidic even at depth after long exposure to highly acidic waters (60,63). Nevertheless, both denitrification and sulfate reduction seem relatively insensitive to low pH. Short-term experimental acidification of sediments to pH 4.5 does not reduce the rate of either process (63), although inhibition of nitrification at low pH may alter N cycling within sediments (25).

3. EFFECT OF ACIDIFICATION ON ORGANIC MATTER DECOMPOSITION

Although the organisms responsible for decomposition may vary with pH and the anionic content of a natural water, the overall rate of decomposition of organic matter appears to be affected minimally by pH above approximately 5.5. Decomposition is certainly less affected by pH declines than are processes accomplished by higher forms of life. However, changing patterns of decomposition may in turn affect alkalinity, since organic acids contribute substantially to ANC in some waters (73–75). Unfortunately, the degradation pathways for organic matter in natural waters and sediments are not well understood, nor is the influence of changes in these pathways on alkalinity.

Organic matter may build up at the sediment surface in acidified waters. However, this is thought to be due to pH-induced chemical flocculation of dissolved organic carbon (DOC) rather than decreased rates of decomposition in most ecosystems (8). The concentration of DOC in acidifying surface waters generally declines with pH, a process that has been seen in experimentally and atmospherically acidified lakes and streams (12,13,76). In an experimental whole lake study (ELA Lake 223), the total rate of organic matter decomposition in the lake, as measured by methane plus carbon dioxide accumulation in the hypolimnion or under ice, was unaffected by a pH as low as 5.1 (15). Part of the reason for this lack of decline is probably self-buffering of sediments, as discussed above. However, sediments and detritus adjusted to pH values as low as 5 to 5.5 do not demonstrate decreased decomposition rates (see below), suggesting that the microbial community can readily adapt to a pH above 5.

Nevertheless, microbial organic matter degradation is affected at higher acidity. Although a decrease in bacterial cell numbers in lakes is seldom seen, numbers of respiring bacteria or bacterial activity may decline. For example, bacterial sulfate reduction in sediments may decline below pH 4.5 in pore waters (63), although ANC generation by sulfate reduction generally holds pore pH closer to neutral except near the sediment surface. Decomposition has often been studied experimentally by examining colonization patterns and weight loss of leaf litter or more defined substrates, using litter bags placed in sediments or in the water column. In this type of study, decomposition is not generally slowed at a pH above about 5 (77,78). An example of typical data, in this case metabolic activity on leaf litter, is shown in Figure 2.5. Other examples include the following. Cellulose decomposition (as measured by weight loss of cellophane strips in litter bags) was lower in surface (0.5 cm depth) sediments of a pH 4.6 lake than in surface sediments of similar lakes with water a pH of 5.5 to 6.6 (79,80); and decomposition of newly sedimented organic carbon in experimentally acidified sediments declined below about pH 5, whereas decomposition of older organic matter was not decreased until a pH of 4.0 was reached (15).

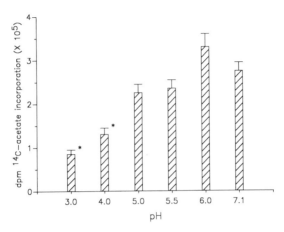

Fig. 2.5. Metabolic activity of the *Carex* litter microbiota measured as the rate of acetate incorporation into lipids after incubation for 2 days in water from arctic Toolik Lake initially adjusted to various pH values. $*P < 0.01$ compared with ambient (pH 7.1) controls. (Reprinted from McKinley and Vestal [16], with permission of the publisher.)

Epilithic communities may be more sensitive to acidity than sediment communities, since they are relatively unbuffered because of their shallow depths (17). In one stream study, epilithic bacterial communities were more sensitive to pH declines than epilithic algae; among the autotrophs, cyanobacteria were most affected by pH (81).

Acidity may shift the balance of decomposition between bacteria and fungi. It has been suggested that extracellular cellulases produced by fungi function better at low pH than bacterial cellulases, which may be associated with the cell surface (80,82). However, the importance of bacteria compared with fungi in low pH decomposition appears to vary among ecosystems and potentially with substrate. During the acidification (to pH 4) of a stream in the Hubbard Brook Experimental Forest, aquatic fungal hyphae covered most of the streambottom directly downstream of the acidification area, but were not noticeable in a reference area of the stream. Leaves collected from the reference stream had greater microbial species diversity and a larger number of spores than did leaves from the acidified stream (76). In acidic (pH 4.6) Bat Lake in Ontario, Hoeniger (49) found that cellophane was first colonized by fungi, then by *Cytophaga*-like gliding bacteria. However, in Harp Lake (pH 6.6), colonization was entirely bacterial. Conversely, colonization of leaf litter in pH-adjusted experimental microcosms from an arctic lake was not dominated by fungi (16).

Bacteria are more important in lignocellulose degradation in the naturally acidic Okefenokee Swamp than are fungi. Microbial lignocellulose degradation rates were lower in the acidic swamp (83) than in nearby circumneutral marshes, but total heterotrophic activity and production were not (84), suggesting that under acidic conditions bacteria rapidly process labile substrates, while less labile organics accumulate as peat. The authors (83,84) suggest that macromolecular degradative processes that rely on extracellular enzymes are depressed at low pH, while utilization of low molecular weight DOC is not (85). Lignocellulose degradation is often N limited, and might therefore be stimulated by N deposition (83).

4. CHANGES IN METAL BIOGEOCHEMISTRY

Acidification of natural waters and watersheds alters the biogeochemistry of a number of metals (86). Acid-induced mobilization or solubilization of toxic metals (*e.g.*, Al) can be a major contributor to the deleterious effects of acidification (*e.g.*, 87). Seasonal bacterial mediation of pH can influence the cycles of these metals. ANC generation by bacterial processes may decrease the potential for mobilization; however, ANC generation below the hypolimnion in summer may also lead, for example, to formation of Al hydroxides, to which some fish species may be particularly sensitive (88). Redox changes as a result of bacterial activity may affect the solubility of metals. Addition of sulfate or nitrate to lakes may increase the duration or extent of an anaerobic hypolimnion (14), leading to increased release of soluble reduced metals species (*e.g.*, Al, Fe, Mn, Zn, Cu) from sediments. The solubility and speciation of metals is also influenced by the concentration of organic ligands (86) and by the availability of solid phases for adsorption (89), both of which may be influenced by bacterial metabolism.

While many changes in metal speciation and solubility are induced directly by pH, changes in the relative activities of various bacterial groups that occur because of acid deposition can also alter the biogeochemical cycling of metals. The cycle of Hg in particular is a subject of considerable current interest because of the increased bioaccumulation of methylmercury in fish from lakes impacted by acid deposition. Methylmercury is a potent human neurotoxin, to which the fetus is particularly sensitive (90). Methylmercury concentrations in fish from acidified freshwater lakes are often elevated above guidelines for human consumption, and health hazard advisories have been released for a large number of lakes in eastern Canada, the north–central and northeast United States, Sweden, and Finland. Although atmospheric deposition of mercury has increased since preindustrial times (91), increased mercury deposition alone does not account for increased mercury bioaccumulation in acid lakes. Experimental lake acidification without increased mercury deposition has resulted in higher mercury burdens in fish (92). Although almost all of the mercury deposited from the atmosphere is

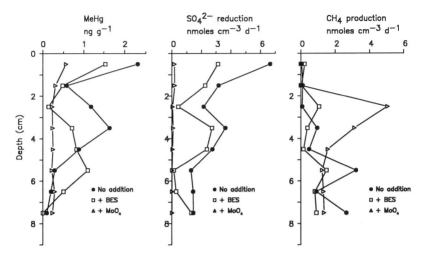

Fig. 2.6. Sediment depth profiles of methylmercury (MeHg) produced from 1 μg added Hg (as HgCl₂) cm⁻³, sulfate reduction and methane production in the presence and absence of specific inhibitors of dissimilatory sulfate reduction and methanogenesis. Inhibitors were injected at 1 cm depth intervals into intact cores to final calculated pore water concentrations of 40 mM BES or 20 mM Na₂MoO₄. Rate measurements were begun 12 hours after addition of inhibitors to allow for diffusion. Hg was injected at 1 cm intervals, and MeHg concentrations in 1 cm sediment depth horizons were measured after 12 hours incubation at about 23°C. MeHg data are averages of two analyses, with standard error between analyses of <0.2 ng g⁻¹. Sulfate-reduction rates were estimated by injection of Na₂³⁵SO₄ into intact cores followed by measurements of acid-volatile sulfides in each depth horizon after 12 hours incubation. Methane production was estimated by the increase over time in headspace CH₄ in vials containing sediment subsamples from each depth.

in inorganic form (93), mercury accumulates in the biota predominantly as methylmercury (94).

The production of methylmercury has long been considered primarily a biological process (95), occurring within lakes via the bacterial methylation of inorganic mercury. However, the cause of acceleration of net methylmercury production or bioaccumulation in these lakes is poorly understood. Although many microorganisms display mercury methylating activity in pure culture (95), field studies have indicated that metal methylation occurs most rapidly in anoxic sediments, in the presence of active microbial sulfate reduction (96,97). Specific inhibition of sulfate reduction, using molybdate, blocks mercury methylation in both estuarine and freshwater sediments (94,98). This is illustrated in Figure 2.6. Isolated sulfate-reducing species are capable of active methylation of both mercury and tin (97,99). In freshwater sediments, where sulfate reduction is generally sulfate limited, experimental additions of sulfate to either sediment slurries

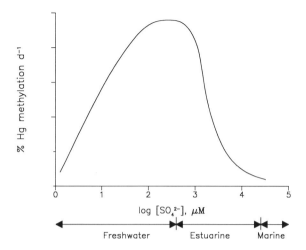

Fig. 2.7. Theoretical relationship between water column sulfate concentration and mercury-specific methylation rate in sediments. Methylation rate is unitless since *in situ* rates have not been measured. (Reproduced from Gilmour and Henry [94], with permission of the publisher.)

or intact lake sediments results in increased microbial production of methylmercury (94). However, high pore-water sulfide concentrations appear to inhibit methylation, effectively limiting methylmercury production in saline sediments (99,100).

The methylmercury concentration in a sediment represents the equilibrium between methylmercury production and degradation. Demethylation of methylmercury is also a microbial process in aquatic systems (101). Unlike the mechanism of methylation, which remains largely unknown, demethylation is well understood as an inducible, enzymatic process (95,102). It appears to be maximal near the sediment–water interface, favored at higher redox potentials than methylation, and less affected by acidification than methylation (103).

Because sulfate-reducing bacteria appear to be important metal methylators, it has been postulated that increased levels of sulfate in freshwater ecosystems will also result in increased rates of transformation of inorganic mercury to methylmercury (94). Based on our work and on that of other investigators, we have suggested that there is an optimal sulfate concentration for mercury methylation by sulfate-reducing bacteria in sediments (Fig. 2.7). Production of sulfide would inhibit methylation above this optimal sulfate concentration, whereas at lower sulfate levels microbial sulfate-reduction and hence mercury methylation would be limited by available sulfate. The optimal sulfate level for methylmercury production would vary somewhat among sediments as a function of other factors that affect sediment sulfate-reduction rates and sulfide concentrations, particularly the average sediment temperature, organic carbon, and iron content. This hypothesis provides an explanation for the relationship between acid deposition to freshwaters and bioaccumulation of mercury in the biota and suggests that sulfate deposition may adversely affect human health by increasing biomethylation of mercury.

5. SUMMARY

Biogeochemical processes mediated by microorganisms are not adversely affected by the acidification of natural waters to the same extent as are the life cycles of higher organisms. Basic processes, *e.g.*, primary production and organic matter decomposition, are not slowed in moderately acidified systems and do not generally decline above a pH of 5. More specifically, the individual components of the carbon, nitrogen, and sulfur cycles are, with few exceptions, also acid resistant.

The influence of acid deposition on microbial processes is more often stimulation of nitrogen and sulfur cycling, often leading to alkalinity production, which mitigates the effect of strong acid deposition. Bacterial sulfate reduction and denitrification in sediments are two of the major processes that can be stimulated by sulfate and nitrate deposition, respectively, and result in ANC generation.

One of the negative effects of acid deposition is increased mobilization and bioaccumulation of some metals. Bacteria appear to play an important role, especially in mercury cycling, with acidification leading to increased bacterial methylation of mercury and subsequent bioaccumulation in higher organisms.

REFERENCES

1. Neary BP, Dillon PJ: Effects of sulphur deposition on lake water chemistry in Ontario, Canada. Nature 333:340–343, 1988.

2. Wright RF, Henriksen A: Chemistry of small Norwegian lakes, with special reference to acid precipitation. Limnol Oceanogr 23:487–498, 1978.

3. Henriksen A, Brakke DF: Increasing contributions of nitrogen to the acidity of surface waters in Norway. Water Air Soil Pollut 42:183–201, 1988.

4. Dillon PJ, Lusis M, Reid R, Yap D: Ten-year trends in sulfate, nitrate and hydrogen deposition in central Ontario. Atmos Environ 22:901–905, 1988.

5. Galloway JN, Schofield CL, Peters NE, Henry GR, Altwicker ER: Effect of atmospheric sulfur on the composition of three Adirondack lakes. Can J Fish Aquat Sci 40:799–806, 1983.

6. Likens GE, Wright RF, Galloway JN, Butler TJ: Acid rain. Sci Am 241:43–51, 1979.

7. National Research Council: Acid Deposition: Long-Term Trends. Washington, DC: National Academy Press, 1986.

8. Turner RS, Cook RB, Van Miegroet H, Johnson DW, Elwood JW, Bricker OP, Lindberg SE, Hornberg GM: Watershed and Lake Processes Affecting Surface Acid-Base Chemistry. NA-PAP Report 10. In: Acidic Deposition: State of Science and Technology. Washington, DC: National Acid Precipitation Program.

9. Cowling EB: Acid precipitation in historical perspective. Environ Sci Technol 16:110A–123A, 1982.

10. Gorham E, Underwood JK, Martin FB, Ogden III JG: Natural and anthropogenic causes of lake acidification in Nova Scotia. Nature 324:451–453, 1986.

11. Huckabee JW, Goodyear CP, Jones RD: Acid rock in the Great Smokies: Unanticipated impact on aquatic biota of road construction in regions of sulfide mineralization. Trans Am Fish Soc 104:677–684, 1975.

12. Dillon PJ, Reid RA, deGrosbois E: The rate of acidification of aquatic ecosystems in Ontario, Canada. Nature 329:45–48, 1987.

13. Schindler DW, Turner MA: Biological, chemical and physical response of lakes to experimental acidification. Water Air Soil Pollut 18:259–271, 1982.

14. Schindler DW, Wagemann R, Cook RB, Ruszczynski T, Prokopowich J: Experimental acidification of Lake 223, Experimental Lakes Area: Background data and the first three years of acidification. Can J Fish Aquat Sci 37:342–354, 1980.

15. Kelly CA, Rudd JWM, Furutani A, Schindler DW: Effects of lake acidification on rates of organic matter decomposition in sediments. Limnol Oceanogr 29:687–694, 1984.

16. McKinley VL, Vestal JR: Effects of acid on plant litter decomposition in an Arctic Lake. Appl Environ Microbiol 43:1188–1195, 1982.

17. Palumbo AV, Bogle MA, Turner RR, Elwood JW, Mulholland PJ: Bacterial communities in acidic and circumneutral streams. Appl Environ Microbiol 53:337–344, 1987.

18. Cook RB, Kelly CA, Schindler DW, Turner MA: Mechanisms of hydrogen ion neutralization in an experimentally acidified lake. Limnol Oceanogr 31:134–148, 1986.

19. Schafran GC, Driscoll CT: Comparison of terrestrial and hypolimnetic sediment generation of acid neutralizing capacity for an acidic Adirondack lake. Environ Sci Technol 21:988–993, 1987.

20. Schindler DW, Turner MA, Hesslein RH: Acidification and alkalization of lakes by experimental addition of nitrogen. Biogeochemistry 1:117–133, 1985.

21. Baker LA, Brezonik PL: Dynamic model of in-lake alkalinity generation. Water Resource Res 24:65–74, 1988.

22. Kelly CA, Rudd JWM, Cook RB, Schindler DW: The potential importance of bacterial processes in regulating rate of lake acidification. Limnol Oceanogr 27:868–882, 1982.

23. Wetzel RG: Limnology. Philadelphia: WB Saunders, 1975.

24. Redfield AC: The biological control of chemical factors in the environment. Am J Sci 46:206–226, 1958.

25. Rudd JWM, Kelly CA, Schindler DW, Turner MA: Disruption of the nitrogen cycle in acidified lakes. Science 240:1515–1517, 1988.

26. Schindler DW: The significance of in-lake production of alkalinity. Water Air Soil Pollut 30:931–944, 1986.

27. Schindler DW: Confusion over the origin of alkalinity in lakes. Limnol Oceangr 33:1637–1640, 1988.

28. Kelly CA, Rudd JWM, Hesslein RH, Schindler DW, Dillon PJ, Driscoll CT, Gherini SA, Hecky RE: Prediction of biological acid neutralization in acid-sensitive lakes. Biogeochemistry 3:129–140, 1987.

29. Baker LA, Urban NR, Brezonik PL, Sherman LA: Sulfur cycling in an experimentally acidified seepage lake. In Saltzman E, Cooper W (eds): Biogenic Sulfur in the Environment. Washington, DC: American Chemical Society, 1989, pp 79–100.

30. Cook RB, Schindler DW: The biogeochemistry of sulfur in an experimentally acidified lake. Ecol Bull 35:115–127, 1983.

31. Rudd JWM, Kelly CA, Schindler DW, Turner MA: A comparison of the acidification efficiencies of nitric and sulfuric acids by two whole-lake addition experiments. Limnol Oceanogr 35:663–679, 1990.

32. Schiff SL, Anderson RF: Limnocorral studies of chemical and biological acid neutralization in two freshwater lakes. Can J Fish Aquat Sci 44:173–187, 1987.

33. Cook RB: The Biogeochemistry of Sulfur in Two Small Lakes. Ph.D. dissertation, Columbia Univeristy, New York.

34. Schiff SL, Anderson RF: Alkalinity production in epilimnetic sediments: Acidic and non-acidic lakes. Water Air Soil Pollut 31:941–948, 1986.

35. Carpenter EJ, Capone DG: Nitrogen in the Marine Environment. New York: Academic Press, 1983.

36. Paerl HW: Enhancement of marine primary production by nitrogen-enriched acid rain. Nature 326:747–749, 1985.

37. Paerl HW, Rudeck J, Mallin MA: Stimulation of phytoplankton production in coastal waters by natural rainfall inputs: Nutritional and trophic implications. Mar Biol 107:247–254, 1990.

38. Fisher D, Ceraso J, Oppenheimer M: Polluted Coastal Waters: The Role of Acid Rain. Review. New York: Environmental Defense Fund, 1988.

39. Smith RA, Alexander RB, Wolman MG: Water quality trends in the nation's rivers. Science 235:1607–1615, 1987.

40. Gottschalk G: Bacterial Metabolism. New York: Springer-Verlag, 1979.

41. Capone DG, Kiene RP: Comparison of microbial dynamics in marine and freshwater sediments: Contrasts in anaerobic carbon metabolism. Limnol Oceangr 33:725–749, 1988.

42. Froelich PN, et al.: Early oxidation of organic matter in pelagic sediments of the eastern equatorial Atlantic: Suboxic diagenesis. Geochim Cosmochim Acta 43:1075–1090, 1979.

43. Mountford DO, Asher RA, Mays EL, Tiedje JM: Carbon and electron flows in mud and sandflat intertidal sediments at Delaware Inlet, Nelson, New Zealand. Appl Environ Microbiol 39:686–694, 1980.

44. Winfrey MR, Zeikus JG: Effect of sulfate on carbon and electron flow during microbial methanogensis in freshwater sediments. Appl Environ Microbiol 33:275–281, 1977.

45. Berner RA: Early Diagenesis: A Theoretical Approach. Princeton: Princeton University Press, 1980, 241 pp.

46. Holdren GR, Jr., Brunelle TM, Matisoff G, Whalen M: Timing the increase in atmospheric sulfate deposition in the Adirondack Mountains. Nature 311:245–247, 1984.

47. Kelly CA, Rudd JWM: Epilimnetic sulfate reduction and its relationship to lake acidification. Biogeochemistry, 1:63—77, 1984.

48. Carignan R: Quantitative importance of alkalinity flux from the sediments of acid lakes. Nature 317:158–160, 1985.

49. Kuivila KM, Murray JW: Organic matter diagenesis in freshwater sediments: The alkalinity and total CO_2 balance and methane production in the sediments of Lake Washington. Limnol Oceanogr 29:1218–1230, 1984.

50. Smith RL, Klug MJ: Reduction of sulfur compounds in the sediments of a eutrophic lake basin. Appl Environ Microbiol 41:1230–1237, 1981.

51. Lovely DR, Klug MJ: Model for the distribution of sulfate reduction and methanogenesis in freshwater sediments. Geochim Cosmochim Acta 50:11–18, 1986.

52. Fry B: Stable sulfur isotope distributions and sulfate reduction in lake sediments of the Adirondack Mountains, New York. Biogeochemistry 2:239–343, 1986.

53. Nriagu JO, Soon YK: Distribution and isotopic composition of sulfur in lake sediments of northern Ontario. Geochim Cosmochim Acta 49:823–834, 1985.

54. Brezonik PL, Baker LA, Perry TE: Mechanisms of alkalinity generation in acid-sensitive soft water lakes. In Hites RH, Eisenreich SJ Sources and Fates of Aquatic Pollutants. Washington DC: American Chemical Society, 1987, pp 229–260.

55. Carignan R, Tessier A: The co-diagenesis of sulfur and iron in lake sediments of southwest Quebec. Geochim Cosmochim Acta 52:1179–1188, 1988.

56. Giblin A, Likens GE, White D, Howarth RW: Sulfur storage and alkalinity generation in New England lake sediments. Limnol Oceanogr 35:852–869, 1990.

57. Owen JS, Mitchell MJ: Sulfur biogeochemistry of an acidic lake in the Adirondack region of New York. In Rao S (ed): Stress and Aquatic Microbial Interactions. Boca Raton, FL: CRC Press, 1988, pp 59–68.

58. Landers DH, Mitchell MJ: [35]Sulfate incorporation in the sediments of three New York lakes. Hydrobiologia 160:85–95, 1988.

59. David MB, Mitchell MJ: Sulfur constituents and cycling in waters, seston and sediment of an oligotrophic lake. Limnol Oceanogr 30:1196–1207, 1985.

60. Rudd JWM, Kelly CA, Furitani A: The role of sulfate reduction in the long term accumulation of organic and inorganic sulfur in lake sediments. Limnol Oceanogr 31:1281–1291, 1986.

61. Brown KA: Formation of anaerobic sulfur in an anaerobic peat. Soil Biol Biochem 18:131–140, 1986.

62. Weider RK, Lang GE: Cycling of inorganic and organic sulfur in peat from Big Run Bog, West Virginia. Biogeochemistry 5:221–242, 1988.

63. Rudd JWM, Kelly CA, St. Louis V, Hesslein RH, Furutani A, Holoka MH: Microbial consumption of nitric and sulfuric acids in acidified north temperate lakes. Limnol Oceanogr 31:1267–1280, 1986.

64. Erlich HL: Geomicrobiology. New York: Marcel Dekker, 1990.

65. Seitzinger SP: Denitrification in freshwater and coastal marine ecosystems: Ecological and geochemical significance. Limnol Oceangr 33:702–724, 1988.

66. Likens GE, Borman FH, Pierce RS, Eaton JS, Johnson NM: Biogeochemistry of a Forested Ecosystem. New York: Springer, 1977.

67. Lindberg SE, Lovett GM, Richter DD, Johnson DW: Atmospheric deposition and canopy interactions of major ions in a forest. Science 231:141–145, 1986.

68. Grennfelt P, Hultberg H: Effects of nitrogen deposition on the acidification of terrestrial and aquatic ecosystems. Water Air Soil Pollut 30:945–963, 1986.

69. Driscoll CT, Newton RM: Chemical characteristics of Adirondack lakes. Environ Sci Technol 19:1018–1024, 1985.

70. Driscoll CT, Yatsko CP, Ungast FJ: Longitudinal and temporal trends in the water chemistry of the north branch of the Moose River. Biogeochemistry 3:37–62, 1987.

71. Kelly CA, Rudd JWM, Schindler DW: Acidification by nitric acid—Concern for the future. Water Air Soil Pollut 50:49–62, 1990.

72. Herlihy AT, Mills AL: The pH regime of sediments underlying acidified waters. Biogeochemistry 2:377–381, 1986.

73. Driscoll CT, Bisogni JJ: Weak acid/base systems in dilute acidified lakes and streams of the Adirondack region of New York State. In Schnoor JL (ed): Modelling of Total Acid Precipitation Impacts. Boston: Butterworth, 1984, pp 53–72.

74. Eschleman KN, Hemond HF: The role of organic acids in the acid–base status of surface waters at Bickford Watershed, MA. Water Resource Res 21:1503–1510, 1985.

75. LaZerte BD, Dillon PJ: Relative importance of anthropogenic versus natural sources of acidity in lakes and streams of central Ontario. Can J Fish Aquat Sci 41:1664–1677, 1984.

76. Hall RJ, Likens GE, Fiance SB, Hendrey GR: Experimental acidification of a stream in the Hubbard Brook Experimental Forest, New Hampshire. Ecology 61:976–989, 1980.

77. Burton TM, Stanford RM, Allan JW: Acidification effects on stream biota and organic matter processing. Can J Fish Aquat Sci 42:669–675, 1985.

78. Carpenter J, Odum WE, Mills AL: Leaf litter decomposition in a reservoir affected by acid mine drainage. Oikos 41:165–172, 1983.

79. Hoeniger JFM: Microbial decomposition of cellulose in acidifying lakes of south-central Ontario. Appl Environ Microbiol 50:315–322, 1985.

80. Hoeniger JFM: Decomposition studies in two central Ontario lakes having surficial pHs of 4.6 and 6.6. Appl Environ Microbiol 52:489–497, 1986.

81. Mulholland PJ, Elwood JW, Palumbo AV, Stevenson RJ: Effect of stream acidification on periphyton composition, chlorophyll, and productivity. Can J Fish Aquat Sci 43:1846–1858, 1986.

82. Ljungdahl LG, Eriksson KE: Ecology of cellulose degradation. Adv Microbial Ecol 8:237, 1985.

83. Benner R, Moran MA, Hodson RE: Effects of pH and plant source on lignocellulosic biodegradation rates in two wetland ecosystems, the Okefenokee Swamp and a Georgia salt marsh. Limnol Oceanogr 30:489–499, 1985.

84. Murray RE, Hodson RE: Microbial biomass and utilization of dissolved organic matter in the Okefenokee Swamp ecosystem. Appl Environ Microbiol 47:685–692, 1984.

85. Benner R, Lewis DL, Hodson RE: Biogeochemical cycling of organic matter in acidic environments: Are microbial degradative processes adapted to low pH? In Rao SS (ed): Acid Stress and Aquatic Microbial Interactions. Boca Raton, FL: CRC Press, 1989, pp 33–46.

86. Nelson WO, Campbell PGC: Effects of acidification on the geochemistry of Al, Cd, Pb, Hg. Environ Pollut 72, 71:91–130, 1991.

87. Cronan CS, et al.: Aluminum toxicity in forests exposed to acid deposition: The ALBIOS results. Water Air Soil Pollut 48:181–192, 1989.

88. Schafran GC, Driscoll CT: Spatial and temporal variations in aluminum chemistry of a dilute acidic lake. Biogeochemistry 3:105–119, 1987.

89. Tessier A, Rapin F, Carignan R: Trace metals in oxic lake sediments: Possible adsorption onto iron hydroxides. Geochim Cosmochim Acta 49:183–194, 1985.

90. Clarkson TW: Human health risks from mercury in fish. Environ Toxicol Chem 9:957–961, 1990.

91. Rada RG, Wiener JG, Winfrey MR, Powell DE: Recent increases in atmospheric deposition of mercury in north-central Wisconsin lakes inferred from sediment analyses. Arch Environ Contam Toxicol 18:175–181, 1989.

92. Wiener JG, Fitzgerald WF, Watras CJ, Rada RG: Partitioning and bioavailability of mercury in an experimentally acidified Wisconsin lake. Environ Toxicol Chem 9:909–918, 1990.

93. Fitzgerald WF: Atmospheric and oceanic cycling of mercury. In Riley JP, Chester P (eds): Chemical Oceanography, Vol 10. London: Academic Press, 1989, pp 151–186.

94. Gilmour CC, Henry EA: Mercury methylation in aquatic systems affected by acid deposition. Environ Pollut 71:131–169, 1991.

95. Robinson JB, Tuovinen OH: Mechanisms of microbial resistance and detoxification of mercury and organomercury compounds: Physiological, biochemical and genetic analyses. Microbiol Rev 48:95–124, 1984.

96. Compeau G, Bartha R: Methylation and demethylation of mercury under controlled redox, pH and salinity conditions. Appl Environ Microbiol 48:1203–1207, 1984.

97. Gilmour CC, Tuttle JH, Means JC: Anaerobic microbial methylation of inorganic tin in estuarine sediments slurries. Microbial Ecol 14:233–242, 1987.

98. Compeau G, Bartha R: Sulfate-reducing bacteria: Principle methylators of mercury in anoxic estuarine sediment. Appl Environ Microbiol 50:498–502, 1985.

99. Compeau GC, Bartha R: Effect of salinity on mercury-methylating activity of sulfate-reducing bacteria in estuarine sediments. Appl Environ Microbiol 53:261–265, 1987.

100. Craig PJ, Moreton PA: Total mercury, methyl mercury and sulphide in River Carron sediments. Mar Pollut Bull 14:408–411, 1983.

101. Winfrey MR, Rudd JWM: Environmental factors affecting the formation of methylmercury in low pH lakes: A review. Environ Toxicol Chem 9:853–869, 1990.

102. Walsh CT, DiStefano MD, Moore MJ, Shewchuk LM, Verdine GL: Molecular basis of bacterial resistance to organomercurial and inorganic mercuric salts. FASEB J 2:124–129, 1988.

103. Korthals ET, Winfrey MR: Seasonal and spatial variations in mercury methylation and demethylation in an oligotrophic lake. Appl Environ Microbiol 53:2397–2404, 1987.

EFFECT OF ACID RAIN ON SOIL MICROBIAL PROCESSES

DAVID D. MYROLD

Department of Crop and Soil Science, Oregon State University, Corvallis, Oregon 97731

G.E. NASON

Alberta Environment, Lethbridge, Alberta, Canada, T1J HC7

1. INTRODUCTION

Concern about the environmental impact of acidic precipitation arose about 20 years ago when Scandinavian studies documented the changing chemical composition of precipitation (1). Research since that time has provided voluminous amounts of data documenting the effects of acid inputs on vegetation, soil chemistry, and soil biology. These data have been summarized in numerous books (2–8).

Our purpose in this chapter is to focus on how acidic precipitation affects soil biological processes. We will not make an extensive review of the literature, because that can be found elsewhere (*e.g.*, 9–16), but rather we will provide selected examples that illustrate the current understanding of the interactions between soil organisms and acidification processes. This will be done within the context of how soil organisms and their metabolic processes respond and adapt to environmental perturbations, both as individuals and as aggregated communities.

The framework for our discussions is illustrated by a conceptual model (Fig. 3.1) that shows that there are a number of intervening steps or interactions that occur before the effects of acidic deposition can be monitored by changes in

Environmental Microbiology, pages 59–81, © 1992 Wiley-Liss, Inc.

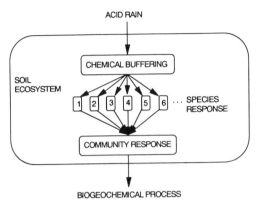

ACID RAIN

Fig. 3.1. Conceptual model of how acid rain inputs interact with soil components and the soil biota to produce changes in soil microbial processes.

microbial processes. First, the chemical composition of the acidic input must be known. Once acidic precipitation enters the soil ecosystem, it reacts with the soil matrix through a number of chemical buffering reactions, which in turn give rise to an altered soil solution chemistry. Different species of soil organisms are affected by the modified soil solution in varying ways. It is the collective responses of these individual species responses that determine how the soil community as a whole reacts to acid precipitation. It is the community response that is finally translated into changes in rates of biologically mediated nutrient cycle processes.

2. ACID RAIN COMPOSITION

Even in pristine environments, precipitation is not pure; it contains a variety of dissolved ions. Water in equilibrium with atmospheric concentrations of CO_2 has a pH of about 5.6 because of bicarbonate buffering. Additional dissolved anions (*e.g.*, Cl^-, NO_3^-, and SO_4^{2-}) and cations (*e.g.*, Na^+, K^+, NH_4^+, Ca^{2+}, and Mg^{2+}) are present in amounts that vary by location. The ionic composition of rainwater collected along a west-to-east transect across the United States illustrates these points (Fig. 3.2). Although the pHs of rainwater collected in Oregon and South Dakota are the same (range 5.36–5.34), there are significant differences in the concentrations of other ions. In Oregon, at the Alsea Ranger Station, which is in the Coast Range about 25 km from the Pacific Ocean, higher concentrations of Na^+, Mg^{2+}, and SO_4^{2-} are found than at the H.J. Andrews Experimental Forest, which is in the Cascade Range about 100 km from the Pacific Ocean. The higher salt concentrations nearer the coast arise from sea-

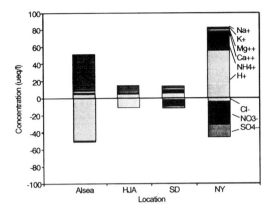

Fig. 3.2. Concentrations of important anions and cations in precipitation collected at locations along a west-to-east transect across the United States. Alsea and HJA (H.J. Andres Experimental Forest) are located in Oregon; SD is Cottonwood, South Dakota; and NY is Aurora Research Farm, New York. Data are from NADP/NTN (76,77).

water aerosolization. Concentrations of Na^+, Mg^{2+}, and Cl^- drop off even more in rainwater collected in South Dakota. Such variations in ion speciation and concentration in rainwater, particularly in NO_3^- and SO_4^{2-}, affect rainwater pH. Based on natural, local variations in N and S cycling, Charlson and Rodhe (17) have estimated that pH 4.5 is a reasonable lower limit for the pH of natural rainfall.

Besides natural variations in rainwater ion composition, there are demonstrated effects of anthropomorphic inputs. Human-related inputs manifest themselves as increased concentrations of H^+, NH_4^+, NO_3^-, and SO_4^{2-}. The increase in H^+ is brought about by the dissolution of N oxides, produced largely by internal combustion engines, and S oxides, largely released by fossil fuel burning. Much of the NH_4^+ seems to come from volatilization of N in animal excrement. Consequently, acidity, N, and S concentrations in rainfall increase as population densities increase. This trend is demonstrated by data for Oregon, South Dakota, and New York (Fig. 3.2).

Although low rainwater pH values spawned the term *acid rain* and produce eye-catching headlines, evaluating the potential impact of atmospheric deposition is probably best done by considering the annual inputs of all its components. In doing so, it is important to remember that, in many locations, dry deposition (particulates and aerosols) can account for greater inputs of N and S, and consequently a greater load of H^+, than wet deposition (rain, fog, and snow). For example, Höfken and Gravenhorst (18) estimated that wet and dry depositions of NH_4^+ were about equal and that dry deposition accounted for three times more

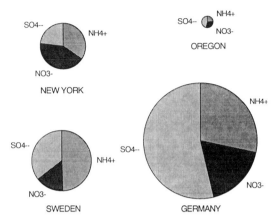

Fig. 3.3. Ranges in annual H^+ deposition from sites throughout the world, partitioned according to the source of acidity. The area of the pie charts is proportional to the total input, which ranges from 0.055 kmol H^+ ha^{-1} yr^{-1} for Alsea, OR, to 2.79 kmol H^+ ha^{-1} yr^{-1} for Solling, Germany. Data are from Höfken and Gravenhorst (18), Grennfelt et al. (78), and NADP/NTN (76,77).

NO_3^- and four times more SO_4^{2-} than wet deposition. Data representing the range of acid inputs and the relative amount associated with NH_4^+, NO_3^-, and SO_4^{2-} are presented in Figure 3.3. To calculate these data, it was assumed that 1 mol of NH_4^+ would generate 1 mol of H^+ as it was utilized, 1 mol of H^+ was associated with every 1 mol of NO_3^-, and that 2 mol of H^+ was associated with every 1 mol of SO_4^{2-}. In addition to the input of <1 to >3 kmol H^+ ha^{-1} yr^{-1}, N inputs range from 0.4 kg N ha^{-1} yr^{-1} (Oregon) to >18 kg N ha^{-1} yr^{-1} (Germany), and S inputs range from 0.4 kg S ha^{-1} yr^{-1} (Oregon) to >24 kg S ha^{-1} yr^{-1} (Germany) for wet deposition. Totals would be greater if dry depositions were included.

3. SOIL BUFFERING REACTIONS

The microbial response to the components of acidic deposition is modified by the soil environment. Soil properties important to an understanding of microbial responses include texture, mineralogy, aeration status, and organic matter content, quality, and distribution. Taken together, these properties determine the buffering and sorption capacities of soil with respect to components of acidic deposition.

Protons, NH_4^+, and oxides of N and S added to the soil solution enter into dynamic equilibria with the solid phase. Amounts of these chemical species associated with the solid phase depend primarily on surface area. The relative masses of solids in sand (2–0.05 mm), silt (0.05–0.002 mm), and clay (<0.002 mm) diameter classes regulate surface area. Material in the sand fraction typically has a specific surface of about 1 m^2 g^{-1}, whereas certain clay minerals have specific surfaces ranging to 800 m^{2}g^{-1}. Therefore, the reactive surface of fine-textured soils greatly exceeds that of coarse-textured soils.

Soil colloids carry electrical charge, which confers the capacity to hold and exchange ions. Negative charges on colloids arise mainly from isomorphous substitution in clay mineral lattices and dissociation of carboxyl, enol, and phenol groups. The cation exchange capacity (CEC) thus generated is critical in the retention of "basic" cations—Ca^{2+}, Mg^{2+}, Na^+, and K^+—but may also retain acidic ions—hydrated forms of Al, Fe, and Mn—in acidic soils. Positive charges on soil colloids are mainly due to protonation of 1) Al groups at clay edges, 2) crystalline and paracrystalline Al and Fe oxyhydroxides, and 3) amine groups on organic matter. Capacity for nonspecific anion retention is thus pH dependent and is generally of lower magnitude than cation exchange capacity.

Acidification of a soil system involves removal, by leaching, of basic cations, and their replacement on the exchange complex by hydrated forms of Al, which are released by weathering of aluminosilicate minerals. Leached basic cations must be accompanied by an electrical equivalent of anions. Because anion exchange capacity is small in relation to cation exchange capacity, the nature and amount of anions in the soil solution may be viewed as a critical control on acidification. In soils unaffected by acidic deposition, concentrations of the strong anions SO_4^{2-} and NO_3^- are kept low by competing biological processes; and leaching of cations is regulated by HCO_3^- and organic acid anions. Bicarbonate leaching is limited by equilibria with CO_2 and carbonate minerals, whereas organic acids may be mobile in surface horizons but precipitated by complexation with Fe and Al at depth. In contrast, soils affected by acidic deposition may receive total (wet plus dry) inputs as high as 100 kg SO_4^{2-}–S ha^{-1} yr^{-1} and 35 kg NO_3^-–N ha^{-1} yr^{-1} (18) which can leach basic cations.

Ulrich (19) organized information on various acid–base reactions in soils under the concept of buffer ranges. In order of decreasing pH, these are as follows.

1. Carbon buffer range. In calcareous soils, $CaCO_3$ is decomposed by acids. In the soil solution carbonic acid is formed by respiratory CO_2 inputs and reacts with $CaCO_3$:

$$CaCO_3 + H_2O + CO_2 \rightleftharpoons Ca^{2+} + 2HCO_3^-. \tag{1}$$

Thus HCO_3^- will be regulated by the partial pressure of CO_2 and by the availability of $CaCO_3$. At CO_2 partial pressures typical of soil atmospheres the soil pH will be near 7.8 as long as $CaCO_3$ remains. After $CaCO_3$ has been consumed, protonation of HCO_3^- provides additional buffering,

$$H^+ + HCO_3^- \rightleftharpoons H_2CO_3 \rightleftharpoons H_2O + CO_2. \tag{2}$$

Maximum buffer capacity occurs at pH 6.4, but essentially all capacity is exhausted at pH 5, where the system is about 95% carbonic acid.

2. Exchange buffer range. Acidity can be neutralized by exchange of protons with basic (nonhydrolytic) metal cations:

$$mH^+ + \underline{\underline{BC}} \rightleftharpoons \underline{\underline{mH}} + BC^{m+}, \tag{3}$$

where BC is base cation and a double underline indicates the adsorbed phase. True exchange occurs where the adsorbent is a clay mineral, whereas organic matter "exchange" involves protonation of discrete functional groups such as carboxyl. Ulrich (19) proposed that this mechanism was important primarily between pH 4 and 5.2, but recent work by De Vries *et al.* (20) suggests a major contribution from pH 4 to 7.

3. Al buffer range. Below a pH of about 4.5, hydrolysis of Al becomes important in regulating soil pH:

$$Al(OH)_3 + 3H^+ \rightleftharpoons Al^{3+} + 3H_2O. \tag{4}$$

De Vries *et al.* (20) estimated a typical Al buffer capacity of 975 kmol kg^{-1} based on a soil depth of 50 cm, bulk density of 1.3 Mg kg^{-1}, and reactive (oxalate extractive) Al oxide/oxyhydroxide content of 150 mmol kg^{-1}. Reactive Al would include monomeric and polymeric hydroxy complexes in the soil solution, sorbed to mineral surfaces, or associated with organic matter. This capacity is large, yet conservative, because it does not consider the approximately 7% by mass of Al contained in primary minerals of the earth's crust.

4. Fe buffer range. Trivalent Fe hydrolyzes in the same way as Al:

$$Fe(OH)_3 + H^+ \rightleftharpoons Fe(OH)_2^+ + H_2O. \tag{5}$$

The first hydrolysis has a pK of about 3. Therefore Fe should not contribute to soil buffering until Al has been consumed or dissolution rates of Al minerals become limiting. The equilibria of protons with the various sinks described above may be gathered under a general concept of intensity and capacity factors. Soil pH, or active acidity, is the intensity factor, whereas "exchangeable acidity" is the capacity factor. Properties of two Alberta soils receiving acidic inputs from regional sour gas processing that contrast in these factors are given in Table 3.1. The Fort McMurray soil has a lower pH but also a lower exchange acidity and buffer capacity because of a paucity of exchangeable bases. Low exchange

TABLE 3.1. Properties of Two Alberta Soils Contrasting in
Susceptibility to Acidification[a]

Property[b]	Soil Location	
	Fort McMurray	Twin Butte
Classification	Dystric Cryochrept	Agric Cryoboroll
Texture	Sand	Sandy clay loam
Organic matter (%)	1.17	13.8
pH in 0.01 M $CaCl_2$	4.3	5.7
Cation exchange capacity[c] ($cmol_c$ kg^{-1})	0.94	41.97
Exchangeable cations ($cmol_c$ kg^{-1})		
Ca	0.50	34.1
Mg	0.04	6.8
Na	0.01	0.02
K	0.04	1.06
Al	0.23	0.01
Exchange acidity[d] ($cmol_c$ kg^{-1})	3.54	14.06

[a]Soils drawn from Alberta Environment's Long-Term Acidification Monitoring Program.
[b]Weighted average of properties in the top 10 cm.
[c]$BaCl_2$-triethanolamine method.
[d]Determined at ambient pH.

capacity, in turn, may be explained by coarse texture and low organic mater content. In contrast, the Twin Butte soil has both a higher pH and a greater buffer capacity by virtue of abundant exchangeable bases and organic matter.

De Vries *et al.* (20) compared long-term responses to high and low acid loads of two hypothetical soils differing in CEC and in some other covarying parameters. Their simulation showed that acid inputs near a global mean resulted in a stable pH of about 5.5, determined by the exchange buffer mechanism (Fig. 3.4). This plateau was reached because inputs of base cations from weathering exceeded removal by leaching. In contrast, an acid load typical of affected areas of Europe, defined as a sum of SO_2, NO_x, and NH_3 of 5 kmol ha^{-1} yr^{-1}, caused a complex temporal response reflecting different buffer mechanisms (Fig. 3.4). The effect of differing amounts of base cations is seen where the soil of low CEC has acidified to pH 4 in less than 100 years, whereas the high CEC soil requires over 250 years. While reactive Al is available both soils remain at pH 4; thereafter pH declines more or less quickly to the Fe buffer range at pH 3. It is important to realize, from a biological standpoint, that the onset of the Fe buffer range means that solution concentrations of Al^{3+} are now in the 100 mM range.

Very few mineral soils affected by acidic deposition have been reported to have a pH value below 4. However, forest soils of pH 4 (Al buffer range) are

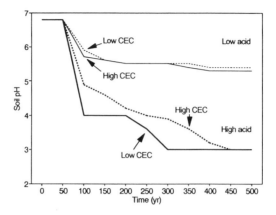

Fig. 3.4. Responses of two different hypothetical soils to high and low acid inputs. Acid input defined as the sum of NH_3, NO_3^-, and SO_4^{2-} equal to 0.5 (low) or 5.0 (high) kmol H^+ ha^{-1} yr^{-1}. Cation exchange capacity of 15 (low) or 100 (high) $cmol_c$ kg^{-1}. After De Vries et al. (20).

common in northeastern United States, eastern Canada (21), and southern Scandinavia (22). Whether soils in this pH range will attain levels of Al^{3+} sufficiently toxic to affect biotic communities qualitatively depends on the balance among rates of leaching of Al and base cations and their replenishment through mineral weathering. As long as base cations remain, their displacement from the exchange complex by Al will hold down solution concentrations of Al. Reuss and Johnson (3) emphasized that leaching losses of base cations will be mediated primarily by NO_3^- and SO_4^{2-} and that the biological sink for the former (plant uptake) exceeds the latter by approximately 15-fold. Nitrate inputs for the most part appear to be within system abilities to assimilate, but questions remain regarding 1) synchrony between addition and uptake processes and 2) effects on natural N supplying processes (16). Assuming NO_3^- leaching does not regulate base cation leaching, retention of SO_4^{2-} becomes the critical consideration.

Sulfate is retained in soils by both biological and chemical mechanisms. Biological mechanisms include plant uptake and immobilization into microbial tissues. Biomass S added as detritus may be stabilized as soil organic matter in either C-bonded or ester forms. Chemical retention of SO_4^{2-} may be through adsorption or precipitation in secondary minerals. Controversy exists as to what proportion of each is typically operative, because precipitated phases may be poorly crystalline and difficult to isolate and identify. In general, a soil's capacity to sorb SO_4^{2-} increases with increasing solution concentration, and this capacity is correlated to amount of Al and Fe oxyhydroxides (3). However, Harrison et al. (23) showed that SO_4^{2-} retention was a much stronger function of oxalate-extractable Al in Ultisols than in Spodosols or Inceptisols. Generally, SO_4^{2-} retention was poor in high elevation forest soils of the U.S. northeast (23).

4. RESPONSE AND ADAPTATION OF SOIL MICROORGANISMS TO ACID RAIN

Survival and growth of microorganisms are dependent on several environmental factors, including substrate and nutrient availability, pH, aeration, and water potential. Typically, each microbial species will grow optimally at one combination of environmental factors. Consequently, acid rain, with increased inputs of H^+, N, and S, has the potential to affect a microorganism either positively or negatively depending on whether the inputs shift it toward or away from the optimum. This short-term, immediate response is what has been measured most often experimentally and is discussed in the following sections on H^+, N, and S effects.

Over time, it is also possible that a particular microbial species can adapt to environmental changes brought about by acid rain and actually change its growth optima with respect to pH, nutrients, and so forth. Few data specific to acid rain exist with respect to this type of long-term adaptive response. The potential for microbial adaptation to stress is discussed in the last section.

4.1 Hydrogen Ion Effects

Acidification of the soil environment can affect microorganisms directly by the increased concentration of H^+ or indirectly through changes in the solution concentrations of other ions, such as PO_4^{3-}, Al, and heavy metals, which result from reduced soil pH. Important implications of increased H^+ concentrations are discussed in depth first, followed by information about Al toxicity. The reader is referred to other sources for detailed explanations of the effects of heavy metals on microorganisms and their activities (24). Heavy metals are also discussed by Gilmour, chapter two of this volume.

4.1.1. Direct pH Effects

An optimum pH for growth exists for each microbial species. This pH is typically close to that of the microorganism's environment. For many soil bacteria the optimum is near pH 7 (growth between pH 5 and 8), consequently these organisms are known as *neutrophiles*. For many soil fungi, the pH range for growth and the pH of optimal growth are about 1–2 pH units lower than for soil bacteria (Fig. 3.5). There are, however, many examples of acidophilic microorganisms that grow optimally at pH 4 or less. By understanding how neutrophiles and acidophiles respond to pH, one can better interpret the empirical results obtained in artificial acidification studies.

The H^+, or proton, is central to many metabolic reactions in microbial cells, including energy generation and ion transport (25). Consequently, all microbes

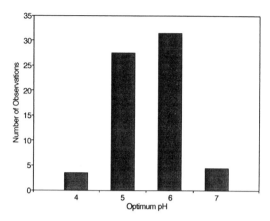

Fig. 3.5. Distribution of pH optimum for 67 strains of ectomycorrhizal fungi. Data are from Dennis (43).

have mechanisms to maintain their cytoplasmic pH (internal pH) at a level consistent with the functioning of cellular enzymes (26,27). Solution H^+ concentrations (external pH) thereby affect microorganisms by forcing them to maintain a favorable cytoplasmic pH (pH homeostasis). In addition, H^+ in the soil solution may interact with external cell structures or secreted substances, such as extracellular enzymes (28).

Internal pH values for neutrophiles range from 7.5 to 8.0, which is slightly more alkaline than the external pH range at which they exist. Acidophiles also maintain an internal pH near neutrality that is more alkaline (pH 6.5 to 7.0) than that of their normal environment (27). Because both neutrophiles and acidophiles maintain an internal pH near neutrality, it is likely that their cytoplasmic enzymes are quite similar with respect to their pH optima. Consequently, microbial adaptation to an acidic environment is most likely to reflect adjustments to those cellular features that function to maintain a near-neutral internal pH.

Several mechanisms exist to provide pH homeostasis in microbial cells: modified cell membrane permeability, cytoplasmic buffering, acid or base production in the cytoplasm, and active H^+ transport (27,28). Booth (27) states that there is little evidence that cytoplasmic buffering (most bacteria have similar cytoplasmic buffering capacities) or metabolic production of acids or bases are major factors in pH homeostasis.

It is known that the archaebacterial acidophiles have unusual branch-chained, ether-linked membrane lipids, but Cobley and Cox (26) suggest that this is unlikely to be an adaptation to low pH because neutrophilic methanogens and halobacteria also possess these membrane constituents. Some acidophilic eubacteria have been found to have unique membrane lipids, however, and it is known that media pH can affect the lipid content of *Staphylococcus aureus,* a neutrophile (29). If such alterations in membrane composition confer greater acid tolerance, it may be possible that soil bacteria could use this mechanism to adapt to a gradually acidified environment.

Most evidence suggests that proton pumping is the dominant method that bacteria use to maintain a neutral internal pH. Extrusion of H^+, typically via a K^+–H^+ antiport, is the specific mechanism that neutrophiles and acidophiles use for pH homeostasis (27). In acidophiles, however, the much larger difference between external and internal pHs requires a small or negative electrical potential gradient to maintain an effective proton motive force. Such negative electrical potential gradients have been measured in acidophiles (*cf.* 26) and may be produced by an ATP-coupled cation extrusion mechanism (27). Booth (27) suggests that the basic proton pumping mechanisms are similar for neutrophiles and acidophiles, but that their regulation or expression may differ in degree. If this is so, then it would seem that neutrophilic bacteria would be able to make phenotypic or genotypic adaptations to lowered soil pH.

Fungi apparently adjust to external pH using many of the same mechanisms as bacteria. The somewhat lower pH optima shown by many fungi may be partly a result of their different cell wall structure, and Padan (25) makes the interesting point that fungi, being eukaryotes, have the ability to sequester unwanted materials in vacuoles. So in addition to extruding H^+ into the environment, fungi could presumably dump excess H^+ into their vacuoles. The relative resistance of fungal spores probably plays an additional role as a survival mechanism at unfavorable soil pH.

Low soil pH has the potential for affecting microbial structures or products external to the organism (28). Certain cell surface constituents may become protonated as the pH drops below the pK_a of the ionized cell wall component. This would have the effect of reducing the generally net negative charge on the microbial cell, which could affect its interactions with soil minerals or organic matter. Protonation might affect specific cell surface components, such as lectins and other compounds implicated in plant–microbe interactions. These interactions could also be influenced by effects that soil pH might have on the ionization of signal molecules, such as luteolin and NodRm-1 (30). There is currently no evidence for such effects, however. Changes in soil pH are known to affect the activities of extracellular enzymes (31). For example, there is both acid (pH 4–6) and alkaline (pH 8–10) phosphatase activity in many soils, but urease typically has a single pH optimum (pH 6.5–7).

4.1.2. Aluminum Toxicity

Robson and Abbott (14) point out several shortcomings of experiments done to determine the effects of soil acidification on microbial activity, one of which is that solution concentrations of potentially toxic cations, such as Al^{3+}, typically increase as soil pH decreases. Compared with plants, however, relatively little research has focused directly on the effects of Al on microorganisms. The complex chemistry of Al, which polymerizes and interacts with phosphate and organic acids, makes interpretation of experimental results difficult.

Wood (32) summarizes the effects of Al toxicity on rhizobia, which are probably the most studied soil bacteria. Concentrations of Al <50 μM were found to be inhibitory for growth of rhizobia in pure cultures. Such Al concentrations are typical for soils of pH 4. Although the direct mode of action for Al toxicity is not know, Johnson and Wood (33) demonstrated that Al can bind directly to DNA, which may inhibit replication, although other mechanisms may be operative (32). Wood (32) notes that strain differences in Al tolerance are not related to soil pH, which could mean that adaptation of *Rhizobium* to high Al concentrations has not occurred in nature.

Studies of mycorrhizal fungi have shown that they have a greater tolerance for Al than *Rhizobium*. Paulus and Bresinsky (34) found no effect of Al concentrations up to 3.7 mM on the growth of *Suillus variegatus*. The sensitivity of mycorrhizal fungi to Al concentration varies by species and isolate; however, Thompson and Medve (35) found growth inhibited in five of six strains tested at 1.9 mM Al (the lowest concentration used). Interestingly, the unaffected strain was *Suillus luteus,* which showed no adverse effect up to 11 mM Al.

4.1.3. Results of Acidification Experiments

Most studies have collected data on numbers of total bacteria or fungi or on selected physiological groups of microorganisms such as starch hydrolyzers and spore formers. Generally, total fungal biomass has shown little response to inputs up to 10 kmol H^+ ha^{-1} yr^{-1}, whereas total bacterial populations have often been observed to decrease (15). Shifts in the abundance of particular bacterial groups following acid inputs are common; for example, Mancinelli (36) found actinomycetes to increase but lipolytic bacteria to decrease. Results of shifts of physiological groups vary greatly among studies, however. For example, the population of starch hydrolyzers has been observed to increase (37), decrease (38), or stay unchanged (36). But data of this type do not provide information about reaction of individual microbial species to acid rain.

The most specific information available is at the level of functional groups (*e.g.,* nitrifiers and denitrifiers) or about symbiotic microorganisms, such as mycorrhizal fungi and rhizobia. Bacteria responsible for specific N cycle processes are discussed later under NH_4^+ and NO_3^- effects because the added N may be a more important factor than the change in acidity. Mycorrhizae are addressed in this section.

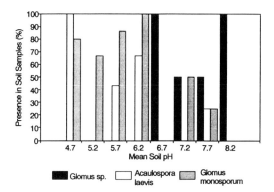

Fig. 3.6. Tolerance of different mycorrhizae species to specific pH conditions. Data are from Robson and Abbott (14).

Because of the concern about acid rain damage to forest systems, much research has been done on the effects of acid rain on the formation of ectomycorrhizae and on the response of ectomycorrhizal fungi. In most cases, ectomycorrhizae have been found to be sensitive to acid inputs at pH 3.0–3.5 (15), which is in the Fe buffer range, but exceptions exist (39–41), and liming to pH 7 also decreases ectomycorrhizal formation (42). Perhaps of greater interest is that, as with the pure culture work on the effect of pH on ectomycorrhizal fungi (43), the response of mycorrhizal fungi in soil media and in symbiosis varies by species (41,42,44), with some species more prevalent under acidic conditions and others dominant at more neutral pH. Surveys of ectomycorrhizal fruiting bodies in polluted areas normally show far less species diversity than in uncontaminated areas (45,46). It is unknown, however, whether adaptation to low pH occurs within species of ectomycorrhizal fungi.

Robson and Abbott (14) provided an excellent review of the effects of soil acidity on vesicular–arbuscular mycorrhizae (VAM). Species of VAM respond to soil pH like their ectomycorrhizal counterparts: There is a shift in which VAM species are most prevalent as soil pH changes (Fig. 3.6). Unlike ectomycorrhizae, however, acidification does not appear to affect mycorrhizal infection (14,47). This may be more a factor of the nearly ubiquitous presence of VAM host plants than any inherent acid tolerance of VAM.

4.2. Ammonium and Nitrate Effects

Nitrogen is an essential nutrient for all microorganisms and is typically the most abundant cell component after C, H, and O. It is a major constituent of amino acids that are the building blocks of structural and enzymatic proteins.

Nucleic acids, amino sugars (important in cell walls), and heterocyclic compounds also contain significant amounts of N. With the exception of N_2-fixing bacteria, microorganisms assimilate N from their environment as NH_4^+, NO_3^-, and some simple organic forms. Acquisition of N does not appear to depend heavily on extracellular enzymes, at least within the normal range of soil C availability. Consequently N for microbial growth can be considered a by-product of organic C metabolism.

Individual heterotrophic microorganisms, which are generally C-limited in soils, would not be expected to be strongly affected by NH_4^+ and NO_3^- inputs, but some exceptions exist. It is well known that the high concentrations of soil inorganic N can inhibit the nodulation of legumes by rhizobia and also depress rates of N_2 fixation (48). In pure culture, these inhibitory effects begin to occur at 2–5 mM NO_3^-, which is roughly equivalent to 9–23 mg NO_3^-–N kg^{-1} soil. Although the NO_3^- concentration of precipitation is much lower (1–100 μM), soil NO_3^- concentrations could reach levels inhibitory to symbiotic N_2-fixing bacteria.

Recent studies of N fertilization in forest ecosystems have shown that added N, albeit at rates up to 10 times those of annual precipitation inputs for highly impacted regions, can influence microbial populations. In Sweden, for example, shifts in species abundance of both saprophytic and ectomycorrhizal fungi were observed in response to N fertilization (49,50). Despite these shifts in fungal species, no effects on mycorrhizal infection were observed. However, other studies at this site suggest that N fertilization may have decreased heterotrophic activity (51). This latter observation is discussed in greater detail in section 5.

Autotrophic nitrifiers, especially ammonia oxidizers that use NH_4^+ as an energy source, are often stimulated by additions of NH_4^+. Usually, this has been documented indirectly by measuring nitrification activity (52,53), although Martikainen (54) measured nitrifier numbers. In a similar manner, denitrifying bacteria may respond positively to NO_3^- inputs in soils with low NO_3^- availability provided that other controlling factors are not limiting (*cf.* 55). Laboratory experiments have shown denitrification potentials, which may be interpreted as being proportional to denitrifier populations, to increase after the addition of simulated acid rain containing NO_3^- (56).

4.3. Sulfur Effects

In addition to nitrogen, S is also an important nutrient required by microorganisms. It is needed in lesser amounts than N, but is still considered a macronutrient. Sulfur is present in the amino acids cysteine and methionine, which are important in the stabilization and activity of proteins and enzymes, and it is present in several vitamins and cofactors. All of these S compounds contain S directly linked to a C-skeleton (C–S bonds). Ester sulfates (C–O–S or C–N–S bonds) are also present in microorganisms. In fact, some fungi are able to store excess S as ester sulfates (57). In times of S deficiency, many microorganisms are able to produce extracellular sulfatase enzymes, which can selectively cleave the SO_4^{2-} moiety from ester sulfates.

Because most soil heterotrophs are not S-limited, growth responses to added S are uncommon. Decreased populations of heterotrophic bacteria and fungi have been observed in soils in areas receiving large amounts of particulate S or SO_2 (58), but no responses to SO_4^{2-} have been reported. Conversely, populations of S-oxidizing bacteria and fungi increase when reduced forms of S are added (59,60). In anaerobic soils, there might be a positive response of SO_4^{2-}-reducing bacteria to additions of SO_4^{2-}, which is used as an electron acceptor by these anaerobic organisms.

4.4. Microbial Adaptation Mechanisms

Microorganisms can adapt to environmental changes by phenotypic or genetic changes. A phenotypic change could result from an altered gene (genotypic change), but it could also arise from a change in gene expression. Regulation of transcription, e.g., derepression of the *lac* operon in the presence of lactose, is one way that microbes can respond phenotypically to an external stimulus. Sporulation by bacteria and fungi might represent a phenotypic change that could occur as a result of acidification. In fact, increases in viable numbers of spore-forming bacteria have been observed in soils exposed to SO_2 (61) or low amounts of acidity (37). Bååth et al. (37) also observed that bacteria in acid-treated soils were generally smaller in size. Except for these observations, no other phenotypic adaptations by bacteria or fungi have been noted. These results can be interpreted in several ways: 1) microorganisms are limited in their physiological (phenotypic) response to altered soil pH, 2) acidification treatments have not been severe enough to evoke phenotypic responses, or 3) responsive groups of microorganisms have been overlooked. The first and last explanations appear to be most likely, because many experiments have used quite concentrated (e.g., pH 2.5–3.5) acid solutions.

Genetic changes in response to soil acidification, through either mutation and selection or horizontal gene transfer, have not been demonstrated or even studied. That such adaptation has occurred is likely for two reasons. First, microbial diversity is presumed to have arisen as a result of mutation and subsequent selection in different habitats (62–65). Second, this type of adaptation has been shown to occur in the laboratory under artificially induced stresses or selection pressures (66–68).

The best example of genetic adaptation to low pH by soil bacteria is the work of O'Hara et al. (69) with *Rhizobium meliloti*. They found a direct correspondence between acid tolerance and the ability of *R. meliloti* strains to maintain an

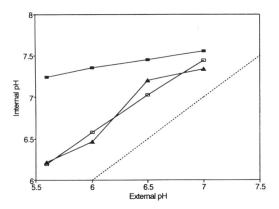

Fig. 3.7. Ability of acid-tolerant (closed squares), acid-sensitive (closed triangles), and acid-sensitive (open squares) mutants of *R. meliloti* to maintain internal pH at various levels of external pH. The acid-sensitive mutants were derived by Tn5 mutagenesis of an acid-tolerant strain. Data points represent the mean response of six acid-tolerant, six acid-sensitive, and four acid-sensitive mutants. Data are from O'Hara et al. (68).

alkaline internal pH over a range of external pH of 7.0 to 5.6 (Fig. 3.7). By subjecting acid-tolerant strains to Tn5 mutagenesis, they were able to isolate acid-sensitive mutants that were unable to maintain an alkaline internal pH (Fig. 3.7). This suggests that acid tolerance is genetically determined, and presumably acid-sensitive strains could undergo genetic changes via mutation or other means and become more acid tolerant.

5. RESPONSE AND ADAPTATION OF SOIL MICROBIAL COMMUNITIES TO ACID RAIN

How soil microbial processes respond to acid rain is a reflection of the composite response of all soil microorganisms. The response, or lack of response, may be due to the resilience of individual organisms, as discussed in the previous section, or may represent functional redundancies as different species assume a dominant role in a particular niche. Both resiliency and redundancy are likely to operate simultaneously and are therefore difficult to separate. It is similarly difficult, and may be neither desirable nor realistic, to separate acid rain effects into those of acidity and those of N and S fertilization. Examples illustrating these issues will be considered in the following sections on community adaptation, resiliency and redundancy, and the interactive effects of acid rain components.

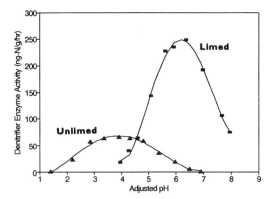

Fig. 3.8. pH profiles of denitrifier enzyme activity of limed and unlimed soil, which demonstrate adaptation to a long-term change in soil acidity. Data are from Parkin et al. (69).

5.1. Community Adaptation

One of the best examples of adaptation is in a study by Parkin et al. (70). They found different pH optima for denitrification enzyme activity in soils maintained for over 15 years at either pH 4 or pH 6 (Fig. 3.8). Unfortunately, it is not possible to distinguish whether this was the result of genotypic or phenotypic change in acid tolerance within acid-sensitive denitrifying bacteria or if it represented selection of preexisting acid-tolerant denitrifiers. It should also be pointed out that despite the development of an acid-tolerant population of denitrifiers, denitrification rates in the low pH soil were lower than those in the unlimed soil. This study suggests that at least some microbial communities may be capable to adapting to significant changes in soil acidity, but comparable investigations of adaptation to acid rain are needed.

5.2. Resiliency and Redundancy

The principles of resiliency and redundancy can be applied to artificial acidification studies that have examined the response of several N cycle processes. In reviewing the literature, Myrold (15) found that net N mineralization was relatively unaffected by inputs of acid, whereas net nitrification was more sensitive and tended to decrease as the result of acid inputs (Fig. 3.9). The differing effects of acidification on N mineralization compared with nitrification are likely because of differences in redundancy between these two processes, although nitrifiers may also be less resilient to acidification. Net N mineralization is mediated by virtually all heterotrophic microorganisms, by both bacteria and fungi. In

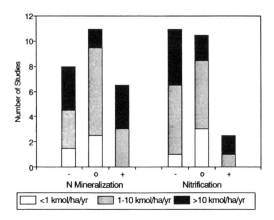

Fig. 3.9. Summary of studies that examined the response of N-cycle processes to artificial acidification. Each investigation measured the response of net N mineralization, a process mediated by diverse genera of bacteria and fungi, and net nitrification, a process mediated primarily by a restricted group of autotrophic bacteria. Responses were classified as negative ($-$), positive ($+$), or no change (O). Responses are grouped according to ranges of H^+ input. Data are from Myrold (15).

contrast, nitrification is mediated mainly by relatively few genera of autotrophic bacteria. Thus there is probably less functional redundancy available to act as a buffer for nitrification than there is for N mineralization.

Other examples of the importance of redundancy of function could be given. For example, Myrold and Nason (16) used soil respiration as an example of a microbial process that has a great degree of redundancy and has often been found to be minimally affected by artificial acidification. They found that soil respiration remained unchanged or actually increased in response to acid inputs in more than 70% of the studies surveyed. In contrast, symbiotic associations, particularly ectomycorrhizae, appear to be more sensitive to acidification. This may be more the result of a lack of resiliency or redundancy of the host plant than to the inability of the mycorrhizal fungi to adapt to changing soil pH. Nevertheless, it is consistent with the concept that more specialized organisms or processes are likely to be more vulnerable to environmental changes.

5.3. Interactions Among Constituents of Acid Rain

The soil respiration studies surveyed by Myrold and Nason (16) are complicated by the potential for interactive effects of H^+ and inputs of N and S. It is possible that the additional N and S could act either to mitigate or to exacerbate the effect of increased acidity. For example, N fertilization studies have given mixed results with respect to the effects of additional N on microbial activity. In high C/N ratio litter layers, microbial activity responded positively to N additions (71). However, numbers of specific microbial genera were not measured. In many cases, however, N fertilization has depressed soil respiration (51,72) or litter decomposition (73). The reason for these negative effects on C cycling is not well understood, but excess N inhibits ligninolytic activities of several white-rot fungi (74), apparently because of the repression of the genes coding for ligninolytic enzymes (75).

6. CONCLUSIONS

Acid rain is real; the pH of precipitation in many areas of the world is below its normal equilibrium value, and concentrations of inorganic N and S are elevated above background. The impact of acid rain on soil microbial processes is less clear. This is largely because of the chemical buffering of the soil ecosystem and the inherent resiliency and redundancy of soil microorganisms.

Microorganisms have an amazing capacity to adapt to new situations, which is enhanced by their ability to evolve under selection pressure. Their resilience is a function of both the large number of microorganisms present in a given volume of soil and their high growth rate relative to macroorganisms. This suggests that microorganisms are likely to be able to adapt more quickly to acidification than plants or animals, which may be one reason why symbiotic associations, such as ectomycorrhizae, are more susceptible to acid inputs than their saprophytic counterparts.

The relative stability of many microbial populations and processes in acidification stresses is also a result of functional redundancy. The large number and great diversity of heterotrophic microorganisms is probably a major reason why microbial processes of C and N mineralization are less affected by acid inputs than are processes that rely on a more restricted community, such as nitrification and autotrophic nitrifiers.

ACKNOWLEDGMENTS

This is Technical Paper No. 9587 from the Oregon Agricultural Experiment Station. We acknowledge the financial support of the National Science Foundation through the Presidential Young Investigator Award, BSR-8657269, to D.D.M.

REFERENCES

1. Bolin B, Granat L, Ingelstom L, Johannesson M, Mattsson E, Oden S, Rodhe H, Tamm CO: Air Pollution Across National Boundaries: The Impact on the Environment of Sulfur in Air and Precipitation. Stockholm: Norstedt and Soner, 1972.

2. Kennedy IR: Acid Soil and Acid Rain: The Impact on the Environment of Nitrogen and Sulfur Cycling. New York: Wiley, 1986.

3. Reuss JO, Johnson DW: Acid Deposition and the Acidification of Soils and Waters. Ecological Studies 59. New York: Springer-Verlag, 1986.

4. Mathy P: Air Pollution and Ecosystems. Dordrecht: D. Reidel, 1988, 981 pp.

5. Binkley D, Driscoll CT, Allen HL, Schoeneberger P, McAvoy D (1989): Acidic Deposition and Forest Soils: Context and Case Studies of the Southeastern United States, Ecological Studies 72. New York: Springer-Verlag, 149 pp.

6. Schulze E-D, Lange OL, Oren R: Forest Decline and Air Pollution: A Study of Spruce (*Picea abies*) on Acid Soils. Ecological Studies 77. New York: Springer-Verlag 1989, 475 pp.

7. Lucier AA, Haines SG: Mechanisms of Forest Response to Acidic Deposition. New York: Springer-Verlag, 1990.

8. Smith WH: Air Pollution and Forests: Interaction Between Air Contaminants and Forest Ecosystems, ed 2. New York: Springer-Verlag, 1990, 618 pp.

9. Alexander M: Effects of acidity on microorganisms and microbial processes in soil. In Hutchinson TC, Havas M (eds): Effects of Acid Precipitation on Terrestrial Ecosystems. New York: Plenum, pp 363–374, 1980.

10. Cook RB: The impact of acid deposition on the cycles of C, N, P, and S. In Bolin B, Cook RB (eds): The Major Biogeochemical Cycles and Their Interactions. New York: Wiley, 1983, pp 345–364.

11. Tabatabai MA: Effect of acid rain on soils. CRC Crit Rev Environ Control 15:65–110, 1985.

12. Hågvar S: Atmospheric deposition: Impact via soil biology. In Schneider T (ed): Acidification and Its Policy Implications. Amsterdam: Elsevier, 1986, pp 153–160.

13. De Vries W, Breeuwsma A: The relation between soil acidification and element cycling. Water Air Soil Pollut 35:293–310, 1987.

14. Robson AD, Abbott LK: The effect of soil acidity on microbial activity in soils. In Robson AD (ed): Soil Acidity and Plant Growth, Sydney, Australia: Academic Press, pp 139–165.

15. Myrold DD: Effects of acidic deposition on soil organisms. In Lucier AA, Haines SG (eds): Mechanisms of Forest Response to Acidic Deposition. New York: Springer-Verlag, 1990, pp 163–187.

16. Myrold DD, Nason GE: Microorganisms as mediator of acid deposition effects in forest soils. Trends Soil Sci (in press).

17. Charlson RJ, Rodhe H: Factors controlling the acidity of natural rainwater. Nature 295:683–685, 1982.

18. Höfken KD, Gravenhorst G: Deposition of atmospheric aerosol particles to beech and spruce forest. In Georgii HW, Pankrath J (eds): International Symposium: Deposition of Atmospheric Pollutants. Dordrecht: D. Reidel, 1981, pp 191–194.

19. Ulrich B: Soil acidity and its relation to acid deposition. In Ulrich B, Pankrath J (eds): Effects of Accumulation of Air Pollutants on Forest Ecosystems. Boston: Reidel, 1983, pp 127–146.

20. De Vries WD, Posch M, Kämäri: Simulation of the long-term soil response to acid deposition in various buffer ranges. Water Air Soil Pollut 48:349–390, 1989.

21. Tan B: Extent and effect of acid precipitation in northeastern United States and eastern Canada. Arch Environ Contam Toxicol 18:55–63, 1989.

22. Hallbäcken L, Tamm CO: Changes in soil acidity from 1927 to 1982–1984 in a forest area of south-west Sweden. Scand J For Res 1:219–232, 1986.

23. Harrison RB, Johnson DW, Todd DE: Forest soil sulfur pools and SO_4^{2-} adsorption and desorption capacity following elevated inputs of SO_4^{2-}. In Olsen RK, Lefohn AS (eds): Effects of Air Pollution on Western Forests. Pittsburgh: Air Waste Management Association, 1989, pp 529–546.

24. Bååth E: Effects of heavy metals in soil on microbial processes and populations (a review). Water Air Soil Pollut 47:335–379, 1989.

25. Padan E: Adaptation of bacteria to external pH. In Klug MJ, Reddy CA (eds): Current Perspectives in Microbial Ecology. Washington, DC: American Society for Microbiology, 1984, pp 49–55.

26. Cobley JG, Cox JC: Energy conservation in acidophilic bacteria. Microbiol Rev 47:579–595, 1983.

27. Booth IR: Regulation of cytoplasmic pH in bacteria. Microbiol Rev 49:359–378, 1985.

28. Krulwich TA, Guffanti AA: Physiology of acidophilic and alkalophilic bacteria. Adv Microbial Physiol 24:173–213.

29. Haest CWM, DeGier J, Op Der Kamp JAF, Bartels P, Van Deenen LLM: Changes in permeability of *Staphylococcus aureus* and derived liposomes with varying lipid composition. Biochim Biophys Acta 255:720–733, 1972.

30. Lerouge P, Roche P, Promé J-C, Faucher C, Vasse J, Maillet F, Camut S, de Billy F, Barker DF, Dénarié J, Truchet G: *Rhizobium meliloti* nodulation genes specify the production of an alfalfa-specific sulfated lipo-oligosaccharide signal. In Gresshoff PM, Roth LE, Stacey G, Newton WE (eds): Nitrogen Fixation: Achievements and Objectives. New York: Chapman and Hall, 1990, pp 177–186.

31. Burns RG: Soil Enzymes. New York: Academic, 1978, 380 pp.

32. Wood M: Aluminum toxicity to rhizobia. In Meguśar F, Gantar M (eds): Perspectives in Microbial Ecology. Ljubljana: Slovene Society for Microbiology, 1986, pp 659–663.

33. Johnson AC, Wood M: DNA, a possible site of action of aluminum in *Rhizobium* spp. Appl Environ Microbiol 56:3629–3633, 1990.

34. Paulus W, Bresinsky W: Soil fungi and other microorganisms. In Schulze E-D, Lange OL, Oren R (eds): Forest Decline and Air Pollution, Ecological Studies 77. New York: Springer-Verlag, 1989, pp 110–120.

35. Thompson GW, Medve RJ: Effects of aluminum and manganese on the growth of ectomycorrhizal fungi. Appl Environ Microbiol 48:556–560, 1984.

36. Mancinelli RL: Alpine tundra soil bacterial responses to increased soil loading rates of acid precipitation, nitrate, and sulfate. Front Range, Colorado, USA. Arctic Alpine Res 18:269–275, 1986.

37. Bååth E, Berg B, Lohm U, Lundgren B, Lundkvist H, Rosswall T, Söderström B, Wiren A: Effects of experimental acidification and liming on soil organisms and decomposition in a Scots pine forest. Pedobiologia 20:85–100, 1980.

38. Shafer SR: Influence of ozone and simulated acidic rain on microorganisms in the rhizosphere of *Sorghum*. Environ Pollut 51:131–152, 1988.

39. Shafer SR, Grand LF, Bruck RI, Heagle AS: Formation of ectomycorrhizae on *Pinus taeda* seedlings exposed to simulated acidic rain. Can J For Res 15:66–71.

40. McAfee BJ, Fortin JA: The influence of pH on the competitive interactions of ectomycorrhizal mycobionts under field conditions. Can J For Res 17:859–864.

41. Meier S, Robarge WP, Bruck RI, Grand LF: Effects of simulated rain acidity on ectomycorrhizae of red spruce seedlings potted in natural soil. Environ Pollut 59:315–351, 1989.

42. Erland S, Söderström B: Effects of liming on ectomycorrhizal fungi infecting *Pinus sylvestris* L. I. Mycorrhizal infection in limed humus in the laboratory and isolation of fungi from mycorrhizal roots. New Phytol 115:675–682, 1990.

43. Dennis JJ: Effect of pH and Temperature on *In Vitro* Growth of Ectomycorrhizal Fungi. Information Report BC-X-273. Victoria: Pacific Forestry Center, 1985, 19 pp.

44. Erland S, Söderström B, Andersson S: Effects of liming on ectomycorrhizal fungi infecting *Pinus sylvestris* L. II. Growth rates in pure culture at different pH values compared to growth rates in symbiosis with the host plant. New Phytol 115:683–688, 1990.

45. Kowalski S, Wojewoda W, Bartnik C, Rupik A: Mycorrhizal species composition and infection patterns in forest plantations exposed to different levels of industrial pollution. Agric Ecosyst Environ 28:249–255, 1989.

46. Schaffers AP, Termorshuizen AJ: A field survey on the relations between air pollution, stand vitality and the occurrence of fruitbodies of mycorrhizal fungi, in plots of *Pinus sylvestris*. Agric Ecosyst Environ 28:449–454, 1989.

47. Killham K, Firestone MK: Vesicular–arbuscular mycorrhizal mediation of grass response to acidic and heavy metal deposition. Plant Soil 72:39–48, 1983.

48. Streeter J: Inhibition of legume nodule formation and N_2 fixation by nitrate. CRC Crit Rev Plant Sci 7:1–23, 1989.

49. Arnebrandt K, Söderström B: The influence of nitrogen fertilization on ectomycorrhizal mycelium colonization. Agric Ecosyst Environ 28:21–25, 1989.

50. Arnebrandt K, Bååth E, Söderström B: Changes in microfungal community structure after fertilization of Scots pine forest soil with ammonium nitrate or urea. Soil Biol Biochem 22:309–312, 1990.

51. Nohrstedt H-Ö, Arnebrandt K, Bååth E, Söderström B: Changes in carbon content, respiration rate, ATP content, and microbial biomass in nitrogen-fertilized pine forest soils in Sweden. Can J For Res 19:323–328, 1989.

52. Vitousek PM, Matson PA: Causes of delayed nitrate production in two Indiana forests. For Sci 31:122–131, 1985.

53. Fyles JW, McGill WB: Nitrogen mineralization in forest soil profiles from central Alberta. Can J For Res 17:242–249, 1987.

54. Martikainen PJ: Numbers of autotrophic nitrifiers and nitrification in fertilized forest soil. Soil Biol Biochem 17:245–248, 1985.

55. Davidson EA, Myrold DD, Groffman PM: Denitrification in temperate forest ecosystems. In Gessel SP, Lacate DS, Weetman GF, Powers RF (eds): Sustained Productivity of Forest Soils. Vancouver: University of British Columbia, 1990, pp 196–220.

56. McColl JG, Firestone MK: Cumulative effects of simulated acid rain on soil chemical and microbial characteristics and conifer seedling growth. Soil Sci Soc Am J 51:794–800, 1987.

57. Saggar S, Bettany JR, Stewart JWB: Measurement of microbial sulfur in soil. Soil Biol Biochem 13:493–498, 1981.

58. Bewley RJ, Parkinson D: Effects of sulfur dioxide pollution on forest soil microorganisms. Can J Microbiol 30:179–185, 1984.

59. Wainwright M: Microbial S-oxidation in soils exposed to heavy atmospheric pollution. Soil Biol Biochem 11:95–98, 1979.

60. Gupta VVSR, Lawrence JR, Germida JJ: Impact of elemental sulfur fixation on agricultural soils. I. Effects on microbial biomass and enzyme activities. Can J Soil Sci 68:463–473, 1988.

61. Prescott CE, Bewley RJF, Parkinson D: Litter decomposition and soil microbial activity in a forest receiving SO_2 pollution. In Stone EL (ed): Forest Soils and Treatment Impacts. Knoxville: University of Tennessee, 1984, pp 448.

62. Clarke PH: Evolution of new phenotypes. In Klug MJ, Reddy CA (eds): Current Perspectives in Microbial Ecology. Washington, DC: American Society for Microbiology, 1984, pp. 71–78.

63. Freter R: Factors affecting conjugal plasmid transfer in natural bacterial communities. In Klug MJ, Reddy CA (eds): Current Perspectives in Microbial Ecology. Washington DC: American Society for Microbiology, 1984, pp 105–114.

64. Slater JH: Genetic interactions in microbial communities. In Klug MJ, Reddy CA (eds): Current Perspectives in Microbial Ecology. Washington DC: American Society for Microbiology, 1984, pp 87–93.

65. Young JPW: The population genetics of bacteria. In Hopwood DA, Chater KF (eds): Genetics of Bacterial Diversity. New York: Academic Press, 1989, pp 417–438.

66. Hall BG: Adaptation by acquisition of novel enzyme activities in the laboratory. In Klug MJ, Reddy CA (eds): Current Perspectives in Microbial Ecology. Washington, DC: American Society for Microbiology, 1984, pp 79–86.

67. Kurihara Y: Interaction and stability of microbial communities in experimental model systems. In Hattori T, Ishida Y, Maruyama Y, Morita RY, Uchida A (eds): Recent Advances in Microbial Ecology. Tokyo: Japan Scientific Societies Press, 1989, pp 11–20.

68. Dykhuizen DE: Experimental studies of natural selection in bacteria. Annu Rev Ecol Syst 21:373–398, 1990.

69. O'Hara GW, Goss TJ, Dilworth MJ, Glenn AR: Maintenance of intracellular pH and acid tolerance in *Rhizobium meliloti*. Appl Environ Microbiol 55:1870–1876, 1989.

70. Parkin TB, Sexstone AJ, Tiedje JM: Adaptation of denitrifying populations to low soil pH. Appl Environ Microbiol 49:1053–1056, 1985.

71. White CS, Moore DI, Horner JD, Gosz JR: Nitrogen mineralization–immobilization response to field N-perturbations or C-perturbations—An evaluation of a theoretical model. Soil Biol Biochem 20:101–105, 1988.

72. Söderström B, Bååth E, Lundgren B: Decrease in soil microbial activity and biomasses owing to nitrogen amendments. Can J Microbiol 29:1500–1506, 1983.

73. Titus BD, Malcolm DC: The effect of fertilization on litter decomposition in clearfelled spruce stands. Plant Soil 100:297–322, 1987.

74. Leatham GF, Kirk TK: Regulation of ligninolytic activity by nutrient nitrogen in white-rot basidiomycetes. FEMS Microbiol Lett 16:65–67, 1983.

75. Kirk TK, Fenn P: Formation and action of the ligninolytic system in basidiomycetes. In Frankland JC, Hedger JN, Swift MJ (eds): Decomposer Basidiomycetes. British Mycological Society Symposium 4. New York: Cambridge University Press, 1982, pp 67–90.

76. NADP/NTN: NADP/NTN Data Report, Precipitation Chemistry, 1 January–30 June, 1988. Vol XI, No. 1. Fort Collins: NADP/NTN Coordination Office, 1989.

77. NADP/NTN: NADP/NTN Data Report, Precipitation Chemistry, 1 July–31 December, 1988. Vol. XI, No. 2. Fort Collins: NADP/NTN Coordination Office, 1989.

78. Grennfelt P, Larsson S, Leyton P, Olsson B: Atmospheric deposition in the Lake Gårdsjön area, SW Sweden. Ecol Bull (Stockh) 37:101–108, 1985.

MICROBIAL TRANSPORT
OF TOXIC METALS

TIM FORD
RALPH MITCHELL
Division of Applied Sciences, Harvard University, Cambridge, Massachusetts 02138

1. INTRODUCTION

Disposal of toxic metal wastes to surface and subsurface waters poses unacceptable health risks. Contamination of drinking water supplies and concentration of metals in edible fish are of particular concern. However, we are unable to assess accurately the extent of the risk. Available models of metal transport and transformation are grossly inadequate. The deficiencies of transport models are due in part to failure to take into account the complexity of biological interactions with metals. This chapter reviews current information about microbiological processes that affect cycling, transport, and transformation of toxic metals. Hopefully, a better understanding of the role of microorganisms will permit the development of more accurate metal transport models.

2. MICROBE–METAL INTERACTIONS

The chemical reactions between microorganisms and metals can be divided into six distinct processes.

1. Intracellular accumulation. Concentration of metals within bacterial and other microbial cells can result from interaction with surface ligands followed by

Environmental Microbiology, pages 83–101, © *1992 Wiley-Liss, Inc.*

slow transport into the cell. This may be an important form of detoxification or a means of incorporating specific metals into enzymes (*e.g.*, Cu and Zn). Whatever the biochemical function, mobility of the metal is directly dependent on the mobility and physiology of the cell.

2. Cell wall–associated metals. Many metals readily bind to cell walls as a result of the presence of specific functional groups, which readily exchange protons for divalent cations. In a similar manner to intracellular accumulation, there is a close association between the fate of the metal and the cell.

3. Metal–siderophore interactions. Siderophores are chelating agents excreted by many microorganisms to facilitate uptake of ferric ions. In general, they have extremely strong binding affinities for ferric ions and do not appear to react strongly with other metals. However, there is some evidence that specific siderophores bind copper and molybdenum. Siderophore interaction with other metals cannot be ruled out, especially considering the ease with which they can be modified chemically to bind metals other than iron (1).

4. Extracellular mobilization/immobilization of metals by bacterial metabolites. Preferential leaching of metals from natural ores has considerable economic implications. Leaching can result from production of acidic metabolites or release of toxic metals bound to iron or manganese oxides through microbial reduction. Immobilization occurs because of formation of insoluble salts, *e.g.*, sulfides as a biproduct of sulfate-reducing bacterial activity.

5. Extracellular polymer–metal interactions. Many microorganisms produce extracellular polysaccharides that strongly bind metals. The role of these polymers in mobilization or immobilization of toxic metals is now well understood. However, there is increasing evidence that both processes may be important in metal cycling. Quantification and rate estimates of these interactions are extremely difficult to evaluate. Most of the available information has been obtained in waste treatment research.

6. Transformation and volatilization of metals. Examples include formation of elemental mercury (Hg^0) by demethylation and formation of methylated and, possibly, ethylated forms of mercury, selenium, tin, tellurium, arsenic and lead (2). In addition, toxic metal oxides can be used as electron acceptors under anaerobic conditions. The reduced form is frequently less toxic and may be either more volatile or precipitated.

2.1. Intracellular Accumulation

Bacterial assimilation of metals may be important in detoxification, enzyme function, and physical characteristics of the cell. Potential mechanisms of toxic metal flux across a membrane are shown in Figure 4.1. Sigg (4) presented a probable scenario for intracellular accumulation. Extracellular or cell wall–attached ligands are thought to bind toxic metals. These ligands transport the

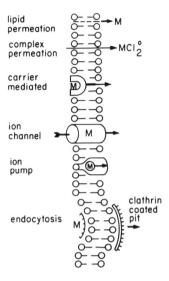

Fig. 4.1. Suggested mechanisms to explain the flux of metal ions (M) into a bacterial cell. (Reproduced from Simkiss and Taylor (3).)

complexed metals through the cell wall in a slow transport step. The metals are released inside the cell, incorporated into biochemical pathways, or trapped in an inactive form by complexation with another high-affinity ligand (5). This process is illustrated in Figure 4.2. The protoplasm of *Escherichia coli* typically contains 0.3% trace elements, including manganese, cobalt, copper, zinc, and molybdenum (8). Of these, both copper and zinc are considered toxic metals. Yet these are essential for the activity of several enzymes. In addition, nickel is a component of hydrogenase in many organisms, and its addition has been found to stimulate chemolithotrophic growth (9,10).

Cadmium is accumulated by a large number of organisms. Research by Macaskie *et al.* (11) on *Citrobacter* suggests that it is accumulated as cell-bound cadmium phosphate. This is presumably a detoxification mechanism and is similar to the accumulation of lead as $PbHPO_4$ by a different *Citrobacter* species suggested by Aickin *et al.* (12). Fate and transport of the metal is therefore directly related to the fate and transport of the bacterial cell. When the cell dies the metal is presumably released for further complexation following a similar model proposed for exopolymer–metal interactions (13).

A number of metals are enzymatically altered intracellulary. This does not usually result in accumulation. Rather, transformation occurs to a less toxic and, in certain cases, volatile form. This will be discussed further in the section on

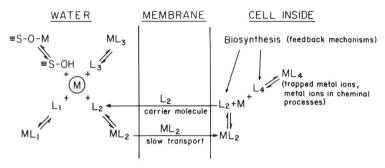

Fig. 4.2. A simplified uptake model showing the relationship between bacterial ligands (L) and metal ions (M), either as detoxification mechanisms or as incorporation into biochemical pathways. (Adapted from Sigg [4], Wood and Wang [5], and Williams [6,7].)

transformation of metals. The ability of certain bacterial cells to accumulate metals intracellularly has been exploited in mining practices, particularly in management of effluent treatment lagoons (14). Uranium has been shown to accumulate rapidly in cells of *Saccharomyces cerevisiae* and *Pseudomonas aeruginosa* (15).

2.2. Cell Wall Interactions

Binding of metals to cell surfaces plays a dominant role in the distribution of metals in natural waters (4,16). In addition, sorption of metals to living or dead cells is considered a practical solution to many metal contamination problems (4). Algal surfaces contain functional groups that have been shown to bind metals competitively with many dissolved ligands (16). Xue et al. (16) suggest that carboxylic amino, thio, hydroxo, and hydroxy–carboxylic groups on the surface of phytoplankton cells interact coordinatively with metal ions.

Bacteria possess lipopolysaccharides (LPS) in their outer membranes. Figure 4.3 shows diagrammatically the central role that LPS play in bacterial membrane structure. These chemicals are extremely complex, consisting of a hydrophobic, phosphorylated section, known as *lipid A*; a core oligosaccharide; and variable O-specific side chains consisting of a number of unusual sugars. The side chains project out from the cell membrane and contain different functional groups capable of binding metals.

Phosphoryl groups of LPS and phospholipids are the most abundant electronegative sites available for metal binding (18,19). The polyvalent toxic metals are primarily bound to LPS molecules because of the presence of closely opposed reactive sites (19). It has been suggested that this may provide a mechanism to immobilize toxic metals and prevent their entry into the cell (20). The fate of the

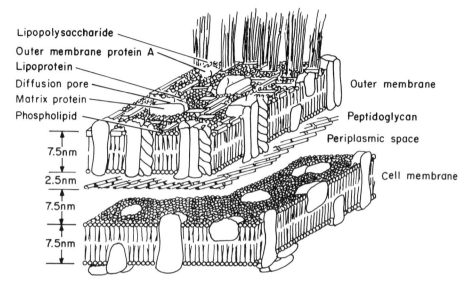

Lipopolysaccharide

Outer membrane protein A

Lipoprotein

Diffusion pore

Matrix protein

Phospholipid

7.5nm

2.5nm

7.5nm

7.5nm

Outer membrane

Peptidoglycan

Periplasmic space

Cell membrane

Fig. 4.3. The possible molecular configuration of the *E. coli* cell envelope. LPS side chains are only shown partially covering the cell wall for clarity. (Reproduced from Ingraham *et al.* [8], as modified from DiRienzo [17].)

metal is closely tied to the fate of the cell, and we may therefore assume that distribution of bacteria as well as algal populations (4,16) can influence the distribution of metals.

2.3. Siderophores

Siderophores are iron-complexing, low-molecular-weight organic compounds. Two major types are generally considered, the hydroxamate and catecholate siderophores. Figure 4.4 shows typical chemical structures of these complexing agents. Their function depends on their ability to concentrate iron in environments where the concentration is low and to facilitate transport into the cell. This is accomplished by a very strong affinity of the molecule for iron.

Siderophores vary chemically. One class is characterized by the presence of hydroxamate groups that strongly bind ferric iron (22). This activity is a function of the size and charge of the ion. It follows that analogs may also be strongly bound by these siderophores. For example, aluminium, gallium, and chromium form trivalent metal ions of similar size to iron (25). Aluminium may be a specific competitor for catecholate siderophores. In his review paper, Hider (24) reported stability constants of Fe^{3+} and Al^{3+} for catechol of 10^{44} and 10^{46},

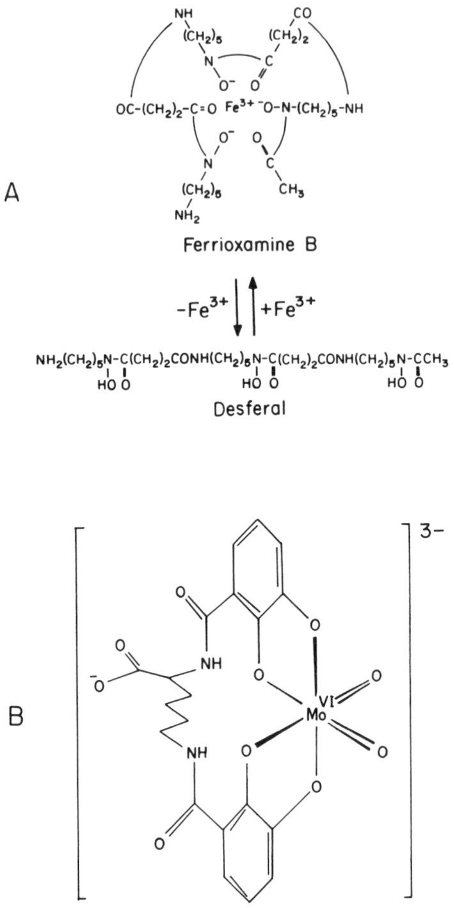

Ferrioxamine B

$$-Fe^{3+} \Big\updownarrow +Fe^{3+}$$

$$NH_2(CH_2)_5N-C(CH_2)_2CONH(CH_2)_5N-C(CH_2)_2CONH(CH_2)_5N-CCH_3$$

Desferal

Fig. 4.4 Typical siderophore structures. **A:** Structure of ferrioxamine B and its iron-free form Desferal. (Reproduced from Bickel *et al.* [21], as reported in Davis and Byers [22].) **B:** Structure of a molybdenum complexing dicatechol. (Reproduced from Emery and Neilands [23], as reported in Hider [24].)

respectively. Molybdenum and copper have been shown to form strong complexes with siderophores (24). In particular, catecholate siderophore–molybdenum complexes may provide an important uptake mechanism for intracellular molybdenum accumulation, required for nitrogenase enzymes. Ochiai (26) reports that a number of nitrogen-fixing bacteria secrete catecholate siderophores under molybdenum-deficient conditions.

Cu^{2+} complexation with both hydroxamate and catecholate siderophores has been reported and may be important in sequestering copper for production of tyrosinase (27). In addition, McKnight and Morel (28) suggest that hydroxamate siderophores may play an important role in reduction of copper toxicity to cyanobacteria. They did not provide direct evidence but speculated that the geometry of the copper siderophore complex made it unlikely that it would be assimilated by cyanobacterial cells. It would not be recognized by the microbial iron transport system.

2.4. Extracellular Processes

Research into industrial applications of microbial–metal interactions, especially in mining, has provided insights into the mechanisms of metal solubilization and transport. Organic or inorganic acids produced by microorganisms, including *Thiobacillus, Serratia, Pseudomonas, Bacillus, Penicillium,* and *Aspergillus*, are able to extract metals from solid substrates (29). Schinner and Burgstaller (29) used citric acid production by a *Penicillium* to extract zinc selectively from industrial waste. Toxic metals can be mobilized under anaerobic conditions as coprecipitates with iron oxides (30). This may be an important process, since coprecipitation of toxic metals with ferric iron is a widely used treatment for high metal content waste streams (31).

Francis and Dodge (30) used highly defined pure culture systems in their studies using a N_2-fixing *Clostridium* sp. They concluded that metals closely associated with iron (Cd and Zn) were solubilized by enzymatic reduction of ferric iron, whereas others (particularly Pb) are solubilized by the indirect action of bacterial metabolites.

Very little is known about microbial mobilization of metals under environmentally realistic conditions. Probably the closest approximation has been the work of Chanmugathas and Bollag (32–34), who studied the transport of cadmium in soil columns, with particular emphasis on biological mobilization processes. Cadmium was added to sterile (gamma-irradiated) and nonsterile soil columns and monitored for mobilization and speciation in the column eluates. Figure 4.5 shows that 36% of sorbed cadmium was mobilized in nonsterile columns. In contrast, <16% was mobilized in sterile columns. A significant amount of the mobilized cadmium was present as a hydrophilic organic complex.

In our own laboratory, we have studied microbial mobilization of manganese by bacterial exopolymers in soil (13). Manganese bound to bacterial exopolymers and free manganese were circulated through soil columns. After 24 hours a final eluate was removed and manganese assayed in the high- and low-molecular-weight fractions. More total manganese (high plus low molecular weight) was released from polymer metal application (7%) than from free metal application (3%). In addition, a considerably higher percentage of manganese was released in the high-molecular-weight fraction (45%) in the nonsterile than in the sterile (28%) controls.

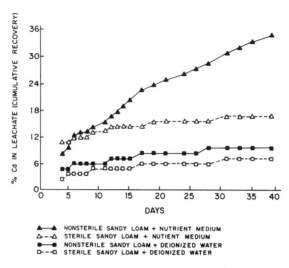

Fig. 4.5. Recovery of sorbed cadmium in soil column leachates, showing increased mobilization of cadmium in the presence of nutrient-stimulated bacterial activity. (Reproduced from Chanmugathas and Bollag [34].)

Many metal salts are insoluble, and their formation results in immobilization of toxic metals by sequestering to sediments or adsorption to soil particles. Frequently metals form insoluble complexes with, for example, hydroxides, carbonates, phosphates, and sulfides. Probably the best known microbial immobilization process is sulfide production by the sulfate-reducing bacteria, characteristic of anoxic sediments. As a result, these sediments will often contain high concentrations of lead and mercuric sulfides.

2.5. Extracellular Polymer–Metal Interactions

Interaction between bacterial exopolysaccharides and metals has been the subject of increasing research interest. Most of the research stems from the importance of biological flocculation of metals in waste water treatment processes (35–38). As a result of extensive chemical studies of extracellular polysaccharides by Sutherland (39–41) and metal-binding studies by Geesey and Ford and their coworkers (42–50), we now have a clearer picture of the interactions between bacterial exopolymers and metal ions. Interaction with metal ions is generally considered a direct consequence of the presence of negatively charged functional groups on the exopolymer. These groups include pyruvate, phosphate, hydroxyl, succinyl, and uronic acids. A pH-dependent binding of

TABLE 4.1. Maximum Binding Abilities and Conditional Stability
Constants of Exopolymer–Metal Complexes[a]

Exopolymer sample	Metal	MBA (nmol mg^{-1})	K_c ($\times 10^8$)
Thermus sp.			
Suspended	Cu	9	0.7
Attached	Cu	85	0.09
Deleya marina	Cu	263	24
	Mn	556	9
	Fe	39	1
	Ni	435	14
Pedomicrobium manganicum	Fe	13	750
	Mn	184	19
Pedomicrobium ferrugineum	Mn	409	2.9

[a]Data are from refs. 13, 47, and 48. MBA, maximum binding ability; K_c, conditional stability constant.

positively charged cations can rapidly occur, with stability constants in excess of those generally measured for humic substances and other naturally occurring ligands (51). Table 4.1 shows a summation of some binding characteristics of purified bacterial polysaccharides and different cations.

The role of bacterial exopolymer–metal interactions in the environment is a question that has only recently been addressed. For example, Mittelman and Geesey (51) measured a maximum binding affinity of 489 nmol Cu/mg carbohydrate for exopolymer from a freshwater sediment bacterium. Black *et al.* (13) proposed a simple model to explain the role of microbial polysaccharides in metal cycling in soil (Fig. 4.6). Within marine systems (*e.g.*, 52–54), freshwater systems (*e.g.*, 55–57), and soils (13,30,32), the importance of mixed bacterial/organic/inorganic aggregates to metal cycling is thought to be considerable. In marine systems, biogeochemical cycling of metals is strongly mediated by large, rapidly sinking organic particles and their associated bacteria (54). Freshwater and soils may be more dependent on clay/organic aggregates (56), of which the biogenic fraction may be particularly important in binding specific metals (4,16). Adsorption of dissolved toxic metals to organic or inorganic aggregates may result in the following processes.

1. Immobilization as an insoluble inorganic salt.

2. Adsorption to bacterial cell walls with subsequent transformation and release.

3. Binding to extracellular material that may subsequently be sloughed from an aggregate/soil particle with subsequent retention in the mobile, aquatic phase as a dissolved or colloidal organic metal complex.

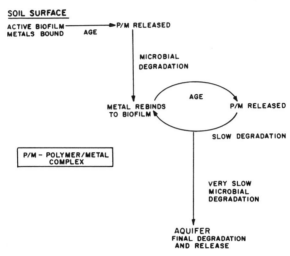

Fig. 4.6. A potential model for the cycling and transport of toxic metals in the subsurface. (Adapted from Black *et al.* [13].)

3. VOLATILIZATION OF METALS BY TRANSFORMATION

Mercuric ion resistance has been well studied. The mechanism is generally accepted to involve intracellular reduction of Hg^{2+} to Hg^0 by mercuric reductase, with a subsequent volatilization. Mercuric reductases have been isolated from a number of microorganisms, including *E. coli* (58,59), *Thiobacillus ferrooxidans* (60), *Streptomyces*, *Streptococcus* (61), and *Caulobacter* (62).

Evidence exists that certain metal-tolerant bacteria use toxic metal species as electron acceptors. Selenate has been shown to be reduced by anaerobic bacteria (63–65). Chromate reduction under anaerobic conditions has also been reported (*e.g.*, 66), and chromate reductase activity has been associated with a soluble protein (67).

Undoubtedly the most studied environmentally mediated transformation of toxic metals is the methylation of mercury. This arose from research on the mercury-polluted sediment of Minemata Bay, Japan. Extensive mercury poisoning from fish consumption was traced to mercury in industrial waste disposal. Subsequent biological transformation to methylmercury, an extremely lipid-soluble form of the metal, resulted in accumulation in fish. Although there is now a considerable literature describing the methylation of mercury (see refs. 68–70 for reviews), it is still unclear as to whether microbial mercury methylation is a significant factor in mercury accumulation throughout the biota. Pure culture experiments have shown that many bacteria, including *Clostridium*, *Neurospora*, *Pseudomonas*, *Bacillus*, *Mycobacterium*, *E. coli*, *Aerobacter aerogenes*, *Bacillus megaterium*, and a number of fungi have the capability to methylate mercury (70). However, the sulfate-reducing bacteria (SRB) are the most significant (71). The mechanism of mercury methylation has been, at least partially, elucidated (68). The metal is methylated intracellularly by nonenzymatic transfer of methyl groups from methylcobalamin (vitamin B12). The reader is referred to the review article by Robinson and Tuovinen (68) for further information.

Methylation of mercury by sulfate-reducing bacteria may provide an important pathway for mercury accumulation by fish (71–73). It is also apparent that lake acidification increases levels of methylmercury in fish. However, there is considerable debate as to the mechanism. Winfrey and Rudd (69) report decreased methylation in sediments after acidification. Wiener *et al.* (74) suggest that acidification results in increased gill permeability to methylmercury. In contrast, Gilmour and Henry (70) postulate that there is an optimal degree of lake acidification for bacterial methylation. They propose that above a certain concentration of sulfate, methylation is inhibited because of the rapid formation of the insoluble mercuric sulfide. However, some acidification will stimulate SRB activity and hence mercury methylation. The importance of SRB in mercury methylation within a particular aquatic system will depend to a large degree on concentrations of methylmercury that enter from the surrounding catchment and in precipitation (74,75). This subject is treated in more detail by Gilmour, this volume.

Volatilization of other metals by microbial methylation has received less attention, as have other forms of transformation. These processes are reviewed by Thayer and Brinckman (2), who suggest that rates of other transformations, *e.g.*, ethylation and phenylation, are unlikely to be significant because of the large size of the organic groups. Methylation has, however, been shown for arsenic, lead, selenium, tellurium, tin, thallium, and antimony. As a result of the use of organo-tins in anti-fouling paints, there is some information about biomethylation of tin by estuarine bacteria. Gilmour *et al.* (76) correlated production of monomethyl tin in whole sediment samples with numbers of sulfate-reducing and sulfide-oxidizing bacteria. In addition, these authors showed that *Desulfovibrio* isolated from sediments methylate tin in culture medium at similar rates to natural methylation at sediment–water interfaces.

Methylation of arsenic by fungi has been studied extensively as a result of human poisoning by fungal transformations of arsenic in paints (reviewed in ref. 77). More recent work has been concerned with transformation and mobilization in sediments and soils (*e.g.*, 78). The importance of biomethylation cannot be overstressed because of the potential health consequences of high methylation rates.

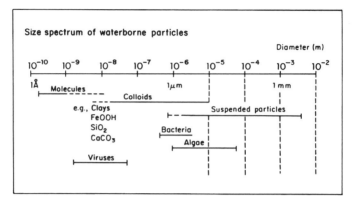

Fig. 4.7. Size spectrum of waterborne particles. Reproduced from Stumm [79].

4. MODELS OF BIOLOGICAL FATE AND TRANSPORT OF METALS

With limited biological information, it is extremely difficult to construct predictive models of fate and transport of toxic metals in either terrestrial or aquatic systems. If we are prepared to make assumptions based on the limited data available, then we may treat microorganisms and their products as colloids. Using the size spectra for particles from Stumm (79) shown in Figure 4.7, we can see that bacteria, viruses, and algae fit easily into the colloidal size range. Transport of bacterial products such as exopolymers can be described directly using colloidal transport theory.

A microbial polysaccharide–metal complex can be expected to be mobile under certain soil conditions, primarily for two reasons: 1) hydrophobicity (surface charge) and 2) size exclusion from soil pores. Mobility of a colloidal particle has been described in terms of changes in mass concentrations as a function of dispersive and convective flux and mass transfer between adjacent phases (80). The relationship between retardation of a chemical and the fluid velocity is described in the equation:

$$ R\frac{dC}{dt} = D\frac{dC}{dx^2} - V\frac{dC}{dx}, $$

where R is the retardation factor, C is the mass concentration of the mobile phase, t is the time (in days), D is the dispersion coefficient ($m^2\ d^{-1}$), x is the flow path distance (m), and V is the interstitial velocity of mobile phase (80).

Enfield and Bengtsson (80) used blue dextran, to study this relationship. The movement of this dye in soil columns was compared with tritiated water as a conservative tracer. The theoretical relationship suggested that the retardation factor for blue dextran would be 1.009. However, in practice the dye eluted from experimental columns slightly before the water, suggesting that average interstitial velocity is higher for blue dextran than for water. The authors suggest that the apparent anomalous results occur because of size exclusion, with effective pore volume for blue dextran approximately 13% less than for tritiated water. This type of information has been used to argue that hydrophilic organic macromolecules may be responsible for mobilization of hydrophobic compounds, e.g., DDT (81). This hypothesis depends on the ability of macromolecules to sorb hydrophobic contaminants preferentially over the stationary soil particles (80,82). Transport of toxic metals by bacterial products must therefore be implied. Active binding, often in excess of binding constants with humic material and other soil constituents, may result in retention of metals in solution over long time periods (13). Transport of toxic metals in association with colloids has been shown in a number of cases (83,84).

Transport models applied to microorganisms are much more complex. Mobility, stickiness, replication, metabolite production, aggregate formation, growth, and death should all be considered. However, understandably, we are still at the stage of applying simplistic models. Yates and Yates (85) attempted to address the problem by adding a decay term to the advection, dispersion, and adsorption terms usually included in a transport model. However, as they correctly point out, decay rates are organism specific, as are the other metabolic factors mentioned above. To model a natural population, all characteristics of each component organism need to be factored in. Clearly this is impossible. We are only able to identify a fraction of the organisms present in the environment, many of which are nonculturable with current technology.

When traditional models based on soil hydraulic parameters are used, it appears that movement of microorganisms in soils is minimal (86). Field research suggests that this is not the case. Viruses and bacteria have been shown to move considerable distances (87,88). Germann *et al.* (89) used a kinematic wave model to investigate movement of *E. coli* through soil columns. The kinematic wave theory treats flow as a boundary-layer phenomenon and not as flow in a potential field (90). Parameters were derived from percolation of bacterial suspensions through soil columns. Although the approach may be useful in modeling gravity-driven percolations, it is unclear whether any advantage is provided for modeling environmental systems. Size exclusion chromatography and cell surface characteristics are the mechanisms proposed for transport behavior of bacteria in a sandy aquifer (91). Harvey *et al.* (91) compared transport rates between cells, latex microspheres, and bromide as a conservative tracer. Although breakthrough of bacteria occurred well in advance of the tracer, as expected with the size exclusion model, it also occurred in advance of similarly sized microspheres. A charge affect was indicated, since carboxylated micro-scopheres were retarded more than neutral latex microspheres.

Recently, Miralles-Wilhelm *et al.* (92) presented a model for biodegradation of organic contaminants. It combines hydrological and chemical considerations with a biodegradation term based on growth of the microorganisms with respect to the contaminant and oxygen. The model is complex, and the reader is referred to the original paper. However, it serves to illustrate the necessary direction that transport models must take if we are to predict subsurface fate and transformations of toxic metals accurately.

5. MODELS OF MICROBIAL TRANSPORT OF METALS IN SURFACE WATER

Metal mobility in surface waters is dependent on numerous chemical and physical characteristics of the system. These factors all affect speciation of the given metal and hence its affinity for biological material. Numerous models have been applied to speciation of dissolved metals, based on equilibrium chemistry. However, bacteria/algae/viruses and their products present an extremely complex equilibrium! Studies on particle-based systems in freshwater are limited. Sigg *et al.* (93) calculated the removal rate of metals by settling particles:

$$fp \times k_s = \frac{F_m}{h \times C_t} \ (d^{-1}),$$

where fp is the fraction of element in particulate form, k_s is the rate constant characterizing sedimentation (d^{-1}); F_M is the flux of metal $(mg \ m^{-2} \ d^{-1})$, h is the height of water column (m), and C_t is the total concentration in the water column $(mg \ m^{-3})$.

As with groundwater models, we are limited by the lack of a biological component. Microbial parameters are essential to our understanding of metal transport in surface waters. We must take into account exudate production and sloughing rates from surface films, as well as growth and death of the microbial community.

6. CONCLUSIONS

1. Microorganisms play an important role in metal transport.
2. Mechanisms include
   a. Intracellular accumulation
   b. Cell surface binding

   c. Siderophore interactions

   d. Metabolite solubilization or immobilization

   e. Polymer binding and transport (either as a hydrophilic colloid in soil water or as part of a suspended aggregate in surface waters)

   f. Enzyme-mediated transformations

3. Industrial application of microogranisms for metal mining and treatment of toxic wastes rich in metals is providing new insights into transport processes. This information will facilitate development of more accurate predictive transport models.

REFERENCES

1. Brierley CL, Brierley JA, Davidson MS: Applied microbial processes for metals recovery and removal from wastewater. In Beveridge TJ, Doyle RJ (eds): Metal Ions and Bacteria. New York: Wiley, 1989, p 359.

2. Thayer JS, Brinckman FE: The biological methylation of metals and metalloids. Adv Organomet Chem 20:313, 1982.

3. Simkiss K, Taylor MG: Metal fluxes across the membranes of aquatic organism. Rev Aquat Sci 1:173, 1989.

4. Sigg L: Surface Chemical Aspects of the Distribution and Fate of Metal Ions in Lakes. In Stumm W (ed): Aquatic Surface Chemistry: Chemical Processes at the Particle–Water Interface. New York: Wiley, 1987, p 319.

5. Wood JM, Wang HK: Strategies for microbial resistance to heavy metals. In Stumm W (ed): Chemical Processes in Lakes. New York: Wiley, 1985.

6. Williams, RJP: Physico-chemical aspects of inorganic element transfer through membranes. Philos Trans R Soc Lond [Biol] 294:57, 1981.

7. Williams RJP: The symbiosis of metal ion and protein chemistry. Pure Appl Chem 55:35, 1983.

8. Ingraham JL, Maaloe O, Neidhardt, FC: Growth of the Bacterial Cell. Sunderland, MA: Sinauer Assoc, 1983.

9. Tabillon R, Weber F, Kaltwasser H: Nickel requirement for chemilothotrophic growth in hydrogen-oxidizing bacteria. Arch Microbiol 124:131, 1980.

10. Barraquio WL, Knowles R: Beneficial effects of nickel on *Pseudomonas saccharophila* under nitrogen limited chemolithotrophic conditions. Appl Environ Microbiol 55:3197, 1989.

11. Macaskie LE, Dean ACR, Cheetham AK, Jakeman RJB, Skarnulis AJ: Cadmium accumulation by a *Citrobacter* sp: The chemical nature of the accumulated metal precipitate and its location on the bacterial cells. J Gen Microbiol 133:539, 1987.

12. Aickin RM, Dean ACR, Cheetham AK, Skarnulis AJ: Electron microscope studies on the uptake of lead by a *Citrobacter sp.* Microbios Lett 9:7, 1979.

13. Black JP, Ford TE, Mitchell R: The role of bacterial polymers in metal release into water. In Cullimore R (ed): Proc Int Symp Biofouled Aquifers. AWRA, 1986, p 37.

14. Brierly JA, Brierly CL: Biological methods to remove selected inorganic pollutants from uranium mine wastewater. In Trudinger PA, Walter MR, Ralph BJ (eds): Biogeochemistry of Ancient and Modern Environments, Canberra: Australian Academy of Science, 1980, p 661.

15. Strandberg GW, Shumate SE, Parrott JR: Microbial cells as biosorbents for heavy metals: Accumulation of uranium by *Saccharomyces cerevisiae* and *Pseumonas aeruginosa*. Appl Environ Microbiol 41:237, 1981.

16. Xue H-B, Stumm W, Sigg L: The binding of heavy metals to algal surfaces. Water Res 22:917, 1988.

17. DiRienzo, JM, Nakamura K, Inouye M: The outer membrane proteins of gram-negative bacteria: Biosynthesis, assembly, and functions. Annu Rev Biochem 47:481, 1978.

18. Coughlin RT, Tonsager S, McGroarty EJ: Quantitation of metal cations bound to membranes and extracted lipopolysaccharide from *Escherichia coli*. Biochemistry 22:2002, 1983.

19. Ferris FG: Metallic ion interactions with the outer membrane of gram-negative bacteria. In Beveridge TJ, Doyle RJ (eds): Metal Ions and Bacteria. New York: Wiley, 1989, p 295.

20. Beveridge TJ: Mechanisms of the binding of metallic ions to bacterial walls and the possible impact on microbial ecology. In Reddy CA, Klug MJ (eds): Current Perspectives in Microbial Ecology. American Society of Microbiology, Washington, DC: 1984, p 601.

21. Bickel H, Hall GE, Keller-Schierlein W, Prelog V, Vischer E, Weltstein A: Stoffwechselprodukte von actinomyceten uber die konstitution von ferrioxamin B. Helv Chim Acta 43:2129, 1960.

22. Davis WB, Byers RR: Active transport of iron in Bacillus megaterium: Role of secondary hydroxamic acids. J Bacteriol 107:491, 1971.

23. Emery TF, Neilands JB: Structure of ferrichrome compounds. J Am Chem Soc 83:1626, 1961.

24. Hider RC: Siderophore mediated absorption of iron. Struct Bonding 58:26, 1984.

25. Raymond KN, Muller G, Matzanke BF: Complexation of iron by siderophores. Top Curr Chem 123:49, 1984.

26. Ochiai J: Bioinorganic chemistry of oxygen. J Inorg Nucl Chem 37:1503, 1975.

27. Fekete FA, Emery T, Spence JT: Molybdenum-coordinating compound produced by nitrogen-fixing cells of *Azotobacter vinelandii*. Am Soc Microbiol Abstr No 45, 1982.

28. McKnight DM, Morel FMM: Copper complexation by siderophores from filamentous blue-green algae. Limnol Oceanogr 25:62, 1980.

29. Schinner F, Burgstaller W: Extraction of zinc from industrial waste by a *Penicillium sp.* Appl Environ Microbiol 55:1153, 1989.

30. Francis AJ, Dodge CJ: Anaerobic microbial remobilization of toxic metals coprecipitated with iron oxide. Environ Sci Technol 24:373, 1990.

31. DeCarlo EH, Thomas DM: Removal of arsenic from geothermal fluids by adsorptive bubble flotation with colloidal ferric hydroxide. Environ Sci Technol 19:538, 1985.

32. Chanmugathas P, Bollag J-M: Microbial role in immobilization and subsequent mobilization of cadmium in soil suspensions. Soil Sci Soc Am J 51:1184, 1987.

33. Chanmugathas P, Bollag J-M: Microbial mobilization of cadmium in soil under aerobic and anaerobic conditions. J Environ Qual 16:161, 1987.

34. Chanmugathas P, Bollag J-M: A column study of the biological mobilization and speciation of cadmium in soil. Arch Environ Contam Toxicol 17:229, 1988.

35. Brown MJ, Lester JN: Metal removal in activated sludge: the role of bacterial extracellular polymers. Water Res 13:817, 1979.

36. Brown MJ, Lester JN: Role of bacterial extracellular polymers in metal uptake in pure bacterial culture in activated sludge. I. Effects of metal concentration. Water Res 16:1539, 1982.

37. Brown MJ, Lester JN: Role of bacterial extracellular polymers in metal uptake in pure bacterial culture in activated sludge. II. Effects of mean cell retention time. Water Res 16:1549, 1982.

38. Lester JN, Sterrett RM, Rudd T, Brown MJ: Assessment of the role of bacterial extracellular polymers in controlling metal removal in biological waste water treatment. In Grainger JM, Lynch JR (eds): Microbiological Methods for Environmental Biotechnology. London: Academic Press, 1984, p 197.

39. Sutherland IW: Bacterial exopolysaccharides. Adv Microbiol Physiol 8:143, 1972.

40. Sutherland IW: Biosynthesis of microbial exopolysaccharides. Adv Microbiol Physiol 23:79, 1982.

41. Sutherland IW: Biosynthesis and composition of gram negative bacterial extracellular and wall polysaccharides. Annu Rev Microbiol 39:243, 1985.

42. Geesey GG, Mittelman MW: The role of high-affinity, metal-binding exopolymers of adherent bacteria in microbial-enhanced corrosion. In Corrosion/85. Houston: Natl Assoc Corrosion Eng, 1985, No. 297.

43. Geesey GG, Mittelman MW, Iwaoka T, Griffiths PR: The role of bacterial exopolymers in the deterioration of metallic copper surfaces. Mat Perf 25:37, 1986.

44. Geesey GG, Iwaoka T, Griffiths PR: The characterization of interfacial phenomena occurring during exposure of a thin copper film to an aqueous suspension of an acidic polysaccharide. J Colloid Interface Sci 120:370, 1987.

45. Geesey GG, Jang L: Interactions between metal ions and capsular polymers. Beveridge TJ, Doyle RJ (eds): Metal Ions and Bacteria. New York: Wiley, 1989, p 325.

46. Geesey GG, Jang L, Jolley JG, Hankins MR, Iwaoka T, Griffiths PR: Binding of metal ions by extracellular polymers of biofilm bacteria. Water Sci Technol 20:161, 1989.

47. Ford TE, Maki JS, Mitchell R: The role of metal-binding bacterial exopolymers in corrosion processes. In Corrosion/87. Houston: Natl Assoc Corrosion Eng, 1987, No. 380.

48. Ford TE, Maki JS, Mitchell R: Involvement of bacterial exopolymers in biodeterioration of metals. In Houghton DR, Smith RN, Eggins HOW (eds): Biodeterioration 7. London: Elsevier Applied Science, 1988, p 378.

49. Ford TE, Black JP, Mitchell R: Relationship between bacterial exopolymers and corroding metal surfaces. In Corrosion/90. Houston: Natl Assoc Corrosion Eng, 1990, No. 110.

50. Ford TE, Mitchell R: The ecology of microbial corrosion. In Marshall KC (ed): Advances in Microbial Ecology, vol 11. New York: Plenum, 1990, p 231.

51. Mittelman MW, Geesey GG: Copper binding characteristics of exopolymers from a freshwater sediment bacterium. Appl Environ Microbiol 49:846, 1985.

52. Lion LW, Harvey RW, Young LY, Leckie JO: Particulate matter: Its association with micro-organisms and trace metals in an estuarine salt marsh microlayer. Environ Sci Technol 13:1522, 1979.

53. Cowen JP, Silver MW: The association of iron and manganese with bacteria on marine macroparticulate material. Science 224:1340, 1984.

54. Cho BC, Azam F: Major role of bacteria in biogeochemical fluxes in the ocean's interior. Nature 332:441, 1988.

55. Walker SG, Flemming CA, Ferris FG, Beveridge TJ, Bailey GW: Physiochemical interaction of *Escherichia coli* cell envelopes and *Bacillus subtilis* cell walls with two clays and ability of the composite to immobilize heavy metals from solution. Appl Environ Microbiol 55:2976, 1989.

56. Flemming CA, Ferris FG, Beveridge TJ, Bailey GW: Remobilization of toxic heavy metals adsorbed to bacterial wall-clay composites. Appl Environ Microbiol 56:3191, 1990.

57. Hamilton-Taylor J, Willis M: A quantitative assessment of the sources and general dynamics of trace metals in a soft water lake. Limnol Oceanogr 35:840, 1990.

58. Izaki K, Tashiro Y, Funaba T: Mechanisms of mercuric chloride resistance in microorganisms. III. Purification and properties of a mercuric ion reducing enzyme from *Escherichia coli* bearing R factor. J Biochem (Tokyo) 75:591, 1974.

59. Schottel JL: The mercuric and organomercurial detoxifying enzymes from a plasmid-bearing strain of *Escherichia coli.* J Biol Chem 253:4341, 1978.

60. Olson GJ, Porter FD, Rubinstein J, Silver S: Mercuric reductase enzyme from a mercury-volatilizing strain of *Thiobacillus ferrooxidans.* J Bacteriol 151:1230, 1982.

61. Nakahara H, Schottel JL, Yamaha T, Miyakawa Y, Asakawa M, Harville J, Silver S: Mercuric reductase enzymes from *Streptomyces* species and group B *Streptococcus.* J Gen Microbiol 131:1043, 1985.

62. Guangyong JI, Salzberg SP, Silver S: Cell-free mercury volatilization activity from three marine *Caulobacter* strains. Appl Environ Microbiol 55:523, 1989.

63. Maiers DT, Wichlacz PL, Thompson DL, Bruhn DF: Selenate reduction by bacteria from a selenium-rich environment. Appl Environ Microbiol 54:2591, 1988.

64. Zehr JP, Oremland RS: Reduction of selenate to selenide by sulfate-respiring bacteria: Experiments with cell suspensions and estuarine sediments. Appl Environ Microbiol 53:1365, 1987.

65. Oremland RS, Hollibaugh JT, Maest AS, Presser TS, Miller LG, Culbertson CW: Selenate reduction to elemental selenium by anaerobic bacteria in sediments and culture: Biogeochemical significance of a novel, sulfate-independent respiration. Appl Environ Microbiol 55:2333, 1989.

66. Wang PC, Mori T, Komori K, Sasatsu M, Toda K, Ohtake H: Isolation and characterization of an *Enterobacter cloacae* strain that reduces hexovalent chromium under anaerobic conditions. Appl Environ Microbiol 55:1665, 1989.

67. Ishibashi Y, Cervantes C, Silver S: Chromium reduction in *Pseudomonas putida.* Appl Environ Microbiol 56:2268, 1990.

68. Robinson JB, Tuovinen OH: Mechanisms of microbial resistance and detoxification of mercury and organomercury compounds: physiological, biochemical, and genetic analyses. Micro Rev 48:95, 1984.

69. Winfrey MR, Rudd JWM: Environmental factors affecting the formation of methylmercury in low pH lakes. Environ Toxicol Chem 9:853, 1990.

70. Gilmour CC, Henry EA: Mercury methylation in aquatic systems affected by acid deposition. Environ Pollut (in press).

71. Compeau GC, Bartha R: Sulfate-reducing bacteria: Principal methylators of mercury in anoxic sediments. Appl Environ Microbiol 50:498, 1985.

72. Compeau, GC, Bartha R: Effect of salinity on mercury-methylating activity of sulfate-reducing bacteria in estuarine sediments. Appl Environ Microbiol 53:261, 1987.

73. Berman M, Bartha R: Levels of chemical versus biological methylation of mercury in sediments. Bull Environ Contam Toxicol 36:401, 1986.

74. Wiener JG, Fitzgerald WF, Watras CJ, Rada RG: Partitioning and bioavailability of mercury in an experimentally acidified Wisconsin lake. Environ Toxicol Chem 9:909, 1990.

75. Schroeder WH, Munthe J, Lindqvist O: Cycling of mercury between water, air, and soil compartments of the environment. Water Air Soil Pollut 48:337, 1989.

76. Gilmour CC, Tuttle JH, Means JC: Tin methylation in sulfide bearing sediments. In Sigleo AC, Hattori A (eds): Marine and Estuarine Geochemistry, Chelsea, MI: Lewis, 1985, p 239.

77. Challenger F: Biological methylation. Chem Rev 36:326, 1945.

78. Brannon JM, Patrick WH: Fixation, transformation, and mobilization of arsenic in sediments. Environ Sci Technol 21:450, 1987.

79. Stumm W: Chemical interaction in particle separation. Environ Sci Technol 11:1066, 1977.

80. Enfield CG, Bengtsson G: Macromolecular transport of hydrophobic contaminants in aqueous environments. Groundwater 26:64, 1988.

81. Enfield CG, Walters DM, Carsell RF, Cohen SZ: Approximating transport of organic pollutants in ground water. Groundwater 20:711, 1982.

82. McCarthy JF, Zachara JM: Surface transport of contaminants. Environ Sci Technol 23:496, 1989.

83. Buddemeier RW, Hunt JR: Transport of colloidal contaminants in ground water: Radionucle-otide migration at Nevada test site. Appl Geochem 3:535, 1988.

84. McCarthy JF, Wobber FJ (eds): Mobility of Colloidal Particles in the Subsurface: Chemistry and Hydrology of Colloid-Aquifer Interactions, Washington, DC: U.S. Department of Energy, 1988, DOE/ER-0425.

85. Yates MV, Yates SR: Modeling in microbial transport in soil and groundwater. ASM News 56:324, 1990.

86. Corapcioglu MY, Haridas A: Transport and fate of microorganisms in porous media: A theoretical investigation. J Hydrol 72:149, 1984.

87. Keswick BH, Wang DS, Gerba CP: The use of microorganisms as ground water tracers: A review. Groundwater 20:142, 1982.

88. Bales RC, Gerba CP, Grondin GH, Jensen SL: Bacteriophage transport in sandy soil and fractured tuff. Appl Environ Microbiol 55:2061, 1989.

89. Germann PF, Smith MS, Thomas GW: Kinematic wave approximation to the transport of *Escherichia coli* in the vados zone. Water Res Res 23:1281, 1987.

90. Germann PF: Fluid mechanical considerations on colloidal transport in porous media. In McCarthy JF, Wobber FJ (eds): Mobility of Colloidal Particles in the Subsurface: Chemistry and Hydrology of Colloid–Aquifer Interactions. Washington, DC: U.S. Department of Energy, 1988, DOE/ER-0425.

91. Harvey RW, George LH, Smith RL, LeBlanc DR: Transport of microspheres and indigenous bacteria through a sandy aquifer: Results of natural- and forced-gradient tracer experiments. Environ Sci Technol 23:51, 1989.

92. Mirrales-Wilhelm F, Gelhar LW, Kapoor V: Effects of Heterogeneities on Field-Scale Biodegradation in Groundwater. Paper H12A-04. Proc Am Geophysical Union, 1990.

93. Sigg L, Sturm M, Kistler D: Vertical transport of heavy metals by settling particles in Lake Zurich. Limnol Oceanogr 32:112, 1987.

5

TRANSPORT OF PATHOGENS THROUGH SOILS AND AQUIFERS

GABRIEL BITTON

Department of Environmental Engineering Sciences, University of Florida, Gainesville, Florida 32611

RONALD W. HARVEY

United States Geological Survey, Menlo Park, California 94025

1. INTRODUCTION

Between 1971 and 1980, the use of untreated groundwater was responsible for more than one-third of the waterborne disease outbreaks in the United States (1). This statistic points to the potential for subsurface contamination by pathogenic microorganisms. The major sources of pathogens are waste water effluents, residual sludges from waste treatment (2), and septic tank effluents (3, 4). The number of septic tanks in the United States has been estimated at 22 million units. These on-site treatment systems serve approximately one-third of the U.S. population (4). They are major contributors to the contamination of subsurface environments. The contaminants are household chemicals (nitrate, heavy metals, organic toxicants), pathogenic microorganisms, and parasitic cysts.

The fate of pathogenic microorganisms in soils and aquifer material is primarily governed by their transport and persistence in these environments. The survival and transport of pathogens in soils and aquifers are controlled by four major factors: climate (*e.g.,* temperature, rainfall), type of soil or aquifer ma-

Environmental Microbiology, pages 103–124, © *1992 Wiley-Liss, Inc.*

terial (*e.g.*, texture, pH, water holding capacity, cation exchange capacity), properties of pore fluids (*e.g.*, chemistry, saturation), and type of pathogen (5). In the following sections, we review the major factors controlling the persistence and transport of pathogens and discuss modeling efforts to assess pathogen fate in soil and subsurface environments.

2. PERSISTENCE OF PATHOGENIC MICROORGANISMS IN SOILS AND GROUNDWATER

The topic of enteric bacterial survival in soils has been extensively reviewed (5–9). The major factors that control their persistence in this complex environment are temperature, moisture content, sunlight, pH, organic matter, bacterial type, and antagonistic microflora. Persistence is generally highest at low temperature, high soil moisture content, and abundant organic matter that may allow the growth of certain bacteria. Some pathogens, such as mycobacteria, are extremely hardy and survive for several months in soils. Pathogen survival is generally lower in acidic soils. Sunlight has a detrimental effect at the soil surface. Competition and antagonism from other bacteria, as well as predation by protozoa, can contribute to the decline of pathogenic and indicator bacteria in soils.

Many of the abiotic factors that affect survival of bacteria in the subsurface also influence the persistence of viruses (10). Development in the late 1970s of better methods for virus recovery from soils increased our knowledge of the persistence of viruses in this complex environment. In general, all methods of virus recovery include an elution step followed by a concentration step. The most commonly used eluants include glycine-EDTA, beef extract, casein, and non-fat dry milk (11, 12). These procedures result in a recovery efficiency of approximately 50%. Viruses have also been recovered by eluting soil with deionized water and concentrating the eluant by sorption onto aluminum hydroxide ($Al[OH]_3$) flocs (13).

Soil temperature (14–16) and moisture (15, 17–21) are the primary factors affecting virus survival in soils treated with waste water effluents or sludge. Viruses survive best in moist soils under low temperature. Soil type may also influence survival, although the effects are probably related to the degree of viral adsorption (10). There is considerable variation among virus types in their ability to survive in subsurface environments. Among the enteric viruses, hepatitis A virus (HAV) appears to be among the most resistant to inactivation in soils (22).

Virus persistence in groundwater was studied under laboratory as well as field conditions, using traditional glass containers and McFeters-type diffusion chambers. These investigations indicate that bacteria and viruses persist longer in groundwater than in surface waters (21, 23–25). The decay rate (k) of poliovirus in groundwater was found to be 0.0019 hr^{-1} (23). Similar decay rates were found for enteroviruses (poliovirus 1; echoviruses 6, 11, and 24; coxsackievirus B6) incubated in dialysis tubes under *in situ* conditions (25); decay rates of viruses in well water varied between 0.0004 to 0.0037 hr^{-1}. Virus survival in groundwater appears to be influenced by water temperature, dissolved oxygen, and, possibly, by indigenous microorganisms in groundwater. Yates and Gerba (26, 27) also showed that temperature was important in controlling phage persistence and that survival was also dependent on the groundwater source.

Geostatistical techniques (*e.g.*, kriging, cokriging, and combined kriging and regression) have been considered for estimating virus inactivation in groundwater in order to predict safe septic tank setback distances for drinking water well installation in the vicinity of septic tank drainfields (10, 28, 29). Results of recent laboratory investigations suggest that the relationship between virus inactivation rates and groundwater temperature may be described by a linear equation (10, 30). Combined use of kriging, which can be used to calculate rates of viral inactivation at unsampled locations from values obtained for samples collected nearby, and a linear regression equation relating viral inactivation and temperature allows estimation of setback distances between wells and contamination sources by knowing only the groundwater temperature and flow characteristics (10). Setback distances (D), in meters, can be calculated with the following equation:

$$D = (tKi)/\theta$$

where t is the travel time required (days) for a 10 million-fold decline in number of active viruses, which is calculated from kriged estimates of virus inactivation; K is the the hydraulic conductivity (m/day); i is the hydraulic gradient (m/m); and θ is the effective porosity (29).

3. PATHOGEN TRANSPORT THROUGH SOILS AND AQUIFER MATERIALS

3.1 Factors Controlling the Transport of Pathogenic Microorganisms

3.1.1. Bacterial Transport

Straining, which occurs within pores that are smaller than the limiting dimension of the cell, and adsorption onto particles are the major factors controlling transport of bacteria through soils (5). Bacterial cells are negatively charged biocolloids and, as such, they interact with soil particles (sand, silt, clay minerals, metal oxides). Bacterial adsorption to soils is favored in the presence of cations, clay minerals, low concentrations of soluble organics, and at low pH conditions (5). Bacterial removal by straining is inversely proportional to particle size of soils. Bacteria strained at the soil surface promote the retention of finer particles in soil, a phenomenon occurring during mat formation resulting from the application of septic tank effluents. The biological clogging mat or crust appears to be an effective barrier to bacterial breakthrough. The clogging mat may be due to biological factors such as the presence of microbial polysaccharides (31, 32).

Bacterial transport through absorption fields of septic tanks is controlled by soil porosity and the degree of saturation with water. Under saturated conditions bacteria may be transported over much greater distances than under unsaturated conditions. Studies have shown that pathogen transport through soil is promoted by heavy rainfall, and bacterial contamination of wells was found to coincide with periods of heavy rainfall (33, 34). Rainfall generally lowers the ionic strength of the pore fluid and thus promotes bacterial transport through soils. Laboratory experiments with sludge–soil mixtures challenged with bacterial indicators, total coliforms (TC) and fecal coliforms (FC), have shown that only heavy rainfall (12.3 cm/day) promotes significant downward transport of the bacteria to the bottom of the 8 inch-deep column. Lower amounts of rainfall did not cause significant migration of the bacterial cells. Bacterial breakthrough in the column leachates never exceeded 2.4% of TC and 1.4% of FC inputs to the columns (34).

Agricultural scientists have studied the introduction of beneficial microorganisms in the rhizosphere for the control of soilborne plant pathogens. Transport and survival of these microorganisms (*e.g., Rhizobium*) in the rhizosphere is of great interest to agronomists. These studies have shown that the vertical flow of water is an important transport mechanism for bacteria in the rhizosphere (35). Bacterial transport by percolating water is enhanced by preferred flow path structure (channels) created by plant roots and by the activity of soil invertebrates such as earthworms (36). Growing wheat roots were also found to enhance the downward transport of a genetically engineered microorganism by enhancing water flow through the soil (37).

3.1.2. Virus Transport

Because of their small size, viruses are less subject to straining in sandy aquifer sediments and soils than are bacteria. Therefore, bacteria do not serve as good indicators for virus transport into groundwater. Under field conditions, indicator bacteria, total coliforms, fecal coliforms, and fecal streptococci (TC, FC, and FS) originating from a septic tank system were efficiently retained by a sandy/loamy sand soil and were detected at levels less than 1 count per 100 ml of groundwater. However, polioviruses were transported to groundwater. (38).

The major abiotic impediment to the migration of viruses through the sub-surface is sorption onto solid surfaces. Virus adsorption to solids, including soils, can be explained largely in terms of surface interactions between the amino acids on the capsid and biological or nonbiological surfaces. These interactions include both electrostatic and hydrophobic interactions (39–43). Several factors control virus adsorption to soils. Binding to soils is affected by soil texture. The presence of clays and other minerals such as hematite and magnetite increases virus retention (39, 44–46). Conversely, soil organic matter may decrease adsorption (47–50); organic-rich sediments (''mucks'') have been observed to have a low retention capacity for viruses (50, 51). Increases in hydraulic flow rate through soils facilitate virus transport (52–54) by lessening the degree of adsorption.

Both batch and column experiments have shown that viruses do not readily sorb to or are released from soil particle surfaces when suspended in low ionic strength solutions (44, 49, 55–57). Rainwater, having low ionic strength, can be instrumental in the redistribution and transport of viruses within the soil profile (14, 52, 53). Stools from infants who received the trivalent oral poliovirus vaccine were used to inoculate a septic tank so that poliovirus could be used as a marker and followed downgradient in nearby groundwater (38). It was observed that virus numbers in groundwater increased as the distance from the drainfield increased. It was concluded that the relatively high ionic strength of the septic tank effluent (1,000–3,100 μS cm^{-1}) allowed virus adsorption to take place in the vicinity of the septic system. Virus desorption and movement away from the septic tank was promoted by dilution of the relatively high ionic strength of the septic tank effluent (38).

In contrast, soluble organic compounds in waste water effluents inhibit virus adsorption to surfaces including soils. Their presence would thus promote virus transport through soils (42, 58–60). Humic and fulvic acids also interfere with virus sorption to soils (50, 61).

Several investigations have shown that virus adsorption to solids (*e.g.,* soil, activated sludge flocs, sediments) depends on the type and strain of virus under consideration (55, 60, 62, 63). Virus adsorption to a sandy loam soil varied between 0% for echovirus 1, strain V239, to 99.9% for poliovirus type 1, strain LSc (63), presumably because of differences in surface characteristics of the viruses. The adsorption patterns of HAV and rotaviruses can differ from those of polioviruses and phages (22, 59). Adsorption capacity may also vary between isolates of the same virus type (49). A field study carried out at a groundwater recharge site in Western Australia showed considerable variation with regard to virus adsorption by a sandy soil (64). Viruses such as echoviruses 11, 14, 24, 29, and 30, coxsackievirus B4, and adenovirus type 3 penetrated 3.0 m of soil. Echovirus type 11 was detected in groundwater at 9 m below the recharge basin. Conversely, indigenous and laboratory strains of polioviruses both displayed higher affinity for the soil under study; it was also observed that the laboratory strains adsorbed better to soil than did the indigenous viruses.

Properly designed soil-column experiments offer a relatively rapid means for assessing virus transport patterns in sludge-amended soils (65). Soil column experiments as well as field studies have shown that sludge-associated viruses become immobilized at the soil surface and do not migrate significantly through soils (15, 16, 66–69). Hurst and Brashear (70), using a vacuum filtration technique to study leaching of viruses from sludge, indeed confirmed that viruses will not desorb significantly from sludge particles following land application of sludge. A similar trend was observed for sludge-associated bacteria. Liu (71), investigating the transport of sludge-associated bacteria into groundwater, has concluded that bacterial contamination from on-land sludge disposal was unlikely. Most of the sludge-associated bacteria were retained at the soil surface.

3.2. Use of Tracers To Study Transport in Aquifer Materials

Abiotic tracers have been used in a number of field and laboratory studies involving transport of microorganisms through soils and groundwater. Well-defined particulate and dissolved tracers can be useful in accounting for abiotic processes that contribute to or control movement of bacteria and viruses through subsurface media. Often, the processes controlling transport *in situ* are complicated by a number of geohydrological, chemical, and biological factors (72). Nonpathogenic microorganisms were employed in many of the co-transport studies involving tracers and microorganisms. However, such studies have shed light on a number of processes affecting the migration of pathogens. Dissolved tracers that are assumed to be conservative (nonreactive) were employed in a number of microbial transport experiments performed in aquifers. The tracers provide information about the hydrologic characteristics along the paths of travel. This can be difficult to determine solely from the concentration histories (breakthrough curves) of the organisms themselves, since microorganisms tend to interact with solid surfaces and do not necessarily follow the same mean flow paths as the groundwater in which they are suspended.

Tracer experiments that are employed to examine microbial transport behavior in aquifers are typically of two types: forced gradient and natural gradient (Fig. 5.1). In forced-gradient experiments, an artificial flow field is created by continuous pumping of water into an injection well (divergent tracer tests, Fig. 5.1A) or continuous withdrawal from a sampling well (convergent tracer test, Fig. 5.1B) at a rate that will override ambient flow conditions in the aquifer. The injectate containing the nonindigenous or labeled organisms and the abiotic tracers is thereby forced through the aquifer to the sampling well. One disadvantage of forced-gradient experiments is that the distance over which the introduced microbes can be followed is limited because of a rapid decrease in their

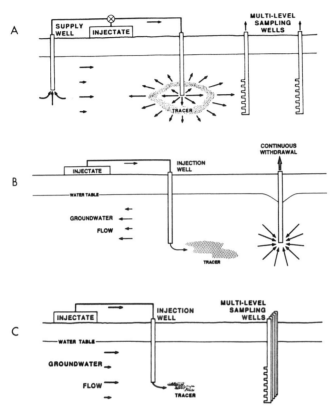

Fig. 5.1. Schematic representations of the design of small-scale groundwater injection experiments involving microorganisms and abiotic tracers. **A:** Divergent, forced gradient. **B:** Convergent, forced gradient. **C:** Natural gradient.

concentrations caused by dilution with groundwater beyond that caused by normal dispersion. The forced dilution is a consequence of the radial flow (convergent or divergent) created by the continuous pumping or injection. The hydrological component of tracer and microbe transport is also more difficult to model in forced-gradient tests, because the flow velocity changes along the flow path. These problems may be ameliorated in natural-gradient experiments, which involve passively injecting the tracers into the aquifer and monitoring the injectate as it moves with the natural flow of groundwater past wells that have been installed along the path of travel (Fig. 5.1C). However, correct placement of downgradient sampling wells can be problematic when the hydrology of the site is poorly understood because of physical heterogeneity, geohydrologic complexity, or lack of instrumentation.

Bromide is often used as the conservative tracer in *in situ* subsurface transport studies with microorganisms, because it is geochemically and biologically non-reactive, generally has a low background concentration, and can be readily assayed in the field. The differences in the concentration histories (breakthrough curves) at the sampling well for bromide and bacteria can provide useful information about the importance of physical factors that influence microbial transport: mean groundwater velocity, porosity, and longitudinal dispersion. The use of dissolved, conservative tracers also allows determination of a retardation factor for the microbial transport, which involves comparison of the time of arrival of the peak concentration of bromide to that of the microorganism that had been coinjected with bromide into the aquifer.

Retardation factors (mean flow velocity of the groundwater/mean transport velocity of the microorganisms) for the transport of microorganisms co-injected with bromide into freshwater aquifers are listed in Table 5.1. For small-scale (2–7 m) and forced- and natural-gradient tracer tests within a sandy aquifer in Massachusetts, U.S.A. (73, 74), arrival time of the maximum abundance of labeled, indigenous bacteria was nearly coincident with that of bromide, resulting in a calculated retardation factor of 1.0. In contrast, retardation factors for the other injection tests employing microorganisms were substantially less than 1.0, indicating that transport of microorganisms not immobilized by the medium was, on average, significantly faster than that of the conservative tracer and, presumably, mean groundwater flow. The apparent enhancement in transport velocity of the unattenuated microorganisms was greatest in the experiment that involved a fractured crystalline-rock aquifer (75). This phenomenon may be caused by preferential transport of organisms along preferred flow paths (large pores, fractures, and channels), because they may be excluded from the smaller pores on the basis of size. The absence of more rapid transport for the indigenous bacteria relative to bromide in the tracer tests on Cape Cod may be caused by the absence of a secondary pore structure; this is consistent with the well-sorted nature of the aquifer sediments at that site.

The results of the various *in situ* transport experiments suggest that the usefulness of halide tracers in studying the transport behavior of bacteria depends on the pore structure of the aquifer. At the Massachusetts site, the patterns of breakthrough for bacteria and bromide were similar (Fig. 5.2). Breakthrough of bromide and stained bacteria each exhibited single peaks and followed similar temporal patterns. In this system, bromide was useful in the construction of the hydrological portion of the overall bacterial transport model (74). In contrast, the breakthrough patterns observed for bacteria in forced-gradient experiments performed in other types of aquifers were quite different from that of bromide (Fig. 5.3). The earlier arrival of peak abundance and shorter duration of breakthrough for the bacteria relative to bromide (75) suggest that there are significant differences in paths of travel between bacteria and dissolved species in fractured or

**TABLE 5.1. Differences in Apparent Transport Velocity Between
Microorganisms and Bromide in Small-Scale Groundwater Tracer Experiments[a]**

Microorganism	Type of Distance Retardation				References
	aquifer	test	(m)	factor[b]	
E. coli	Fractured crystalline rock	Forced gradient	12.7	0.1	Champ and Schroeter (75)
Indigenous isolate (Bacillus sp.)	Layered basalt	Forced gradient	27	0.6	Harvey, Voss and Souza, (unpublished data)
Saccharo- myces cerevisiae	Sand and gravel (with clay and carbonate)	Forced gradient	1.5	0.7	Wood and Ehrlich (77)
Indigenous bacterial population	Well-sorted sand and gravel	Forced gradient	1.7 3.2	1.0 0.6[c]	Harvey et al. (73)
Indigenous bacterial population	Well-sorted sand and gravel	Natural gradient	6.7	1.0[d]	Harvey and Garabedian (7)

[a]Reproduced from Harvey (104).

[b]Calculated as the ratio of transport velocity at peak abundances of bromide to that of the microorganisms.

[c]The reason for the retardation factors substantially less than one for the sampler at 3.2 m is not clear. This was not observed in any of the other Massachusetts tracer tests involving bacteria.

[d]Value is for both 8.5 and 9.1 m depths.

fractured–layered rock aquifers. Therefore a great deal of caution should be used in employing dissolved, nonreactive tracers in bacterial transport tests in such systems.

In addition to bromide, a number of other dissolved organic and inorganic constituents have been used as tracers in studies of microbial transport through subsurface materials. These include rhodamine WT dye used in studies involving *Escherichia coli* migrating through a sandy New Zealand aquifer (76), iodide in studies involving yeast (*Saccharomyces cerevisiae*) introduced in a sand and gravel aquifer in Texas (77), and thiocyanate and pentafluorobenzoic acid in studies involving the bacteriophage f2 moving through fractured rock (column experiment) (78). In the latter experiment, the dissolved, nonreacting tracers were retarded by a factor of 1.2–2.0 relative to fluid flow through the fractures, apparently resulting from their diffusion into the porous matrix of the fractured-

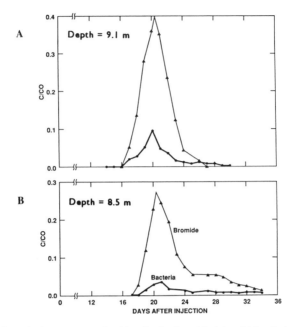

Fig. 5.2. Dimensionless concentration histories for bromide (closed triangles) and fluorochrome-labeled bacteria (closed circles) at a multilevel sampler 6.9 m downgradient from the point of injection. Calculated using data from Harvey and Garabedian (74). The experiment was a natural-gradient tracer test in layered, sandy aquifer sediments. **A:** 9.1 m below land surface; **B:** 8.5 m below land surface.

rock cores. Because the f2 phage remained in the fractures, estimates of dispersion and effective porosity applicable to the phage could not be calculated from data provided by the nonreactive tracer.

The advantage of using microbe-sized microspheres instead of dissolved tracers to delineate abiotic aspects of bacterial transport behavior is that the microspheres and bacteria should follow the same flow paths, even in aquifers with substantial preferred flow-path structure. The disadvantage of using microspheres is that the nature and intensity of interactions with solid surfaces can differ substantively from those experienced by bacteria. The differences in relative transport velocity and breakthrough between bromide or chloride and various types of bacteria-sized microspheres in small-scale, natural-gradient tests are listed in Table 5.2. Differences in transport behavior between microspheres and bacteria can involve the magnitude of retardation. In these experiments, bacteria at peak breakthrough were not retarded (relative to a conservative tracer), whereas most of the microspheres were subject to substantial retardation

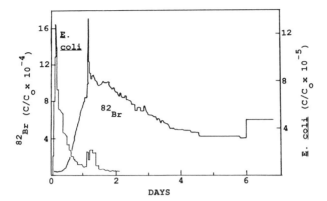

Fig. 5.3. Dimensionless concentration histories for bromide and bacteria (*E. coli*) downgradient from point of injection in the Chalk River Aquifer (Ontario, Canada). The tracer test was a forced gradient in fractured, crystalline rock. (Redrawn from Champ and Schroeter [75].)

TABLE 5.2. Differences in Apparent Transport Velocity and Magnitude of Breakthrough for Bacteria-Sized Colloids Relative to Bromide in Small-Scale, Natural-Gradient Groundwater Tracer Experiments (Mass. U.S.A.)[a]

Colloid	Type	Diameter (μm)	Retardation factor[b]	Relative break-through ($\times\ 10^{-2}$)[c]	References
Bacteria	Indigenous population	0.6	1.0	15–21	Harvey and Garabedian (74)
Plain latex	Uncharged	0.6	~1.0	0.05	Harvey et al. (73)
Polyacrolein	Carbonyl surface gps.	0.8	1.3	3.1	
Carboxylated latex	Carboxyl surface gps.	0.5	1.4	0.04	

[a]Reproduced from Harvey (104).

[b]Calculated as the ratio of transport velocity at peak abundances of bromide to that of the microorganisms or microspheres.

[c]Calculated as the integral of the dimensionless concentration history normalized to that of bromide.

(73). Only the neutral (uncharged) microspheres traveled at approximately the same rate as bromide or chloride. However, the degree of attenuation (immobilization at solid surfaces) for the neutral microspheres was several hundredfold higher than that observed for the bacteria in similar experiments. Although the polyacrolein microspheres exhibited rates of immobilization much closer to those exhibited by bacteria, they were significantly retarded, as were the carboxylated latex spheres (Table 5.2).

All of the particulate or dissolved tracers considered herein appear to have limitations for use in investigations of subsurface microbial transport behavior. In at least one aquifer having well-sorted sediments and little preferred flow-path structure, dissolved, nonreactive tracers, such as bromide, appear useful in delineating the hydrologic aspects of bacterial transport behavior (74). Even in fractured media, halides are useful as indicators of flow and may provide information regarding the potential movement of nutrients that sustain growth of moving bacterial populations. None of the bacteria-sized microspheres examined in the Massachusetts study exhibited the same combination of attenuation, retardation, or apparent dispersion as bacteria (71). However, microspheres appeared to be useful in the delineation of certain abiotic aspects of bacterial transport behavior.

Polioviruses as well as bacterial viruses have been routinely used as model indicator viruses in studies on transport through soils. An attempt was also made to use positively and negatively charged dyes (methylene blue and amaranth) and proteins with various isoelectric points (ferritin, myoglobin, and cytochrome c) to predict virus transport through soils (79). Dye adsorption to soils did not correlate with viral adsorption. However, ferritin (isoelectric point = 4.5) and cytochrome c (isoelectric point = 9.3) adsorption appeared to correlate well with the adsorption of specific types of viruses proposed by Gerba et al. (48).

4. MODELS DESCRIBING PATHOGEN TRANSPORT

4.1. Approach

Increasing contamination of soils and shallow aquifer sediments by microbial pathogens has led to a growing interest in the development of more accurate models that describe their transport. Because of the complexity involved in modeling microbial transport through subsurface environments, the approach can be as important as the mathematical descriptions. As discussed above, the extent of pathogen transport in soils may be affected by factors influencing their survival as well as their movement (5). Consequently, a number of modeling efforts involving pathogenic bacteria and viruses in subsurface environments has emphasized both persistence and transport (34, 72, 80–82). Often, a deterministic approach is employed. However, accurate simulations of microbial transport over distances that are on larger spatial scales than those of the physical heterogeneity in soils or aquifer sediments may require a probabilistic (stochastic) approach.

Both conceptual and data-based approaches have been used in the development of models for simulating subsurface transport of microorganisms. Recently, theoretical models for describing microbial transport through soils and other porous media have been developed (83, 84). However, verification of these conceptual models using field observations is problematic in part because of the complexity of the models and the number of parameters that would need to be determined *a priori*. Without adequate testing, it is difficult to ensure that proper emphasis is given to the various factors controlling transport and that the mathematics are adequate. There is at least some evidence to suggest that the predictive value of the conceptual models may be limited. For example, it has been reported that when flow parameters in the theoretical model described by Corapcioglu and Haridas (84) are stressed to permissible limits, the predicted extent of bacterial transport through unsaturated soil within a 2 week period is only 0.2 m (85). In contrast, Smith et al. (86) demonstrated that *E. coli* could penetrate through a column of undisturbed soil to a depth of 0.3 m within 20 minutes, and transport over several meters through well-sorted aquifer sediments has been observed for labeled bacteria and bacteria-sized microspheres (73). Nevertheless, comprehensive theoretical models provide a valuable conceptual framework for the development of more realistic models that are grounded on field observations.

A number of recent advances have been made in understanding the processes that govern microbial transport through soils and groundwater as a direct result of laboratory microcosm and *in situ* transport studies. Column experiments, which offer a greater degree of control than *in situ* studies, have facilitated improvements in the understanding of several factors governing microbial migration through the subsurface, including bacterial and viral adsorption to grain surfaces (65, 78, 87, 88) and bacterial growth and taxis (89–91). However, it has been shown that the transport behavior of microorganisms in aquifer sediments repacked in columns can be quite different from that observed *in situ,* even when flow velocity, porosity, and physicochemical conditions are similar (92). Consequently, the overall modeling of microbial transport through soils and aquifer sediments may require information derived from *in situ* experiments, since repacking soils and sediments into columns destroys the secondary (preferred flow-path) pore structure (86).

4.2. Bacteria

A number of biotic and abiotic factors can promote or inhibit the subsurface movement of bacterial pathogens. Several important factors that affect their

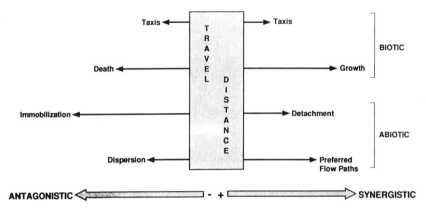

Fig. 5.4. Schematic representation of the role of biotic and abiotic processes in modeling transport of bacteria through the subsurface. "Antagonistic" and "synergistic" refer to the promotion and inhibition, respectively, of microbial transport in the direction of groundwater movement.

transport in the direction of flow are schematically depicted in Figure 5.4. An evaluation of the various factors as they relate to modeling transport of bacteria in aquifers is given by Harvey (93). Bacteria can move through soils and aquifer sediments by several mechanisms, including continuous, discontinuous (intermittent), and chemotactic migration. The relative contribution of each mechanism is not always well understood (93). Much of the modeling effort has focused on continuous transport, which assumes passive, unretarded transport of bacteria. However, discontinuous transport because of reversible sorption onto solid surfaces can be a primary mechanism of bacterial movement through the subsurface, especially where substantial distances are involved.

This discontinuous transport creates an apparent retardation of the bacteria relative to conservative, nonreactive transport. Retardation factors as high as 10 have been reported for bacterial populations traveling through porous aquifers in central Europe (72). The contribution of bacterial motility to overall transport can also be significant, because bacterial movement by taxis is much faster than that caused by random thermal (Brownian) motion. Jenneman et al. (90) reported that the rate of bacterial penetration through Berea sandstone cores in the presence of a nutrient gradient was up to eightfold higher for a motile bacterium (*Enterobacter aerogenes*) than for a nonmotile species (*Klebsiella pneumoniae*). However, for most pathogens in nutrient-limiting soils and aquifer sediments, the extent of transport may be limited largely by the geohydrology, interactions with the solid phase, and survival.

Recent efforts to model transport of enteric bacteria through unsaturated soils have employed data from laboratory column studies. Results of these modeling efforts suggest that pathogens may penetrate to greater depths in soils than was generally believed. Germann et al. (85) used a kinematic wave approximation (employed previously in describing fluid flow through a variety of porous media) to model published data involving migration of *E. coli* through columns containing four different types of soils (94). Their model suggested that few microorganisms will be transported any deeper than 3 m when durations and intensities of naturally occurring precipitation events are considered. However, transport to depths greater than 100 m was predicted for prolonged and/or intensified precipitation events. To predict the likelihood of substantive bacterial migration in coarse soils, Peterson and Ward (95) applied Monte Carlo (probability) simulations to a one-dimensional, transport model that had been verified using earlier observations of FC penetration into a column of fine sand (96). Their simulations suggested that, under reasonable conditions, enteric bacteria can penetrate well beyond the 1.2 m depth in unsaturated soil, which is the minimum standard depth used in the installation of on-site drain fields.

Few attempts have been made to model *in situ* observations of bacterial transport through soils and saturated subsurface sediments. Harvey and Garabedian (74) describe a model that simulates breakthrough curves for nongrowing, fluorochrome-labeled indigenous bacteria populations injected into sandy aquifer sediments 6.7 m upgradient from the sampling well. Their model includes terms for storage, advection, dispersion, reversible and irreversible adsorption and superpositions, and separate solutions for each size class of bacteria and, where appropriate, for each uniformly conductive zone or layer along the path of travel. Because the aquifer sediments are well sorted with a fairly uniform distribution of pores, coinjection of a conservative, nonreactive tracer with the labeled bacteria facilitates construction of the hydrologic portion of the model. This deterministic approach works well for their small-scale test, because the length of the layers having similar conductive properties (as delineated by hydraulic conductivity profiles) appears to be on the order of several meters (97). However, a stochastic approach probably would be required for modeling transport over longer distances or thicknesses of the aquifer because it would be difficult to define aquifer structure deterministically at a larger scale. Both stochastic and deterministic approaches incorporating physical heterogeneity into transport models are facilitated by information concerning the variability of hydraulic and sorptive properties within the aquifer.

A major determinant in the transport of bacteria through subsurface sediments and a major component of most transport models involves their immobilization at grain surfaces. In fine grain sediments, both sorption and straining can result in significant removal of unattached bacteria as they are transported downgradient. It is generally believed that straining becomes an important mechanism for removal of microbes in groundwater when the average cell diameter is greater than 5% of that of the average grain size (98), although the range of pore sizes in the sediments is also important. In some models of subsurface microbial transport, straining and sorption are not differentiated. However, these two mechanisms of microbial immobilization can lead to different results. For example, straining preferentially removes smaller bacteria, whereas the reverse is true for sorption.

The sorptive-filtration theory (99), commonly employed to describe colloid removal during packed-bed filtration in water treatment applications, appears to offer a number of advantages in model descriptions of bacterial immobilization during transport through porous media; it is relatively simple, and it accounts for the abiotic mechanisms by which microbes contact grain surfaces and for the effect of cell size. Results of previously described experiments with different-sized microspheres suggest that sorption and not straining was primarily responsible for immobilization of bacteria traveling in the Cape Cod aquifer (73). Although there are a number of uncertainties in the application of colloid-filtration to microbial immobilization in the subsurface, it appears to be useful in multicomponent descriptions of microbial sorption in transport models (74).

In addition to immobilization, reversible sorption can be an important determinant of bacterial transport. However, it is not always clear how best to represent reversible sorption in models. In the Massachusetts study, it was found that the breakthrough curves could be accurately modeled by using either of two fundamentally different approaches to adsorption (74). The first assumes that adsorption is instantaneous and linear, but that the propensity of the bacteria for solid surfaces differs among segments of the population. The second employs a kinetic approach that assumes that all the bacteria are equally capable of interacting with solid surfaces, but that the rates of adsorption and desorption are different. The column studies described by Hendricks et al. (87) suggest that bacterial sorption onto soils may be modeled in terms of classic chemical thermodynamics.

Since bacteria that are weakly attached to solid surfaces can later desorb, particularly in response to chemical or physical changes, the ultimate limiting control of pathogen migration through the subsurface is the duration of viability. For nonindigenous pathogens in groundwater and soils, mortality can be very significant. Following an initial period of relatively constant abundance, temporal changes in abundance of displaced, nonindigenous bacteria may be modeled by exponential decline. The initial period before exponential decline may vary from days in highly contaminated water to months under oligotrophic conditions (72). In general, decay times for bacterial pathogens in soils and groundwater appear to be slow, and a one order of magnitude reduction in abundance can involve several weeks or longer (23, 100).

4.3. Viruses

Many aspects of modeling subsurface transport of bacteria are also applicable to viruses. One approach taken in modeling virus transport has involved modification of the one-dimensional solution of the classic advection-dispersion equation (101) to include sorption and decay of the virus (*e.g.*, (82). Use of the one-dimensional model is amenable to transport experiments performed in columns and in sandy aquifers where the transverse and vertical dispersivity are negligible compared with that along the direction of flow (longitudinal dispersivity). However, a three-dimensional approach may be required in modeling transport in a number of subsurface environments, particularly where there are sharp gradients in physical and/or chemical properties within the aquifer. In some instances, viruses may be restricted on the basis of size from some of the finer pore structure. For example, it was observed that MS-2 bacteriophages were excluded from 35%–40% of the void volume in a recent study involving a saturated soil column (78).

Sorption of viruses in transport models is generally assumed to be reversible and is often described mathematically by Langmuir or Freundlich isotherms. A retardation of the viruses relative to fluid flow may, therefore, be expected in aquifer sediments and can be calculated from known values of porosity, bulk density of the aquifer material, and the empirical distribution coefficient (81). However, there is a good deal of experimental evidence to suggest that many viruses may move faster through soils and aquifer sediments than conservative tracers of mean fluid flow, because of size-dependent exclusion of the viruses from some of the finer porosity (78). Adsorptive behavior of viruses in the presence of soils varies greatly with virus (48) and soil (47, 88) type.

Model descriptions of viral survival in the subsurface generally involve simple functions of time and temperature (82). Although temperature was found to be the most important (30), a number of other factors can affect the inactivation of viral pathogens in soils and aquifer sediments. These factors include moisture content, temperature, pH, organic matter, antagonism from soil microflora, aggregation, association with particulates, soil properties, virus type, and hydrostatic pressure (5, 102, 103).

REFERENCES

1. Craun GF: Statistics of waterborne outbreaks in the U.S. (1920–1980). In Craun GF (ed): Water Diseases in the United States. Boca Raton, FL: CRC Press, 1986, pp 73–159.

2. Reed SC, Middlebrooks EJ, Crites RW: Natural Systems for Waste Management and Treatment. New York: McGraw-Hill, 1988.

3. Hagedorn C: Microbiological aspects of groundwater pollution due to septic tanks. In Bitton G, Gerba CP (eds): Groundwater Pollution Microbiology. New York: Wiley, 1984, pp 181–195.

4. U.S. EPA: Septic Systems and Groundwater Protection: An Executive Guide. Washington, DC: Office of Groundwater Protection, 1986.

5. Gerba CP, Bitton G: Microbial pollutants: Their survival and transport pattern to groundwater. In Bitton G, Gerba CP (eds): Groundwater Pollution Microbiology. New York: Wiley, 1984, pp 65–88.

6. Foster DH, Engelbrecht RS: In Sopper WE, Kardos LT (eds): Recycling Treated Municipal Wastewater and Sludge Through Forest and Cropland. University Park, PA: Pennsylvania State University Press, 1973, pp 247–259.

7. Gerba CP, Wallis C, Melnick JL: Fate of wastewater bacteria and viruses in soils. J Irrig Drainage Div ASCE 3:157–168.

8. Lehmann DL, Wallis PM: Literature review: Occurrence and survival of pathogenic bacteria in sludge and on soil. In Biological Health Risks of Sludge Disposal to Land in Cold Climates. Calgary, Canada: University of Calgary Press, 1983, pp 115–143.

9. Sepp E: The use of sewage for irrigation: A literature review. Sacramento, CA: California State Department of Public Health, 1971.

10. Yates MV, Yates SR: A comparison of geostatistical methods for estimating virus inactivation rates in ground water. Water Res 21:1119–1125, 1987.

11. Bitton G, Charles MJ, Farrah SR: Virus detection in soils: A comparison of four recovery methods. Can J Microbiol 25:874–880, 1979.

12. Hurst CJ, Gerba CP: Development of a quantitative method for the detection of enteroviruses in soil. Appl Environ Microbiol 37:626–632, 1979.

13. Filip Z, Seidel K, Dizier H: Distribution of enteric viruses and microorganisms in long-term sewage-treated soil. Water Sci Technol 15:129–135, 1983.

14. Duboise SM, Moore BE, Sagik BP: Poliovirus survival and movement in a sandy forest soil. Appl Environ Microbiol 31:536–543, 1976.

15. Bitton G, Pancorbo OC, Farrah SR: Virus transport and survival after land application of sewage sludge. Appl Environ Microbiol 47:905–909, 1984.

16. Damgaard-Larsen S, Jensen KO, Lund E, Nisser B: Survival and movement of enterovirus in connection with land disposal of sludges. Water Res 11:503–508, 1977.

17. Hurst CJ, Gerba CP, Lance JC, Rice RC: Survival of enteroviruses in rapid-infiltration basins during the land application of wastewater. Appl Environ Microbiol 40:192–200, 1980.

18. Farrah SR, Bitton G, Hoffmann E, Lanni O, Pancorbo OC, Lutrick MC, Bertrand GE: Survival of enteroviruses and coliform bacteria in a sludge lagoon. Appl Environ Microbiol 41:459–465, 1981.

19. Sadovski AY, Fattal B, Goldberg D, Katzenelson E, Shuval HI: High levels of microbial contamination of vegetables irrigated with wastewater by the drip method. Appl Environ Microbiol 36:824–830.

20. Sagik BP, Moore BE, Sorber CA: Public health aspects related to the land application of municipal sewage effluents and sludges. In Sopper WE, Kerr SN (eds): Utilization of Municipal Sewage Effluents and Sludge on Forest and Disturbed Land. Philadelphia: Pennsylvania State University Press, 1979, pp 241–253.

21. Yeager JG, O'Brien RT: Enterovirus inactivation in soil. Appl Environ Microbiol 38:694–701, 1979.

22. Sobsey MD, Shields PA, Hauchman FH, Hazard RL, Canton III LW: Survival and transport of hepatitis A virus in soils, groundwater and wastewater. Water Sci Technol 18:97–106, 1986.

23. Bitton G, Farrah SR, Ruskin RH, Butner J, Chou YJ: Survival of pathogenic and indicator organisms in groundwater. Ground Water 213:405–410, 1983.

24. Keswick BH, Gerba CP, Secor SL, Cech I: Survival of enteric viruses and indicator bacteria in groundwater. J Environ Sci Health A17:903–912, 1982.

25. Jansons J, Edmonds LW, Speight B, Bucens MR: Survival of viruses in groundwater. Water Res 23:301–306, 1989.

26. Yates MV, Gerba CP: Virus survival in groundwater. Proceedings of the 1983 Meeting of the Arizona Section of the American Water Research Association and the Hydrology Section of the Arizona–Nevada Academy of Science, Flagstaff, Arizona. 1983, p 115.

27. Yates MV, Gerba CP: Factors controlling the survival of virus in groundwater. Water Sci Technol 17:681–687, 1984.

28. Yates MV, Yates SR: Septic tank setback distances: A way to minimize virus contamination of drinking water. Ground Water 27:202–208, 1989.

29. Yates MV, Yates SR, Warrick AW, Gerba CP: Use of geostatics to predict virus decay rates for determination of septic tank setback distances. Appl Environ Microbiol 52:479–483, 1986.

30. Yates MV, Gerba CP, Kelley LM: Virus persistence in groundwater. Appl Environ Microbiol 49:778–781, 1985.

31. Mitchell R, Nevo Z: Effect of bacterial polysaccharide accumulation on infiltration of water through sand. Appl Microbiol 12:219–223, 1964.

32. Nevo Z, Mitchell R: Factors affecting biological clogging of sand associated with ground water recharge. Water Res 1:231–236, 1967.

33. Lamka KG, Lechevallier MW, Seidler RJ: Bacterial contamination of drinking water supplies in a modern rural neighborhood. Appl Environ Microbiol 39:734–738, 1980.

34. Zyman J, Sorber CA: Influence of simulated rainfall on the transport and survival of selected indicator organisms in sludge-amended soils. J Water Pollut Control Fed 60:2105–2110, 1988.

35. Parke JL, Moen R, Rovira AD, Bowen GD: Soil water flow affects rhizosphere distribution of a seed-borne biological control agent, *Pseudomonas fluorescens*. Soil Biol Biochem 18:583–588, 1986.

36. Madsen FL, Alexander M: Transport of *Rhizobium* and *Pseudomonas* through soil. J Environ Qual 46:557–560, 1982.

37. Trevors JT, van Elsas JD, van Overbeek LS, Starodub M-E: Transport of genetically engineered *Pseudomonas* fluorescens strain through a soil microcosm. Appl Environ Microbiol 56:401–408, 1990.

38. Alhajjar BJ, Stramer SL, Cliver DO, Harkin JM: Transport modelling of biological tracers from septic systems. Water Res 22:907–915, 1988.

39. Bitton G: The adsorption of viruses onto surfaces in soil and water. Water Res 9:473–483, 1975.

40. Bitton G: Adsorption of viruses to surfaces: Technological and ecological implications. In Bitton G, Marshall KC (eds): Adsorption of Microorganisms to Surfaces. New York: Wiley Interscience, 1980, pp 331–374.

41. Gerba CP: Applied and theoretical aspects of virus adsorption to surfaces. Adv Appl Microbiol 30:133–168, 1984.

42. Lipson SM, Stotzky G: Interactions between viruses and clay minerals. In Rao VC, Melnick JL (eds): Human Viruses in Sediments, Sludges, and Soils. Boca Raton, FL: CRC Press, 1987, pp 197–230.

43. Tanford C: The hydrophobic effect and the organization of living matter. Science 200:1012–1018, 1978.

44. Lipson SM, Stotzky G: Adsorption of reovirus to clay minerals: Effects of cation exchange capacity, cation saturation and surface area. Appl Environ Microbiol 46:673–682, 1983.

45. Lipson SM, Stotzky G: Effect of proteins on reovirus adsorption to clay minerals. Appl Environ Microbiol 48:525–530, 1984.

46. Schiffenbauer M, Stotzky G: Adsorption of coliphages T1 and T7 to host and non-host microbes and to clay minerals. Curr Microbiol 8:245–259, 1983.

47. Burge WD, Enkiri NK: Virus adsorption by five soils. J Environ Qual 7:73–76, 1978.

48. Gerba CP, Goyal SM, Cech I, Bogdan GG: Quantitative assessment of the adsorptive behaviour of viruses to soils. Environ Sci Technol 15:940–944.

49. Goyal SM, Gerba CP: Comparative adsorption of human enteroviruses, simian rotavirus, and selected bacteriophages to soils. Appl Environ Microbiol 38:241–247, 1979.

50. Scheuerman PR, Bitton G, Overman AR, Gifford GE: Transport of viruses through organic soils and sediments. J Environ Eng Div 105:629–640, 1979.

51. Sobsey MD, Dean CW, Knuckles ME, Wagner RA: Interactions and survival of enteric viruses in soil materials. Appl Environ Microbiol 40:92–101, 1980.

52. Gerba CP, Lance JC: Poliovirus removal from primary and secondary sewage effluent by soil filtration. Appl Environ Microbiol 36:247–251, 1978.

53. Lance JC, Gerba CP: Poliovirus movement during high rate land filtration of sewage water. J Environ Qual 9:31–34, 1980.

54. Vaughn JM, Landry EF, Beckwith CA, Thomas MZ: Virus removal during groundwater recharge: Effects of infiltration rate on adsorption of poliovirus to soil. Appl Environ Microbiol 41:139–143, 1981.

55. Landry EF, Vaughn JM, Thomas MZ, Beckwith CA: Adsorption of enteroviruses to soil cores and their subsequent elution by artificial rainwater. Appl Environ Microbiol 38:680–687, 1979.

56. Landry EF, Vaughn JM, Penello WF: Poliovirus retention in 75-cm soil cores after sewage and rainwater application. Appl Environ Microbiol 40:1032–1038, 1980.

57. Moore RS, Taylor DH, Reddy RMM, Sturman LS: Adsorption of reovirus by minerals and soils. Appl Environ Microbiol 44:852–859, 1982.

58. Bitton G, Masterson N, Gifford GE: Effect of a secondary treated effluent on the movement of viruses through a cypress dome soil. J Environ Qual 5:370–375, 1976.

59. Dizier H, Nasser A, Lopez JM: Penetration of different human pathogenic viruses into sand columns percolated with distilled water, groundwater, or wastewater. Appl Environ Microbiol 47:409–415, 1984.

60. Schaub SA, Sorber CA: Virus and bacteria removal from wastewater by rapid infiltration through soil. Appl Environ Microbiol 33:609–619, 1977.

61. Bixby RL, O'Brien DJ: Influence of fulvic acid on bacteriophage adsorption and complexation in soil. Appl Environ Microbiol 38:840, 1979.

62. Farrah SR, Goyal SM, Gerba CP, Conklin RH, Smith EM: Comparison between adsorption of poliovirus and rotavirus by aluminum hydroxide and activated sludge flocs. Appl Environ Microbiol 35:360–363, 1978.

63. Gerba CP, Goyal SM, Hurst CJ, LaBelle RL: Type and strain dependence of enterovirus adsorption to activated sludge, soils and estuarine sediments. Water Res 14:1197–1198, 1980.

64. Jansons J, Edmonds LW, Speight B, Bucens MR: Movement of viruses after artificial recharge. Water Res 23:293–299, 1989.

65. Bitton G, Davidson JM, Farrah SR: On the value of soil columns for assessing the transport pattern of viruses through soils: A critical outlook. Water Air Soil Pollut 12:449–457, 1979.

66. Lue-Hing C, Sedita SJ, Rao KC: Viral and bacterial levels resulting from the land application of digested sludge. In Sopper WE, Kerr SN (ed): Utilization of Municipal Sewage Effluents and

Sludge on Forest and Disturbed Land. Philadelphia: Pennsylvania State University, 1979, pp 445–462.

67. Moore BE, Sagik BP, Sorber CA: Land application of sludge: Minimizing the impact of viruses on water resources. In Sagik BP, Sorber CA (eds): Risk Assessment and Health Effects of Land Application of Municipal Wastewater and Sludges. San Antonio, TX: University of Texas, 1978, pp 154–167.

68. Pancorbo OC, Bitton G, Farrah SR, Gifford GE, Overman AR: Poliovirus retention in soil columns after application of chemical- and polyelectrolyte-conditioned dewatered sludges. Appl Environ Microbiol 54:118–123, 1988.

69. Sattar SA: Viruses and land disposal of sewage sludge: A literature review. In Biological Health Risks of Sludge Disposal to Land in Cold Climates. Calgary, Canada: University of Calgary Press, 1983, pp 271–292.

70. Hurst CJ, Brashear DA: Use of a vacuum filtration technique to study leaching of indigenous viruses from raw wastewater sludge. Water Res 21:809–812, 1987.

71. Liu D: Effect of sewage sludge land disposal on the microbiological quality of groundwater. Water Res 16:957–961, 1982.

72. Matthess G, Pekdeger A, Schroeter J: Persistence and transport of bacteria and viruses in groundwater—A conceptual evaluation. J Contam Hydrol 2:171–188, 1988.

73. Harvey RW, George LH, Smith RL, LeBlanc DR: Transport of microspheres and indigenous bacteria through a sandy aquifer: Results of natural and forced-gradient tracer experiments. Environ Sci Technol 23:51–56, 1989.

74. Harvey RW, Garabedian SP: Use of colloid filtration theory in modeling movement of bacteria through a contaminated sandy aquifer. Environ Sci Technol (manuscript in submission), 1991.

75. Champ DR, Schroeter J: Bacterial transport in fractured rock—A field scale tracer test at the Chalk River Nuclear Laboratories. In Proceedings of the International Conference on Water and Wastewater Microbiology, Newport Beach, CA, February 8–11, pp 14:1–14:7.

76. Pyle BH: Bacterial movement-experience at Aeretaugna. In The Quality and Movement of Ground Water in Alluvial Aquifers of New Zealand. Dept. Agric. Microbiol. Tech. Publ. No. 2, Canterbury, New Zealand: Lincoln College, 1979, p 105.

77. Wood WW, Ehrlich GG: Use of baker's yeast to trace microbial movement in ground water. Ground Water 16:398–403, 1978.

78. Bales RC, Gerba CP, Grondin GH, Jensen SL: Bacteriophage transport in sandy soil and fractured tuff. Appl Environ Microbiol 55:2061–2067, 1989.

79. Singh SN, Bassous M, Gerba CP, Kelley LM: Use of dyes and proteins as indicators of virus adsorption to soils. Water Res 20:267–272, 1986.

80. Matthess G, Pekdeger A: Concepts of a survival and transport model of pathogenic bacteria and viruses in groundwater. Sci Total Environ 21:149–159, 1981.

81. Matthess G, Pekdeger A: Survival and transport of pathogenic bacteria and viruses in ground water. In Ward CH, Giger W, McCarty P (eds): Ground Water Quality. New York: Wiley, 1985, pp 472–482.

82. Yates MV, Yates SR, Wagner J, Gerba CP: Modeling virus survival and transport in the subsurface. J Contam Hydrol 1:329–345, 1987.

83. Corapcioglu MY, Haridas A: Transport and fate of microorganisms in porous media: A theoretical investigation. J Hydrol 72:149–169, 1984.

84. Corapcioglu MY, Haridas A: Microbial transport in soils and ground water: A numerical model. Adv Water Resour 8:188–200, 1985.

85. Germann PF, Smith MS, Thomas GW: Kinematic wave approximation to the transport of *Escherichia coli* in the vadose zone. Water Resour Res 23:1281–1287, 1987.

86. Smith MS, Thomas GW, White RE, Ritonga D: Transport of *Escherichia coli* through intact and disturbed soil columns. J Environ Qual 14:87–91, 1985.

87. Hendricks DW, Post FJ, Khairnar DR: Adsorption of bacteria on soils. Water Air Soil Pollut 12:219–232, 1979.

88. Vilker VL, Burge WD: Adsorption mass transfer model for virus transport in soils. Water Res 14:783–790, 1980.

89. Reynolds PJ, Sharma P, Jenneman GE, McInerney JJ: Mechanisms of microbial movement in subsurface materials. Appl Environ Microbiol 55:2280–2286, 1989.

90. Jenneman GE, McInerney MJ, Knapp RM: Microbial penetration through nutrient-saturated Berea sandstone. Appl Environ Microbiol 50:383–391, 1985.

91. Bosma TNP, Schnoor JL, Schraa G, Zehnder AJB: Simulation model for biotransformation of xenobiotics and chemotaxis in soil columns. J Contam Hydrol 2:225–236, 1988.

92. Harvey RW: Transport of Bacteria in a Contaminated Aquifer. U.S. Geological Survey Water-Resources Investigations Report No. 88-4220, 1989, pp 183–188.

93. Harvey RW: Considerations for modeling transport of bacteria in contaminated aquifers. In Abriola L (ed): Groundwater Contamination. Wallingford, Oxfordshire, England: IAHS Press, Publ. No. 185. 1989. pp 75–82.

94. Ritonga MD: Bacterial and Chloride Transport Through Soil Macropores to Ground Water. Internal Report. Lexington, KY: University of Kentucky, 72 pp.

95. Peterson CT, Ward RC: Development of a bacterial transport model for coarse soils. Water Resour Bull 25:349–357, 1989.

96. Dazzo F, Smith P, Hubbell D: Vertical disposal of fecal coliforms in Scranton fine sand. Proc Soil Crop Sci Soc Florida 32:99–102, 1973.

97. Hess KM: Use of a borehole flowmeter to determine spatial heterogeneity of hydraulic conductivity and macrodispersion in a sand and gravel aquifer, Cape Cod, MA. In Molz FJ, Melville JG, Guven O (eds): Proceedings of the Conference on New Field Techniques for Quantifying the Physical and Chemical Properties of Heterogeneous Aquifers. Dallas, TX, March 20–23. Dublin, OH: National Water Well Association, 1989, pp 497–508.

98. McDowell-Boyer LM, Hunt JR, Sitar N: Particle transport through porous media. Water Resour Res 22:1901–1921, 1986.

99. Yao KM, Habibian MT, O'Melia CR: Water and waste water filtration: Concepts and applications. Environ Sci Technol 11:1105–1112, 1971.

100. Crane SR, Moore JA: Bacterial pollution of ground water: A review. Water Air Soil Pollut 22:67–83, 1984.

101. Freeze RA, Cherry JA: Groundwater. Englewood Cliffs, NJ: Prentice-Hall, 1979.

102. Sobsey MD: Transport and fate of viruses in soils. In Microbial Health Consideration of Soil Disposal of Domestic Wastewaters. U.S. EPA Publ. No. EPA-600/9-83-017, 1983.

103. Bitton G, Pancorbo OC, Farrah SR: Effect of hydrostatic pressure on poliovirus survival in ground water. Ground Water 21:756–758, 1983.

104. Harvey RW: Evaluation of particulate and solute tracers for investigation of bacterial transport behavior in groundwater. In Proceedings of the First International Symposium on Microbiology of the Deep Subsurface, Orlando, FL, January 15–18 (in press).

6

DETECTION METHODS FOR WATERBORNE PATHOGENS

AJAIB SINGH

Milwaukee Health Department, Bureau of Laboratories, Milwaukee, Wisconsin 53202

GORDON A. MCFETERS

Department of Microbiology, Montana State University, Bozeman, Montana 59717

1. INTRODUCTION

Microbiological monitoring of drinking water has been practiced in the United States and other countries since the beginning of this century. The primary objective of this practice is to protect the health of the community by preventing the spread of waterborne disease. Potable water systems can become polluted with coliform and pathogenic bacteria from normal, diseased, or carrier human and animal excrements. This can occur by cross connections between a water main and a sewer, especially when the pressure in the water main becomes less than atmospheric pressure (1), or from the entry of sewage water through leaks in damaged pipes. Also, treatment deficiencies may allow the escape of either unharmed or reversibly inactivated (injured) organisms, especially when the source water contains high densities of coliform and pathogenic bacteria. Furthermore, microorganisms can gain access to water from air–water interface in the distribution system such as in a storage tank or may remain unaffected by the chlorine treatment (2). For example *Legionella* species not only survive but multiply in storage tanks and other water systems (3,4).

The risk of illness from water-associated recreation, one of the most popular activities in the United States, has been largely overlooked. Pathogenic micro-

Environmental Microbiology, pages 125–156, © 1992 Wiley-Liss, Inc.

organisms are introduced into recreational water environments via sewage treatment outfalls and nonpoint sources of human and animal wastes. Sewage systems from urban areas that discharge into nearby freshwaters and marine waters have been in existence in the United States and other parts of the world for a long time. Yet it was not until 1951 that guidelines for microbial quality of recreational waters were proposed (5). In 1968, after the lapse of 17 years, the National Technical Advisory Committee recommended guidelines for microbiological quality of such waters to the Federal Water Pollution Control Administration (6). The two standards used by the various states and territories of the United States included a total coliform level of 1,000/100 ml and a fecal coliform limit of 200/100 ml in recreational water.

To ensure safe recreational water and a continued supply of potable water, frequent monitoring of both raw water sources and finished products for the presence of pathogens is very important. This procedure will establish baseline data against which microbiological quality can be compared during waterborne epidemics or other unusual circumstances. Also, periodic examination will help to establish a system or protocol that can be activated, without significant delay, in emergency situations, including sudden disease outbreaks. More importantly, the detection of any organism not previously observed or a marked increase in the number of a specific pathogen in the source water may provide an early indication of possible health risks. The detection of any enteric bacteria, even of low pathogenicity, in the aquatic environment may serve as a warning of unsafe recreational water or a breach in the integrity of disinfection or distribution systems for potable water. However, the detection and enumeration of all pathogenic bacteria that may potentially be present in water is not only expensive and time-consuming but impracticable to perform on a regular basis. Clearly the most effective microbiological monitoring of water sources requires simple, rapid, and relatively inexpensive tests to determine the presence of indicator bacteria on a routine basis as well as periodic examination for pathogenic microorganisms prevalent in a particular geographic area.

The overall microbiological quality of water is determined on a routine basis by detecting the coliform group of bacteria that ferments lactose and constitutes a part of the normal mammalian intestinal flora. However, some members of genera such as *Klebsiella, Enterobacter,* and *Citrobacter* also have an extraintestinal prevalence and may gain entry into water from environmental sources. The fecal coliform bacteria, predominantly *Escherichia coli,* which originate primarily in the mammalian intestine, are detected by their ability to ferment lactose at 44.5°C. These bacteria are not usually long-term occupants of aquatic ecosystems. Thus their presence in water serves as a useful indicator of recent fecal contamination, which is the major source of many enteropathogenic diseases transmitted through water. These "indicator bacteria," which are relatively easy to detect and enumerate, suggest the potential existence of enteropathogens in surface and potable water. The concept of a relationship between coliform bacteria and enteropathogens such as cholera and typhoid bacilli was introduced during the early part of this century and published in 1905 in the first edition of *Standard Methods for the Examination of Water and Wastewater,* American Public Health Association, Washington, DC. The methods described were widely used and were partially responsible for the subsequent reduction of waterborne cases of typhoid.

More recently the U.S. Environmental Protection Agency promulgated the National Primary Drinking Water Regulation (NPDWR) for 83 contaminants. This list also includes maximum contamination levels for five microbiological contaminants: *Giardia lamblia,* viruses, *Legionella,* heterotrophic plate count bacteria, and total coliforms. The revised regulations that became effective in 1990 have introduced several changes for public water systems. Accordingly, filtration and or disinfection processes are required for surface water sources. Individual states would determine which process must be installed to protect against the potential adverse effect of exposure to *G. lamblia,* viruses, *Legionella,* and many other pathogenic bacteria that are removed or inactivated by appropriate treatment. Thus tests for detecting the presence of these and other pathogens are expected to be carried out more often than before. Moreover, it is desirable to adopt relatively easy, sensitive, and highly specific methods for detecting every pathogen that could potentially be present in water sources. The purposes of this chapter are to review some of the methods that can be used for monitoring the microbiological quality of water and to describe the techniques for detecting pathogenic bacteria in contaminated water systems.

2. WATERBORNE OUTBREAKS

The reporting of waterborne disease outbreaks is voluntary in the United States. Information on such occurrences is obtained from scientific and medical reports and through the assistance of state and local health officials. Since 1971, the United States Environmental Protection Agency (USEPA) and the Centers for Disease Control (CDC) have, through joint efforts, published data describing waterborne outbreaks, although records describing outbreaks have been available since 1920.

In the United States, both ground and surface water are important sources of drinking water. About 50% of the U.S. population depends on ground water as the principle source of drinking water (7), and surface water sources are used by an estimated 155 million people (8). Contamination of ground water can occur in local areas by seepage of sewage into aquifers and by improperly developed or poorly protected wells. Although surface water sources are usually protected to some extent from contamination by human sewage discharges, heavy reliance

**TABLE 6.1. Waterborne Disease Outbreaks Reported
During 1971–1988**

Year	No. of outbreaks	Cases of illnesses	Cases/outbreak
1971–75[a]	124	27,838	224
1976–80[a]	202	50,590	251
1981–85[a]	176	32,807	186
1986–88[b]	46	25,852	562

[a]Data are from Craun (9).

[b]Data are from Levine et al. (10).

**TABLE 6.2. Waterborne Disease Outbreaks in Public Community and
Noncommunity Water Systems in the United States During 1986–1988[a]**

Agent	Community		Noncommunity		Totals	
	No. of outbreaks	No. of cases	No. of outbreaks	No. of cases	No. of outbreaks	No. of cases
AGI[b]	4	1,467	18	1,423	22	2,890
Giardia	8	1,146	1	123	9	1,269
Chemical	4	102	0	0	4	102
Shigella	2	1,825	1	900	3	2,725
Norwalk	0	0	3	5,474	3	5,474
Salmonella	2	70	0	0	2	70
Campylobactor	1	250	0	0	1	250
Cryptosporidium	1	13,000	0	0	1	13,000
CGI[c]	0	0	1	72	1	72
Combined total					46	25,852

[a]Data are from Levine et al. (10).

[b]Acute gastrointestinal illness of unknown etiology.

[c]Chronic gastrointestinal illness.

is placed on the disinfection processes to prevent spread of diseases through water.

In spite of the progress made in understanding waterborne transmission of diseases, microbial disease outbreaks do occur. The data in Table 6.1 show the incidence of waterborne disease outbreaks and the number of illnesses recorded in the United States during 1971–1988. There is no evidence of a dramatic increase in the number of waterborne disease outbreaks or in the number of cases per outbreak. However, in a 3 year period (1986–1988) the large number of cases per outbreak (Table 6.1) is attributed to a single outbreak of

Cryptosporidium, which involved 13,000 cases (11). Waterborne disease outbreaks recorded between 1986 and 1988 in both public community and noncommunity water systems (10) are shown in Table 6.2. It is evident from the data that the etiology of the majority of gastrointestinal illnesses originating from noncommunity water systems remains obscure.

3. PUBLIC HEALTH IMPORTANCE

Infectious diarrhea among children (especially under age 5 years), elderly inhabitants of nursing homes, and travellers visiting developing countries has substantial public health significance. According to one estimate, more than 200,000 such children are hospitalized in the United States each year at an annual cost of $1 billion (12). The global impact is enormous since more than 250 million new cases of waterborne disease are reported each year, resulting in more than 10 million deaths (13). Worldwide the most common bacterial diseases transmitted through water are caused by *Shigella, Salmonella,* enterotoxigenic *E. coli* (ETEC), *Campylobacter jejuni,* and *Vibrio cholerae* (Table 6.3). Transmission of *Salmonella typhi* through drinking water is nonexistent, and outbreaks of waterborne cholera are rare in the United States. However, during the last 15 years the public health significance of viral and parasitic agents in diarrheal disease has been increasingly recognized. Viral infections include hepatitis A, gastroenteritis caused by rotavirus, Norwalk-like agents, and virus-like particles. Common parasites include *Giardia lamblia, Cryptosporidium,* and *Entamoeba histolytica* (Table 6.4). The disease caused by these agents, their incubation times, and symptoms are listed in Tables 6.3 and 6.4. Norwalk agents, because of their resistance to high levels of chlorine, are believed to be more commonly transmitted through water (14). It has been reported that in a period from 1976 to 1980, 13 out of a total of 38 serologically confirmed outbreaks were caused by Norwalk virus (15). Giardiasis, caused by a protozoan parasite, is currently the most common cause of waterborne disease that results from the consumption of untreated, inadequately treated, or ineffectively filtered surface water in the United States (16). The first waterborne outbreak caused by *Cryptosporidium* (an enteric coccidian protozoan) occurred in Texas in 1985 (17), and by 1988 this organism was added to the Drinking Water Priority List (18). The significance of viral, parasitic, and some of the bacterial agents in the transmission of disease is vastly enhanced by the observation that many such disease outbreaks were recorded in situations in which residual chlorine levels were acceptable. The inactivation of these agents requires high doses of chlorine and extended periods of contact (14,19–21). Thus the occurrence of disease outbreaks is frequent where heavy reliance is placed on the disinfection of surface water.

**TABLE 6.3. Bacterial Diseases Generally Transmitted by
Contaminated Drinking Water[a]**

Agent	Disease	Incubation time	Symptoms
Shigella	Shigellosis	1–7 days	Diarrhea, fever, cramps, tenesmus, blood in stools
S. typhimurium	Salmonellosis	6–72 hours	Abdominal pain, diarrhea, nausea, vomiting, fever
S. typhi[b]	Typhoid fever	1–3 days	Abdominal pain, fever, chills, diarrhea or constipation, intestinal hemorrhage
Enterotoxigenic *E. coli*	Diarrhea	12–72 hours	Diarrhea, fever, vomiting
Campylobacter jejuni	Gastroenteritis	1–7 days	Abdominal pain suggesting acute appendicitis, fever, headache, malaise, diarrhea, vomiting
Vibrio cholerae[b]	Gastroenteritis	1–3 days	Vomiting, diarrhea, dehydration

[a]Information compiled from various sources.
[b]The incidence is negligible or rare in the United States.

**TABLE 6.4. Viral and Parasitic Diseases
Generally Transmitted by Drinking Water[a]**

Agent	Disease	Incubation period	Symptoms
Hepatitis A	Hepatitis	15–45 days	Fever, malaise, anorexia, jaundice
Norwalk-like agent	Gastro-enteritis	1–7 days	Diarrhea, abdominal cramps, headache, fever, vomiting
Virus-like 27 nm particles	Gastro-enteritis	1–7 days	Vomiting, diarrhea, fever
Rotavirus	Gastro-enteritis	1–2 days	Vomiting followed by diarrhea for 3–8 days
G. lamblia	Giardiasis	7–10 days or longer	Chronic diarrhea, abdominal cramps, flatulence, malodorous stools, fatigue, weight loss
Cryptosporidium	Crypto-spordiosis	5–10 days	Abdominal pain, anorexia, watery diarrhea, weight loss; immunocompromised individuals may develop chronic diarrhea
Entamoeba histolytica	Amebiasis	2–4 weeks	Vary from mild diarrhea with blood and mucus to acute or fulminating dysentery with fever and chills

[a]Information compiled from various sources.

TABLE 6.5. Recreational Water–Associated Disease Outbreaks in the United
States During 1986–1988[a]

Agent	Illness	Source	No. of outbreaks	No. of cases
Pseudomonas	Dermatitis	Whirlpool, tub	10	153
UE[b]	Dermatitis	Whirlpool, hot tub	2	15
Shigella sonnei	Gastroenteritis	Lake, pond, pool	3	280
Giardia	Gastroenteritis	Pool	2	300
Norwalk-like	Gastroenteritis	Lake	1	41
UE	Gastroenteritis	Lake, pond, pool	4	512
Legionella	Pontiac fever	Whirlpool	1	14
Leptospira	Leptospirosis	Stream	1	8
UE	Aseptic meningitis	Creek	1	4
UE	Enterovirus, like infection	Pool	1	26
Total			26	1,363

[a]Data are from Levine *et al.* (10).
[b]Unknown etiology.

All diseases that are spread by the fecal–oral route could potentially be contracted by the unintentional ingestion of polluted recreational water. Swimming-associated outbreaks caused by *Shigella, Giardia,* Norwalk-like viruses, and other enteroviruses have been well documented (22–26). In addition, outbreaks of *Pseudomonas* dermatitis associated with the use of hot tubs, whirlpool baths, and swimming pools have been reported (19). Recreational water–associated disease outbreaks during the years 1986–1988 and reported by Levine *et al.* (10) are shown in Table 6.5.

4. PROBLEMS ASSOCIATED WITH THE DETECTION OF PATHOGENS IN WATER

The increasing demands placed on currently available water resources can raise the potential for contaminating surface water and groundwater by enteric pathogens. Unfortunately, no single procedure is available for the detection of all waterborne pathogens. Also, enteropathogens usually appear intermittently in low concentrations in an aquatic environment that is not ideal for either their growth or extended persistence. Thus, to ensure the good health of a community, there is a need for readily available methods to detect and enumerate pathogens from aquatic sources. Routine or periodic examination of water for pathogens is essential under special circumstances, such as waste water reclamation, water sources with a history of frequent contamination, or during and after actual waterborne outbreaks. However, several limitations render the detection of waterborne pathogens difficult.

The low microbial densities usually found in water necessitate the analysis of large volumes of water to detect pathogens effectively. This may limit the number of samples that could be processed at one time and make the procedure costly. Alternatively, bacteria can be concentrated by filtering water through a glass fiber filter (Whatman, GF/F) as used for *Yersinia enterocolitica* (27). The high "dirt loading" capacity and high degree of bacterial retention capacity of this filter appear to be especially promising for concentrating bacteria from surface and waste waters. Costs associated with sample collection and with the detection of enteric viruses in water are considerably higher than similar tasks with bacteria (28). In potable water, the virus levels are likely to be so low that hundreds or thousands of liters must be sampled to increase the probability of virus detection. Also, the detection of *Giardia* cysts requires filtering a minimum of 380 liters (29) and up to 1,000 liters of water for *Cryptosporidium* (30). Yet another major limitation of these methods is incomplete microbial elution and recovery from the filtering device.

Recovery of pathogens from environmental samples is generally difficult, and methods designed for processing clinical specimens are quite often inadequate. Also, injurious substances or nutrient limitations in aquatic environments induce changes in bacterial physiology and morphology. Injured bacteria have been shown to have reduced intracellular ATP, glucose transport and utilization, aerobic respiration, production of secondary metabolites, resistance to disinfectants, and reduction in cell size (31–34), which necessitate employment of resuscitation steps or use of modified media for growth. In addition, prolonged incubation was extremely important for conditioning cultures isolated from a low nutrient aquatic environment to utilize the concentrated organic substrates found in many media (35). This could be the reason why many investigators who used cultivation procedures described for clinical samples failed to isolate *Legionella* from environmental and potable water samples (36). Also, bacteria in some environments lose their ability to form colonies on agar plates but remain metabolically active and have been referred to as "viable but nonculturable" (37–39). Both metabolically injured and viable but nonculturable pathogenic bacteria fail to grow on standard media but have been shown to be virulent in animal models (40–42).

Preenrichment and selective enrichment of concentrated water samples suspected of being contaminated with *Y. enterocolitica* have been suggested by Schiemann (43) for isolation of this organism. Improvements in the isolation of *Legionella* have been reported after pretreatment techniques, including exposure of water samples to acid (pH 2.2) for 5 minutes (44) or heating to 50°C for 30 minutes (45) prior to plating on selective media.

The detection of viruses of public health importance in water is associated with other problems. Direct examination for viruses in an aquatic environment is not possible because of their small size (10 to 100 nm) and instability. Also, dissolved and suspended material in water and waste water interfere with detection and recovery procedures.

With the present limitations of microbial detection and estimation procedures, a negative finding for a specific pathogen can only be taken as provisional because of the limited sensitivity of the methodology used (46). Thus there is a need for investigation of modifications of the currently used methods and for development of new techniques for detecting waterborne pathogens.

5. DETECTION OF INDICATORS OF PATHOGENIC BACTERIA

The search for an "ideal" indicator bacterial system to determine microbiological water quality has been elusive. An ideal indicator should 1) always be present in the presence of pathogens; 2) be present in a predictable ratio with pathogens; 3) be specific for fecal contamination; 4) be able to resist water treatment and disinfection processes to the same or a slightly greater extent than the pathogens; and 5) be detectable by simple and rapid methods. Detection of coliforms has historically served the purpose to some extent, although coliform-free potable water has been found to be implicated in several waterborne outbreaks (47–49). Thus, the coliform group falls short as an ideal indicator system.

5.1. Total Coliform Determination

5.1.1. Multiple Tube Fermentation Method

This is a technique for determining the most probable number (MPN) of bacteria in water and was first described by McGrady (50) in 1915. Tables for the interpretation were published by him in 1918 (51). This method consists of inoculating known volumes of a water sample into tubes of selective lactose broth. The tubes are examined for growth and production of acid and gas after 24 to 48 hours of incubation. Positive tubes are inoculated into brilliant green lactose bile broth for confirmation of gas production. The resultant MPN is presently determined from the published tables (29).

5.1.2. Membrane Filter Method

In this technique (52), a known volume of a water sample is passed through a membrane filter that is then placed on M-Endo or LES-Endo agar for incubation. All organisms that produce red colonies with a metallic sheen within 24 hours of incubation are considered members of the coliform group. An advantage of the membrane filter method over the MPN technique is that results are available within 24 hours instead of the 3 to 4 days required for confirming MPN results, although verification of colonies observed after membrane filtration is advisable.

Coliforms and other bacteria can become injured in water and waste water from exposure to sublethal levels of stressors such as disinfectants and metals. Such bacteria are incapable of forming colonies on routinely used selective media. However, substantial recovery (10% to greater than 90%) of injured bacteria (53) was observed when m-T7 agar (Difco, Detroit, MI) was used (54) instead of M-Endo agar. The significance of injured bacteria has been recently recognized. As a result, a section describing the recovery of stressed bacteria has been included in recent editions of *Standard Methods for the Examination of Water and Wastewater* (29).

5.2. Presence–Absence Coliform Test

Both MPN and membrane filter methods allow an evaluation of sanitary water quality and the effectiveness of treatment processes. However, according to the new regulations, 100 ml of drinking water should not contain any coliforms. Thus quantitative assessment of coliforms may not be necessary. Instead, a qualitative estimation using the presence–absence (P–A) of coliform concept will be allowed. In one P–A technique, 100 ml water samples are mixed with triple-strength P–A broth and incubated at 35°C. The samples are examined after 24 and 48 hours for the production of acid and gas (29).

5.3. Rapid Detection of Coliforms

A method called the *rapid seven hour fecal coliform test,* similar to the membrane filter procedure, has been described by Reasoner *et al.* (55). After filtering a water sample, the membrane filter is placed on M-7 h FC agar medium (29) and incubated at 41.5°C. Fecal coliform colonies (yellow) are enumerated after 7 hours of incubation. However, a compromise has to be made between sensitivity and speed in detecting and enumerating coliforms.

Another rapid method of fecal coliform detection, described by Reasoner and Geldreich (56), involves concentrating bacteria on a membrane filter which is placed in M-FC broth containing radiolabeled ^{14}C-mannitol. The contents are incubated at 35°C for 2 hours, followed by 2.5 hours at 44.5°C. The $^{14}CO_2$ released is assayed by liquid scintillation spectrometry.

A test that requires only 1 hour of incubation for assessing microbiological quality of drinking water has been reported by Dange et al. (57). The test is based on the correlation of ^{32}P uptake when a sample of water containing coliforms is incubated in a synthetic medium. The major drawback with these methods is related to the use of radioactive materials and to the need for specialized instruments.

5.4. Detection of Coliforms and E. coli by a Defined Substrate Technology

A rapid detection test for total coliform bacteria in water by enzymic hydrolysis of a fluorogenic substrate has been described by Berg and Fiksdal (58). A water sample is filtered through 0.45 μm-pore-size membrane filter, which is aseptically transferred to a sterile buffered medium. The medium consists of 0.02% sodium lauryl sulfate, 0.56% nutrient broth, 0.35% lactose and a fluorogenic substrate 4-methylumbelliferone-β-D-galactoside dissolved in phosphate-buffered saline. The sample is incubated at 35°C for coliform detection. Enzymic hydrolysis of 4-methylumbelliferone-β-D-galactoside is determined by measuring the fluorescence intensity with a fluorimeter. This test is useful for determining coliforms in the range of 60 to 100 CFU/100 ml of water and is completed in 15 minutes.

Three other commercially available test systems, Colisure (Millipore Corp.), Coliquick (Hach), and Colilert (Access Analytical), use defined substrate technology and detect bacteria with specific enzyme substrates. Colilert test was first to receive the USEPA approval in 1989 and uses specific indicator nutrients ortho-nitrophenyl-β-D-galactopyranoside (ONPG) and 4-methylumbelliferyl-β-D-glucuronide (MUG) for the target microbes, total coliforms, and E. coli. A water sample is incubated with the Colilert reagent for 24 hours. If a coliform is present, indicator nutrient is hydrolyzed by the enzyme β-galactosidase of the organism, thereby releasing the indicator portion, ortho-nitrophenyl, from ONPG. The free indicator imparts a yellow color to the solution. E. coli possess an additional constitutive enzyme glucuronidase that hydrolyzes the second indicator nutrient (MUG). As a result of this hydrolysis, MUG is cleaved into a nutrient portion (glucuronide) that is metabolized and an indicator portion, methylumbelliferone, that fluoresces under ultraviolet light. Thus two separate and specific microbial assays are carried out simultaneously with the same sample. These products have been developed to detect low levels (1 CFU/100 ml) of total coliforms and E. coli simultaneously in potable and other waters in 24 hours (59). These systems can be used to determine either MPN using 10 fermentation tubes or in the P–A test format. Detection of enzymatic activity also requires culturing of the bacteria.

5.5. Coliform Detection by Gene Probes Using the Polymerase Chain Reaction

The detection of coliforms from environmental samples without culturing viable cells has been made possible by a genetically based method described by Bej et al. (60). Specific nucleotide sequences associated with coliform bacteria are amplified using the polymerase chain reaction (PCR) and detected with ^{32}P-labeled gene probes. Unlike the conventional method of detecting the activity of a gene product such as β-galactosidase to indicate the presence of coliforms, selected regions of the lacZ gene are detected after amplication. Similarly, PCR is used for amplifying a region of the lamB gene that encodes for a surface protein recognized by the E. coli–specific bacteriophage lambda and is detected by a gene probe. Thus detection of a specific genetic sequence following PCR amplification of lacZ and lamB is useful in detecting most coliforms and low levels of E. coli, respectively. In addition, it is possible to detect Salmonella and Shigella after amplifying lamB by using a nucleic acid probe and a primer annealing temperature of 60° C. Radiolabeled gene probes are used for detecting genomic DNA of one to five viable cells of E. coli in 100 ml of water after amplification. This approach appears very promising for use in water and environmental microbiology but awaits development of relatively simple methods to extract DNA as well as appropriate nonradiolabeled gene probe detection procedures.

6. DETECTION OF INDICATORS OF PATHOGENIC BACTERIA IN RECREATIONAL WATERS

Recreational waters usually include freshwater swimming pools, whirlpools, and fresh or marine surface waters. The suggested indicators of contamination used in surface waters are coliforms, enterococci, Pseudomonas, Streptococcus, and, in rare cases, Legionella (29). There is no single method that can be used for determining the bacteriological quality of recreational water because the presence of one organism does not necessarily indicate the presence or absence of others and the choice of selecting one depends on the type of water to be examined. Methods for determining coliform and other indicator bacteria are described in the Standard Methods for Examination of Water and Wastewaters (29).

6.1. Enterococci

Determination of enterococci is a valuable bacterial indicator for determining the extent of fecal contamination of recreational surface water since the incidence of swimming-associated gastroenteritis was found to be directly related to the concentration of these bacteria (61). A modified membrane filter method described by Dufour (62) is used for detecting enterococci in 24 hours. A water sample is filtered through the membrane filter that is placed on a selective modified mE medium (mE agar, Difco, containing inoxyl-β-glucoside instead of esculin) and incubated for 24 hours at 41°C. Typical blue colonies with clear to black centers are counted under the fluorescent lamp using magnification.

6.2. Bacteriophages

Bacteriophages may also serve as useful indicators of fecal contamination. The presence of coliphage (bacteriophages that infect and replicate in coliform bacteria) in water shows good correlation with water quality and coliform bacteria (63,64) because they are more resistant to water treatments and disinfection (65). The numbers of total and fecal coliforms are determined as follows. Total coliforms:

$$log \ y \ = \ 0.627 \ (log \ x) \ + \ 1.864,$$

where y is total coliforms/100 ml and x is coliphage/100 ml.
Fecal coliforms:

$$log \ y \ = \ 0.805 \ (log \ x) \ + \ 0.895,$$

where y is fecal coliforms/100 ml and x is coliphage/100 ml.

Kott *et al.* (64) described a plaque count method of detecting coliphage in water. However, the current methods for detecting bacteriophages to determine water quality are not standardized. As a result, a great variation in the count of indigenous phage has been reported by Havelaar and Hogeboom (65). This may also be due to differences in the susceptibility of the host bacterial strains to a variety of phages and stress factors in the aquatic environments. Havelaar *et al.* (66) suggested the use of F-specific (pilus- or plasmid-dependent), single-stranded RNA bacteriophages (f2 and MS2) as indicators of pollution in sewage because of their similarity to enteroviruses and their greater resistance to various treatments. The F-specific phages were detected with a laboratory-constructed host strain of *Salmonella typhimurium* (WG 49, phage type 3 Nal^r, F' 42 *lac*: :Tn5). The F'42 lac::Tn5 plasmid induces the production of *E. coli* pili on *S. typhimurium* cells, making them susceptible to F-specific coliphages. The use of the WG 49 host strain eliminates interference caused by somatic coliphages in the detection of F-specific coliphages because the host strain lacks an *E. coli* cell wall. Furthermore, somatic *Salmonella* phages are rare in fecally contaminated water and have less chance of interfering with this detection method (66). Specific plaque counts obtained from natural sewage by using the WG 49 host strain were more reliable and higher than those obtained with *E. coli* host (66).

Sobsey *et al.* (67,68) described methods of detecting fecal and enteric virus contamination of natural and treated waters using F-specific WG 49 host strain. The method involves concentrating F^+ bacteriophage by adsorption to a nitrocellulose membrane filter followed by direct application of the filter to an agar lawn of host bacteria and incubation for plaque development. Recovery of F^+ coliphages from fecally contaminated natural waters was comparable with the membrane filter and other conventional methods.

7. DETECTION AND ENUMERATION OF PATHOGENIC BACTERIA

7.1. *Salmonella* and *Shigella*

Procedures for the detection of *Salmonella* and *Shigella* are not standardized, and protocols may be modified to suit the prevailing circumstances. Thus a brief description of the important steps generally used for their detection in water is given.

7.1.1. *Selective Enrichment and Plating Method*

Concentration of a water sample is generally necessary when testing for pathogenic organisms and can be accomplished using filters of borosilicate glass microfibers bonded with epoxy resin. After filtration the filter is placed directly onto a selective enrichment medium according to the method of Levin *et al.* (69). Membrane filtration (0.45 μm pore size) can be used directly with low turbidity waters and may be applied to turbid waters after precoating with diatomaceous earth. After filtration the membrane is blended with a small amount of peptone broth, which is added to double-strength selective enrichment broth for incubation (70). Tetrathionate broth is considered more efficient than the selenite cystine broth and is commonly used for the recovery of *Salmonella* from water. Suppression of nonpathogenic bacteria can be enhanced with the addition of brilliant green dye (1:50,000) during regular analysis. Sodium lauryl sulfate is added with highly turbid marine waters (29).

For *Shigella*, nutrient broth is used for enrichment. Park *et al.* (71) suggested the use of an autocytotoxic enrichment medium to suppress *E. coli*. The enrichment medium can be prepared by adding 4-chloro-2-cyclopentilephenyl-β-D-galactopyranoside in lactose broth. Enrichment is done by incubating at 35°C for 24 hours. Further differentiation between pathogenic and nonpathogenic bacteria is achieved by streaking the culture on a medium such as xylose lysine deoxycholate (XLD) agar, bismuth sulfite (BS) agar or *Salmonella–Shigella* (SS) agar containing selective agents to inhibit the growth of nonenteropathogenic bacteria.

Typical colonies of *Salmonella* develop a black color with or without metallic sheen on BS agar and black centered red colonies on XLD agar. Colonies of *Salmonella* and *Shigella* on SS agar are colorless. Further identification is based on biochemical and serological tests.

7.1.2. Fluorescent antibody test

The direct fluorescent antibody (FA) procedure can be used to test *Salmonella* in recreational and shellfish-harvesting waters. When the pollution is expected to be heavy, direct plating is done on selective media. Otherwise concentration of samples by filtration is recommended using an absorbent pad on a membrane filter and sterile phosphate buffer or 0.1% peptone water for elution. Enrichment is achieved by incubating plugs with absorbent pads in selenite cystine and tetrathionate broths at 37°C for 24 hours. Plates containing selective media are spot inoculated with enrichment culture medium and incubated for 3 hours at 37°C. Impression smears are made on FA microscope slides and stained with diluted *Salmonella* polyvalent conjugate (72). The use of the FA procedure in combination with a direct method of counting living bacteria (73) facilitated the detection of viable but nonculturable *Salmonella, Shigella,* and other pathogens and injured bacteria in natural waters (42,74,75) using the image analysis technique.

7.1.3. Use of DNA probes

A recently described method using DNA probes (76) for detecting *Salmonella* spp. in clinical samples appears to be promising for use with aquatic samples. The method employs a highly conserved, 1,600 base pair probe cloned from *S. enteritidis* DNA that reacts with all 72 *Salmonella* serotypes tested thus far. Although the method is simple, it uses a radioiodinated probe and requires a gamma-counter for measuring the radioactivity after hybridization.

7.2. Enteropathogenic *E. coli*

7.2.1. Bioassay

The detection of enteropathogenic *E. coli* has been extremely difficult because no biochemical marker is available that would differentiate between pathogenic and nonpathogenic strains. A relationship between serotype and pathogenicity has not been proven, although certain serotypes are more frequently isolated from clinical cases of diarrhea than others (77). To determine the pathogenic nature of isolates, immunological analysis of heat-labile (LT) or heat-stable (ST) toxin assessment of the cytopathogenic effects on tissue culture cells, or accumulation of fluid in an animal bioassay is used. However, these methods are time-consuming, and it is impractical to process large numbers of clinical or environmental samples. Calderon and Levin (78) reported a rapid method for enumerating ETEC from water samples. Water samples are filtered through a membrane filter and placed on a selective medium (mTEC, Difco) for incubation at 44°C for 24 hours followed by incubation on tryptic soy agar at 37°C for 6 hours. The filter is then placed on a monolayer of Y-1 adrenal cells for 15 minutes, followed by further incubation. Finally, the monolayer is examined after 15 to 24 hours for a positive rounding effect of the cells because of the presence of LT toxin producing cultures.

7.2.2. Gene Probes

Gene probes are also used to examine water for ETEC that carry genes for enterotoxin (LT, ST) production (79). Samples are passed through a nitrocellulose filter that is placed on MacConkey medium for overnight incubation at 37°C. Bacterial colonies on the filter paper are lysed, washed, and baked overnight at 65°C. Filters are examined after hybridization with ^{32}P-labeled DNA probes. DNA hybridization assay was found to be 10^4 times more sensitive than conventional tests for *E. coli* in the Y-1 adrenal and suckling mouse assay systems for identifying ETEC in water (79).

7.3. *Campylobacter jejuni*

7.3.1. Culture Method

Standardized methods for isolating *Campylobacter* from water are not available. However, like other waterborne pathogens, concentration of a water sample of low turbidity is achieved by passing several liters of water through a sterile membrane filter (0.45 μm pore size). The filter is directly placed on a selective *Campylobacter* agar containing antimicrobic supplement with 10% defibrinated sheep blood (80) and incubated in an atmosphere containing 10% CO_2, 5% O_2, and 85% N_2 at 42°C. Pressure filtration is necessary for turbid waters. Water is passed through a series of filters of 3.0, 1.2, 0.6, and 0.45 μm. After filtration the 0.45 μm-pore-size filter is placed on a selective medium as described above.

Enrichment is necessary when the bacterial density in the sample ·is low compared with the nonspecific microflora or when the organisms are isolated from stressful environments. Campy thioglycollate medium (80) containing 1.5% ox bile permits multiplication of *Campylobacter* spp. and inhibits the competing microflora. The culture is streaked on a selective medium (Campy agar) after 48 hours of incubation in enrichment. Typical spreading and mucoid colonies of *C. jejuni* on the selective medium are tested further for biochemical reactions. Serological identification utilizes an indirect hemagglutination test (81) or a commercial agglutination assay, Meritec-Campy, manufactured by

Meridian Diagnostics (Cincinnati, OH). The latter assay was found to be 100% sensitive in detecting *C. jejuni* and *C. coli* but had low sensitivity with *C. lardis* (82).

7.3.2. DNA Probes

DNA probes have been employed for the detection of *Campylobacter* spp. in clinical specimens using a radioactive hybridization assay (83) or the hybridization of a commercially available phosphatase-conjugated synthetic oligonucleotide probe (84). A DNA-based AccuProbe (Gen-Probe Inc., San Diego, CA) was demonstrated to have 100% accuracy in identifying thermophilic *Campylobacter* species (85). The assay uses a chemiluminescent labeled, single-strand DNA probe that is complementary to the ribosomal RNA of the target organism. The ribosomal RNA is released from the organism and forms a stable DNA–RNA hybrid that is measured with a chemiluminometer. However, the usefulness of these techniques has not yet been explored in the analysis of water.

7.4. *Yersinia enterocolitica*

7.4.1. Culture Method

The major problem associated with the isolation of *Y. enterocolitica* from water is the lack of selective methods to differentiate pathogenic strains of *Y. enterocolitica* from the commonly occurring nonpathogenic water strains (86). In addition, no specific procedure has been described for the detection of *Y. enterocolitica* in water. However, a procedure developed and subsequently evaluated for food analysis might provide equivalent results in the analysis of water (43). Schiemann (43) suggested the following steps for isolating *Yersinia* from water.

Concentration of water samples. Filtration through Whatman GF/F glass filters is effective (27).

Preenrichment. The filter should be immersed in preenrichment basal medium. The inoculated medium should be incubated at 4°C for 7–9 days or at 15°C for 2 days without losing the effectiveness of the procedure (87).

Selective enrichment. After incubation, 1 ml of the preenrichment medium is transferred to 100 ml of bile–oxalate–sorbose broth. The following filter-sterilized solutions are added to the basal medium: sorbose, 10%; asparagine, 1%; methionine, 1%; yeast extract; sodium pyruvate, 0.5%; metanil yellow; sodium nitrofurantoin; irgasan, 0.4%. The inoculated broth is incubated at 25°C and streaked on plating agar after 3–5 days (88).

Plating agar. *Yersinia* selective agar (Difco) is used for plating and plates are incubated at 32°C. Small colonies (1.0 to 1.5 mm in diameter) will develop after 18 to 24 hours and show a dark red "bull's eye" with an outer transparent zone under a stereomicroscope. Typical colonies are further identified by various biochemical tests. Pathogenic strains can be identified by a negative reaction for fermentation of salicin at 35°C and esculin hydrolysis on bile exculin medium at 25°C. Additional markers of virulence include observing small red colonies on Congo red–magnesium agar medium (89) and negative pyrazinamidase activity (90).

7.5. Legionella

In most cases a 1 liter water sample is adequate for processing, but larger volumes (up to 10 liters) are collected when the bacterial density is expected to be low. In some instances it may be useful to swab suspected sources, such as shower heads and other water fixtures. Samples are refrigerated if delay in processing is expected. Filtration and centrifugation is commonly used to concentrate *Legionella*.

7.5.1. Immunofluorescence Procedure

A rapid presumptive identification is accomplished by fluorescent antibody staining. About 100 ml of the sample is centrifuged at 3,500*g* for 30 minutes, and sedimented material is suspended in a small quantity of filter sterilized water. Smears are prepared from the material, fixed by heat, and covered with a small amount of FITC-labeled globulin directed against the antigen. When the antigen–antibody complex is examined under UV light, *Legionella* cells appear as fluorescent yellow green rods. The direct fluorescent antibody test for *Legionella* is regarded as a presumptive test and should be confirmed by isolation of a pure culture and performing biochemical tests.

7.5.2. Culture Isolation

Samples are concentrated by filtration through a 0.2 μm pore size polycarbonate filter. The filter is removed and placed in a centrifuge tube containing 10 ml of sterile tap water. The organisms are dislodged and dispersed from the filter by vortex mixing (3 × 30 seconds) or in a sonic bath for 10 minutes. The sample is then treated with 1 ml of acid reagent (18 parts of 0.2 M KCl and 1 part of 0.2 M HCl, pH 2.0), allowed to stand at room temperature for 5 to 10 minutes, and neutralized with 1 ml of 0.1 N KOH. Improved recovery of *Legionella* has been obtained after acid treatment (44) or heating to 50°C for 30 minutes (45). Concentrated and acid-treated samples are inoculated onto buffeted charcoal yeast-extract agar supplemented with cysteine, ferric pyrophosphate, and α-ketoglutaric acid (BCYE-α) and a selective agar (BCYE containing glycine anisomycine, vancomycin, and polymyxin B). Plates are incubated at 35°C in a humidified atmosphere, preferably with 2% to 5% CO_2 for 10 days. After incubation, *Legionella* colonies show a characteristic "ground glass" appearance. Typical colonies are transferred onto BCYE-α agar and a BCYE agar plate prepared without L-cysteine. Growth on only BCYE-α agar is presumptive for *Legionella*. Confirmation is done by slide agglutination or direct immunofluorescence methods. Calderon and Dufour (91) examined three buffeted charcoal yeast extract agar-based media containing different selective agents. They concluded that the increased ability of a medium to suppress the growth of nonspecific organisms reduced its capacity to recover *Legionella* spp. and emphasized the need to develop a more suitable medium.

7.5.3. Gas Chromatography

Gilbert (92) compared fatty acid profiles of *Legionella* spp. and a range of other bacteria to establish a specific fatty acid signature of this genus that could be used as a marker for the identification of *Legionella* in water. Computer analysis is required to distinguish the chromatographs of some of the species.

7.5.4. Immune autoradiography

For quantitative enumeration of *Legionella pneumophila* in water samples, Martin et al. (93) suggested the use of an immune autoradiography procedure. Concentrated water samples are inoculated onto selective and nonselective buffered charcoal-yeast extract agar plates and incubated for 72 hours. After incubation, a black-and-white photograph is made. A circular nitrocellulose paper is then placed onto the growth to obtain antigen imprints. The paper is dried, heat fixed for 2 hours in an 80°C vacuum oven, and exposed to polyvalent or monospecific antiserum for 15 minutes. The filters are treated with ^{125}I-labeled *Staphylococcus* protein A for 2 hours at room temperature and exposed to X-ray film for 24 hours at −70°C. The colonies are counted and the radiograph used as a transparent template to locate specific *L. pneumophila* for further characterization.

7.5.5. DNA Amplification and Hybridization

A chromosomal DNA sequence from *L. pneumophila* may be amplified by using the polymerase chain reaction. A water sample containing 35 CFU of *L. pneumophila* is used for amplifying the target sequence (35 cycles), which is hybridized with a radiolabeled oligodeoxynucleotide and detected on dot blots (94). This method is highly sensitive and specific and may facilitate better understanding of *Legionella* epidemiology.

7.6. Giardia lamblia

Detection and identification of *G. lamblia* requires the concentration and purification of the organism or its cysts from a large volume of water. Although several methods for detecting cysts in water are available, studies on the relative efficiency and sensitivity of these methods under different conditions have not been reported.

The method for microscopic examination described in the *Standard Methods for the Examination of Water and Wastewaters* (29) was developed by a consensus panel (95) and is regarded as tentative procedure. *Giardia* cysts are captured by filtering 100 to 1,000 gallons of water through a 1.0 μm polypropylene, cotton, or orlon filter cartridge and stained with iodine for light microscopy.

Alternatively, a fluorescent antibody technique described by Sauch (96) can be used for the detection of *Giardia* cysts. The major disadvantage associated with using these techniques is that cysts are lost during concentration. Also, cysts may get trapped with organic detritus and may not float. *Giardia* cysts, as do algae, diatoms, flagellates, ciliates and crustaceans, have a specific gravity of 1.05 and are difficult to separate from similar-sized particles. The use of monoclonal antibody preparations offers more specificity in the detection of *Giardia* cysts in such samples.

7.7. Cryptosporidium

A large volume of water sample (10 to 1,000 liters) is passed through a yarn-wound cartridge or a polycarbonate disc filter (1.0 μm pore size). The filtration process is reasonably efficient (88%–99%), although water quality determines the volume required (30). The filter is eluted by several washings with distilled water (500–3,000 ml). Backwashing and incorporation of Tween-80 into the wash water enhances recovery of cysts that varies from 16% to 76%. The large volume of eluted sample is concentrated using density gradient centrifugation as described by Musiel et al. (97). To clarify the concentrated sample, sucrose density gradient (1.18–1.29 g/ml) with 1% Tween-80 and sodium dodecyl sulfate is used.

7.7.1. Fluorescent Microscopy

Density gradient centrifugation separates oocysts from the other particulate material found in the sample. The top layer, which is rich in oocysts, is filtered through a membrane filter (1.0 μm pore size) and stained with fluorescein-conjugated polyclonal antibody specific for the oocyst wall (98). The polyclonal antibody was shown to react with human, monkey, and bovine oocysts in fecal samples and did not show any cross reactivity with *Candida* or other enteric organisms. Of the monoclonal antibody (MAb) preparations available, MAb IgG and MAb IgM, developed by Sterling and Arrowood (99), were tested exclusively within the clinical setting by Garcia et al. (100) and did not show cross-reactivity with yeasts and bird species of *Cryptosporidium*. MAb IgM has also been used in conjunction with MAb IgG for *Cryptosporidium* detection from water samples (101).

7.7.2. Identification

Rose et al. (102) have suggested a number of criteria for identification of *Cryptosporidium* oocysts when examined with epifluorescence microscopy. These include 1) characteristic apple-green fluorescence around the oocyst wall, 2) spherical shape, 3) 4 to 6 μm diameter, and 4) characteristic folding in the oocyst wall. A limitation of the presently available methodologies is the inability to determine the viability of the oocysts. However, it may be safe to assume that every oocyst detected in potable water is potentially infectious.

8. METHODS OF VIRAL DETECTION

Viruses are present in extremely low concentrations in the environment compared with clinical specimens. Thus the concentration of viruses is an important first step in their detection in water. Excellent reviews on the methodology of virus concentration in water are available in the literature (103–107). However, the method described in *Standard Methods for the Examination of Water and Wastewaters* (29) is most commonly used. A large volume of water is passed through electronegative or electropositive filters to adsorb virus particles that are eluted with a small volume of a proteinaceous liquid such as beef extract. The development of concentration methods and improved laboratory techniques for the propagation of viruses such as rotavirus and hepatitis A virus have significantly increased virus detection in water during the last several years (106).

There is no universal detection system for all enteric viruses. However, many can be detected from clinical and environmental samples using primate and human tissue cultures (107) or suckling mice (group A coxsackieviruses). These agents can be further recognized by their characteristic cytopathogenic effect on the tissue culture or by counting plaques. For viruses that are difficult to grow in tissue cultures or for those requiring long incubation times, other methods such as immunofluorescence (IF) for rotavirus, enzyme-linked immunosorbent assay (ELISA), and radioimmunofocus assay (RIFA) are used. Unfortunately tissue cultures methods are not presently available for many enteroviruses, including the Norwalk-like agent. Moreover, culture techniques are not well suited for routine identification of viral agents in environmental samples and especially in water. Furthermore, isolation and identification of enteric viruses can require 3 to 21 days.

8.1. Immunofluorescence and Immunoperoxidase Assays

These tests are relatively rapid, and polio and rotaviruses can be detected 6 to 9 hours after cell cultures are infected. The detection procedure of the IF assay consists of infecting tissue culture cells and then detecting the viral-specific antigen in the cells using a specific antibody conjugated with a fluorescent dye such as fluorescein isothyiocyanate. Virus-infected cells fluoresce when viewed under ultraviolet light. The method has been successfully used to detect rotavirus in drinking water (21). This technique offers only a limited advantage, since specific antibody is required for each virus one wishes to detect. A large number of antibodies is required to detect a range of viral contaminants of water.

The immunoperoxidase procedure is more sensitive than the IF assay and does not require a fluorescent microscope. The virus-specific antibody molecules are conjugated with the enzyme horseradish peroxidase using glutaraldehyde. The enzyme is bound to the virus-infected cells by the antigen–antibody reaction. In the presence of substrate the enzyme produces a dark reaction product in the virus-infected cells that appears significantly darker than the surrounding uninfected cells when examined with a light microscope.

8.2. Enzyme-Linked Immunosorbent Assay

8.2.1. Direct ELISA

The direct ELISA uses an enzyme directly linked with the antiviral antibody. Because each antiviral agent must be attached to the enzyme, this assay can be inconvenient when a number of different enzyme immunoassays are to be performed.

8.2.2. Indirect ELISA

In the indirect ELISA, virus antibody is adhered to the wall of a microplate. When the test material is added, antigen forms a complex with the specific antibody adhered to the microplate wall and, unlike the nonspecific reaction, is not washed away. Unconjugated rotavirus antibody produced in a different animal than the first used for coating the microplate well is added. This will react with the antigen complexed with the first antibody. Enzyme-labeled antibody directed against the IgG globulin of the animal source of second antibody is added and the well is washed. Finally a substrate is added, and the enzyme adhered to the well will convert the substrate to a visible form. The intensity of the color is proportional to the amount of antigen in the test material. The greater sensitivity of this assay is explained by a single molecule of antiviral antibody that can react with a number of labeled antiglobulins. ELISA has been used to detect rotavirus in sewage (108,109) in England and in West Germany.

8.3. Radioimmunoassay

Radioimmunoassay (RIA) is based on the specific binding reaction of antigen and antibody. The antigen bound to ^{125}I-labeled antibody is quantitatively measured after separation by centrifugation, filtration, or washing. The test specimen containing antigen is added to specific antibody adsorbed onto a well in a microplate. Antigen is attached to antibody, and unattached antigen is washed away. Antibody labeled with radioactive isotope ^{125}I is added to the well and combines with attached antigen in the well. The well is then washed and contains only antibody–antigen–antibody sandwiches. Radioactivity of the well is measured and is proportional to the amount of antigen present in the sample. This method requires 10^5 to 10^6 virus particles per milliliter, and the major disadvantage is the need for radioactive materials. However, the RIA detection method was successfully employed for assaying Hepatitis A virus in both sewage and well water during an outbreak in Texas (110).

8.4. Nucleic Acid Probes

By applying molecular technology, it has become possible to use nucleic acid probes for viruses in both clinical and environmental samples. Viral nucleic acid is extracted from the specimen by treatment with protease, detergent, and organic solvents. The nucleic acid is applied to a filter matrix and hybridized by placing it in a hybridization solution containing labeled complementary DNA (cDNA) and detected by autoradiography. Nonradioactive nucleic acid probes such as those labeled with biotin (111) appear to be more promising in viral detection, but they are generally 10-fold less sensitive than the radioactive probes and are not readily available commercially.

A cDNA probe was used by Jiang et al. (112) for the detection of hepatitis A virus in estuarine water samples. Also, ^{32}P-labeled cDNA probes have been developed to detect low concentrations (100 to 1,000 particles) of polio virus type 1 and hepatitis type A in aquatic environments (113,114).

A dot-blot hybridization assay has been described for detecting rotavirus RNA sequences in the tissue culture or clinical samples by Dimitrov et al. (115). Low virus concentrations may not allow the direct detection of viral sequence in aquatic environments. Viral concentration techniques or enrichment by growth to attain detectable levels of viral RNA might prove to be highly sensitive methods for measuring viable rotaviruses in environmental samples (116).

8.5. Polymerase Chain Reaction

The PCR can be used for enzymatic amplification of viral genes from their low concentration in samples to detectable levels. The principle of PCR is simple and involves sequential, repeated denaturation cycles of double-strand DNA, annealing of primers, and extension of the annealed primers with a heat-stable DNA polymerase. Since the primer extension products serve as templates, each successive cycle theoretically doubles the amount of target DNA synthesized in the previous cycle. This results in the exponential accumulation of a specific target DNA fragment that can be amplified over 1 million times after 30 cycles. RNA sequences can also be amplified after constructing cDNA using reverse transcriptase. PCR techniques are being actively developed for rotavirus and other viral agents. As with nucleic acid probe assays, a disadvantage of this technique is its inability to distinguish between infectious and noninfectious entities. Unfortunately, data about the efficiency of these assays to detect viruses after disinfection or inactivation in water are not available. However, amplification techniques and gene probe assays for viruses in water appear to have a very promising future.

9. FUTURE DEVELOPMENTS

Waterborne disease outbreaks continue to occur in the United States, involving a wide variety of bacteria, viruses, and protozoans. Improved detection techniques are increasingly employed for the direct isolation and identification of previously unknown etiological agents. However, newer and/or more exhaustive detection techniques rather than the routine microbiological monitoring of water are used to investigate disease outbreaks.

Recent promulgation of national primary drinking water regulations in the United States, which became effective in 1990, will have a significant effect on the analytical methodology routinely employed in the microbiological examination of water. Under this rule, the EPA requires a maximum containment level of zero for total coliforms, including both fecal coliforms and *E. coli*. Also, public water systems must conduct total coliform analysis with one of the methods described in the *Federal Register* (117). The rule also specifies criteria for determining turbidity and five microbiological contaminants for public water systems that use surface water. The contaminants are *G. lamblia,* viruses, *Legionella,* heterotrophic bacteria, and total coliforms.

In order to attain zero levels of indicators and pathogens and low numbers of heterotrophic bacteria in drinking water, changes in the areas of water treatment, distribution, and analytical methods are expected to be introduced. Public systems using a surface or groundwater or under the direct influence of surface water, should install and properly operate water treatment processes that reliably achieve at least 99.9% (3-log) and 99.99% (4-log) removal and/or inactivation of *G. lamblia* cysts and virus particles respectively.

The use of alternative disinfectants such as chloramine, chlorine dioxide, and ozone will need critical evaluation, in terms of both their microbiological efficacy and their potential harmful by-products. Changes in the distribution system including engineering control measures will be based on studies to determine the role of assimilable organic carbon and several other factors involved in the development of biofilms that may cause intermittent but persistent water quality problems. Finally, more efficient and cost-effective analytical methods will need to be developed for the detection of *Giardia* cysts and *Legionella* in the aquatic environment.

The importance of periodic monitoring of water quality for the absence of pathogenic microorganisms will remain even after full compliance with the new rules. The detection of pathogens whose presence or absence in water is not related to fecal contamination or that are more resistant to disinfection treatment necessitates invocation of such protocols. Also, the importance of newly emerged human pathogens that may be transmitted through water (101,118,119) should be recognized. Effective methods for their detection and removal need to be developed. Furthermore, with the prevalence of acquired immunodeficiency syndrome and changing demographics, human population groups may become vulnerable to those opportunistic pathogens that were traditionally regarded as environmental contaminants or merely as transient colonizers at unusual sites.

With improved detection methodology, data about the number and types of pathogens present in the aquatic environment will continue to grow and will greatly enhance our evaluation of the magnitude of microbial contamination of water sources. With better understanding, appropriate control measures can be instituted for prevention of health hazards associated with the presence of pathogens in water. Criteria that can relate recreational water-associated illnesses to some aspect of water quality such as genetically constructed phages or bacteria need to be developed for both freshwaters and marine waters. This is an important goal, because die-off rates of indicator organisms in seawater and freshwater vary considerably (120).

REFERENCES

1. Pipes WO: Microbiological methods and monitoring of drinking water. In McFeters GA (ed): Drinking Water Microbiology: Progress and recent developments. New York: Springer-Verlag, 1990, pp 428–451.

2. Kuchta JM, States SJ, McNamara AM, Wadowsky RM, Yee RB: Susceptibility of *Legionella pneumophila* to chlorine in tap water. Appl Environ Microbiol 46:1134–1139, 1983.

3. Tobin JO'H, Dunnill MS, French M, Morris PJ, Beare J, Fisher-Hoch S, Mitchell RG, Muers MF: Legionaires' disease in a transplant unit: isolation of the causative agent from shower baths. Lancet ii:118–121, 1980.

4. Tobin JO'H, Bartlett CLR, Waitkins SA, Barrows GI, Macrae AD, Taylor AG, Fallon RJ, Lynch FRN: Legionnaires' disease: further evidence to implicate water storage and distribution systems as sources. Br Med J 282:573, 1981.

5. Scott WJ: Sanitary study of shore bathing waters. Connect Health Bull 65:74–85, 1951.

6. National Technical Advisory Committee: Water Quality Criteria. Washington DC: Fed Water Pollut Control Adm, 1968.

7. Zaporozec A: Changing patterns of ground water use in the United States. Ground Water 17:199–204, 1979.

8. Environmental Protection Agency. National primary drinking water regulations; Filtration and Disinfection; Turbidity, *Giardia lamblia,* Viruses, *Legionella,* and Heterotrophic bacteria— Proposed rules. Federal Register 52:42178–42222, Nov. 3, 1987 (40 CFR parts 141, 142).

9. Craun GF: Surface water supplies and health. J Am Water Works Assoc 80:40–52, 1988.

10. Levine WC, Stephenson WT, Craun GF: Waterborne disease outbreaks, 1986–1988. Centers for Disease Control; Morbid and Mortal Wkl Rep 39(SS-1):1–13, 1990.

11. Hayes EB, Matte TD, O'Brian TR, McKinley TW, Logsdon GS, Rose JB, Ungar BLP, Word DM, Pinsky PF, Cummings ML, Wilson MA, Long EG, Hurwitz ES, Juranek DD: Contamination of conventionally treated filtered public water supply by *Cryptosporidium* associated with large community outbreak of cryptosporidiosis. N Engl J Med 320:1372–1376, 1989.

12. Ho M, Glass RI, Pinsky PF, Anderson L: Rotavirus as a cause of diarrheal morbidity and mortality in the United States. J Infect Dis 158:1112–1116, 1988.

13. Snyder JD, Merson MH: The magnitude of the global problem of acute diarrhoeal disease: a review of active surveillance data. Bull World Health Organ 60:605–613, 1982.

14. Keswick BH, Satterwhite TK, Johnson PC, DuPont HL, Secor, SL, Bitsura JA, Gary GW, Hoff JC: Inactivation of Norwalk virus in drinking water by chlorine. Appl Environ Microbiol 50:261–264, 1985.

15. Kaplan JE, Gary GW, Baron RC, Singh N, Schonberger LB, Feldman R, Greenberg HB: Epidemiology of Norwalk gastroenteritis and the role of Norwalk virus in outbreaks of acute nonbacterial gastroenteritis. Ann Intern Med 96:756–761, 1982.

16. Craun GF: Statistics of waterborne outbreaks in the U.S. (1920–1980). In Craun GF (ed) Waterborne Diseases in the United States. Boca Raton, Florida: CRC Press, Inc., 1986 pp 73–159.

17. D'Antonio RG, Winn RE, Taylor JP, Gustafson TL, Current WL, Rhodes MM, Gary Jr GW, Zajac RA: A waterborne outbreak of cryptosporidiosis in normal hosts. Ann Intern Med 103:886–888, 1986.

18. Environmental Protection Agency: Drinking Water: substitution of contaminants and drinking water priority list of additional substances which may require regulation under the Safe Drinking Water Act. Federal Register 53:1892–1902, January 22, 1988.

19. Centers for Disease Control: Outbreak of *Pseudomonas aeruginosa* serotype 0:9 associated with whirlpool. Morbid Mortil Wkl Rep 30:329–331, 1981.

20. Kent GP, Greenspan JR, Herdon JL, Mofenson LM, Harris J-AS, Eng TR, Waskin HA: Epidemic giardiasis caused by a contaminated public water supply. Am J Public Health 78: 139–143, 1988.

21. Rose JB, Gerba CP, Singh SN, Toranzos GA, Keswick B: Isolation of entro- and rotaviruses from a drinking water treatment facility. J Am Water Works Assoc 78:56–61, 1986.

22. Makintubee S, Mallonee J, Istre GR: Shigellosis outbreak associated with swimming. Am J Public Health 77:166–168, 1987.

23. Sorvillo FJ, Waterman SH, Vogt JK, England B: Shigellosis associated with recreational water contact in Los Angeles County. Am J Trop Med Hyg 38:613–617, 1988.

24. Greensmith CT, Stanwick RS, Elliot, BE, Fast MV: Giardiasis associated with the use of a water slide. Pediatr Infect Dis J 7:91–94, 1988.

25. Kappus KD, Marks JS, Holman RC, Bryant JK, Baker C, Gary GW, Greenberg HB: An outbreak of Norwalk gastroenteritis associated with swimming in a pool and secondary person-to-person transmission. Am J Epidemiol 116:834–839, 1982.

26. Turner M, Istre GR, Beauchamp H, Baum M, Arnold S: Community outbreak of adenovirus type 7a infections associated with a swimming pool. South Med J 80:712–715, 1987.

27. Schiemann DA, Brodsky MH, Ciebin BW: *Salmonella* and bacterial indicators in ozonated and chlorine dioxide-disinfected effluent. J Water Pollu Control Fed 50:158–162, 1978.

28. Gerba CP, Rose JB: Viruses in source and drinking water. In McFeters GA (ed): Drinking Water Microbiology: progress and recent developments. New York: Springer-Verlag, 1990, pp 380–396.

29. American Public Health Association-American Water Works Association-Water Pollution control Federation: Standard Methods for Examination of Water and Wastewater, 17th ed. Washington, D.C., APHA, 1990.

30. Rose JB: Occurrence and control of *Cryptosporidium* in drinking water. In Mcfeters GA (ed). Drinking Water Microbiology: progress and recent developments. New York: Springer-Verlag, 1990, pp 294–321.

31. Domek MJ, Robbins JE, Anderson ME, McFeters GA: Metabolism of *Escherichia coli* injured by copper. Can J Microbiol 33:57–62, 1987.

32. Singh A, McFeters GA: Recovery, growth, and production of heat-stable enterotoxin by *Escherichia coli* after copper-induced injury. Appl Environ Microbiol 51:738–742, 1986.

33. Carson LA, Favero MS, Bond WW, Petersen NJ: Factors affecting comparative resistance of naturally occurring and subcultured *Pseudomonas aeruginosa* to disinfectants. Appl Environ Microbiol 23:863–869, 1972.

34. Torrella F, Morita RY: Microcultural study of bacterial size changes and microcolony and ultramicrocolony formation by heterotrophic bacteria in seawater. Appl Environ Microbiol 41:518–527, 1981.

35. MacDonell MT, Hood MA: Isolation and characterization of ultramicrobacteria from a Gulf Coast estuary. Appl Environ Microbiol 43:566–571, 1982.

36. Tompkins LS, Troup NJ, Woods T, Bibb W, McKinney RM: Molecular epidemiology of *Legionella* species by restriction endonuclease and allonezyme analysis. J. Clin Microbiol 25:1875–1880, 1987.

37. Rollins LS, Colwell RR: Viable but not culturable stage of *Campylobacter jejuni* and its role in survival in the natural aquatic environment. Appl Environ Microbiol 52:531–538, 1986.

38. Roszak DB, Grimes DJ, Colwell PR: Viable but nonrecoverable stage of *Salmonella enteritidis* in aquatic systems. Can J Microbiol 30:334–338, 1984.

39. Brayton PR, Colwell RR: Fluorescent antibody staining method for enumeration of viable environmental *Vibrio cholerae* 01. J Microbiol Methods. 6:309–314, 1987.

40. Singh A, Yeager R, McFeters: Assessment of in vitro revival, growth, and pathogenicity of *Escherichia coli* strains after copper- and chlorine-induced injury. Appl Environ Microbiol 52:832–837.

41. Singh A, McFeters GA: Survival and virulence of copper- and chlorine-stressed *Yersinia enterocolitica* in experimentally infected mice. Appl Environ Microbiol 53:1768–1774, 1987.

42. Colwell RR, Brayton PR, Grimes DJ, Roszak, DB, Huq SA, Palmer LM: Viable but non-culturable *Vibrio cholerae* and related pathogens in the environment: implications for release of genetically engineered organisms. Bio/Technology 3:817–820, 1985.

43. Schiemann DA: *Yersinia enterocolitica* in drinking water. In McFeters GA (ed) Drinking Water Microbiology: progress and recent developments. New York: Springer-Verlag, 1990, pp 322–339.

44. Bopp CA, Sumner JW, Morris GK, Wells JG: Isolation of *Legionella* spp. from environmental water samples by low-pH treatment and use of a selective medium. J Clin Microbiol 13:714–719, 1981.

45. Dennis PJ, Barlett CLR, Wright AE: Comparison of isolation methods of *Legionella* spp. In Thornsberry C, Balows A, Feeley J, Jakubowski W (eds) *Legionella*: Proceedings of 2nd International Symposium. Washington DC: American Society for Microbiology, 1984, pp 294–296.

46. Sobsey MD: Quality of currently available methodology for monitoring viruses in the environment. Environ Internat 7:39–52, 1982.

47. Dutka BJ: Coliforms are inadequate index of water quality. J Environ Health 36:39–46, 1973.

48. Berry SA, Noton BG: Survival of bacteriophases in seawater. Water Res 10:323–327, 1976.

49. Geldreich EE: Bacterial populations and indicator concepts in feces, sewage, storm water and soil wastes. In Berg G (ed): Indicators of Viruses in Water and Food. Ann Arbor, Mich, Ann Arbor Science, 1978, pp 51–97.

50. McCrady MH: The numerical interpretation of fermentation-tube results. J Infect Dis 17:183–212, 1915.

51. McCrady MH: Tables for rapid interpretation of fermentation-tube results. Publ Health J (Toronto) 9:201–220, 1918.

52. Thomas Jr. HA, Woodward RL, Kabler PW: Use of molecular filter membranes for water quality control. J Am Water Works Assoc 48:1391–1402, 1956.

53. McFeters GA, Kippin JS, LeChevallier MW: Injured coliforms in drinking water. Appl Environ Microbiol 51:1–5, 1986.

54. LeChevallier MW, Cameron SC, McFeters GA: Comparison of verification procedures for the membrane filter total coliform technique. Appl Environ Microbiol 45:1126–1128, 1983.

55. Reasoner DJ, Blannon JC, Geldreich EE: Rapid seven-hour fecal coliform test. Appl Environ Microbiol 38:229–236, 1979.

56. Reasoner DJ, Geldreich EE: Rapid detection of water-borne fecal coliforms by $^{14}CO_2$ release. In Sharpe AN, Clark DS (eds): Mechanizing Microbiology. Springfield, ILL, Charles C. Thomas Publishers, 1978, pp 120–139.

57. Dange V, Jothikumar N, Khanna P: One hour portable test for drinking waters. Water Res 22:133–137, 1988.

58. Berg JD, Fiksdal L: Rapid detection of total coliforms in water by enzymatic hydrolysis of 4-methylumbelliferone-β-D-galactoside. Appl Environ Microbiol 54:2118–2122, 1988.

59. Edberg SC, Allen MJ, Smith DB, and the National Collaborative Study: National field evaluation of a defined substrate method for the simultaneous detection of total coliforms and *Escherichia coli* from drinking water: comparison with presence-absence techniques. Appl Environ Microbiol 55:1003–1008, 1989.

60. Bej AK, Steffan RJ, DiCesare J, Haff L, Atlas RM: Detection of coliform bacteria in water by polymerase chain reaction and gene probes. Appl Environ Microbiol 56:307–314, 1990.

61. Dufour AP: Health effects criteria for fresh recreational waters. Environmental Protection Agency Report 600/1-84-004. Washington DC: Environmental Protection Agency, 1984.

62. Dufour AP: A 24–hour membrane filter procedure for enumerating enterococci. Abstr 80th Ann Meet Am Society Microbiol, Miami Beach, FL (11–16 May, 1980). Washington, DC: American Society for Microbiology, 1980, p 205.

63. Wentsel RS, O'Neill PE, Kitchens JF: Evaluation of coliphage detection as a rapid indicator of water quality. Appl Environ Microbiol 43:430–434, 1982.

64. Kott Y, Roze N, Sperber S, Betzer N: Bacteriophages as bacterial viral pollution indicators. Water Res 8:165–171, 1982.

65. Havelaar AH, Hogeboom WM: Factors effecting the enumeration of coliphages in sewage and sewage-polluted waters. Antonie Van Leeuwenhoek 49:387–397, 1983.

66. Havelaar AH, Hogeboom WM, Pot R: F-specific RNA bacteriophages in sewage: methodology and occurrence. Water Sci Technol 17:645–655, 1984.

67. Sobsey MD, J, Schwab K, Chung H, Handzel T: A simple membrane filter method for concentration and enumeration of male-specific (F^+) coliphage in water. Abstr 90th Ann Meet Am Society Microbiol. Anaheim, Calif (13–17 May 1990). Washington, DC. American Society for Microbiology, 1990, p 289.

68. Sobsey MD, Kellogg J, Schwab KJ, Handzel TR: A simple membrane filter method to concentrate and enumerate male-specific RNA coliphages. J Am Water Works Assoc 82:52–59, 1990.

69. Levin MA, Fischer JR, Cabelli VJ: Quantitative large-volume sampling technique. Appl Environ Microbiol 28:515–517, 1974.

70. Presnell MW, Andrews WH: Use of the membrane filter and a filter aid for concentrating and enumerating indicator bacteria and *Salmonella* from estuerine waters. Water Res 10:549–554, 1976.

71. Park CE, Rayman MK, Szabo R, Stankiewicz ZK: Selective enrichment of *Shigella* in the presence of *Escherichia coli* by use of 4-chloro-2-cyclopentylphenyl β-D-galactopyranoside. Can J Microbiol 22:654–657, 1976.

72. Thomason BM: Rapid detection of *Salmonella* microcolonies by fluorescent antibody. Appl Environ Microbiol 22:1064–1069, 1971.

73. Kogure K, Simidu U, Taga N: A tentative direct microscopic method for counting living marine bacteria. Can J Microbiol 25:415–420, 1979.

74. Brayton PR, Tamplin ML, Huq A, Colwell RR: Enumeration of *Vibrio cholerae* 01 in Bangladesh waters by fluorescent-antibody direct viable count. Appl Environ Microbiol 53:2862–2865, 1987.

75. Singh A, Yu F-P, McFeters GA: Rapid detection of chlorine-induced bacterial injury by the direct viable count method using image analysis. Appl Environ Microbiol 56:389–394, 1990.

76. Scholl DR, Kaufmann C, Jollick JD, York CK, Goodrum GR, Charache P: Clinical application of novel sample processing technology for the identification of salmonellae by using DNA probes. J Clin Microbiol 28:237–241, 1990.

77. Sack RB: Human diarrheal disease caused by enterotoxigenic *Escherichia coli*. Ann Rev Microbiol 29:333–353, 1975.

78. Calderon RL, Levin MA: Quantitative method for enumeration of enterotoxigenic *Escherichia coli*: J Clin Microbiol 13:130–134, 1981.

79. Echeverria P, Seriwatana J, Chityothin O, Chaicumpa W, Tirapat C: Detection of enterotoxigenic *Escherichia coli* in water by filter hybridization with three enterotoxin gene probes. J Clin Microbiol 16:1086–1090, 1982.

80. Blaser MJ, Berkowitz ID, LaForce FM, Cravens J, Reller LB, Wang W-LL: Campylobacter enteritis: clinical and epidemiologic features. Ann Intern Med 91:179–185, 1979.

81. Penner JL, Hennessy JN: Passive hemagglutination technique for serotyping *Campylobacter fetus* subsp *jejuni* on the basis of heat-stable antigens. J Clin Microbiol 12:732–737, 1980.

82. Nachamkin I, Barbagallo S: Culture confirmation of *Campylobacter* spp. by latex agglutination. J Clin Microbiol 28:817–818, 1990.

83. Tompkins L, Krajden M: Approaches to the detection of enteric pathogens including *Campylobacter,* using nucleic acid hybridization. Diagn Microbiol Infect Dis 4:S78–S88, 1986.

84. Olive DM, Johny M, Sethi SK: Use of an alkaline phosphatase-labeled synthetic oligonucleotide probe for detection of *Campylobacter jejuni* and *Campylobacter coli.* J Clin Microbiol 28:1565–1569, 1990.

85. Tenover FC, Carlson L, Barbagallo S, Nachamkin I: DNA probe culture confirmation assay for identification of thermophilic *Campylobacter* species. J Clin Microbiol 28:1284–1287, 1990.

86. Saari TN, Jansen GP: Waterborne *Yersinia enterocolitica* in the midwestern United States. Contrib Microbiol Immun 5:183–196, 1979.

87. Schiemann DA, Olson SA: Antagonism by gram-negative bacteria to growth of *Yersinia enterocolitica* in mixed cultures. Appl Environ Microbiol 48:539–544, 1979.

88. Schiemann DA: Development of a two-step enrichment procedure for recovery of *Yersinia enterocolitica* from food. Appl Environ Microbiol 43:14–27, 1982.

89. Riley G, Toma S: Detection of Pathogenic *Yersinia enterocolitica* by using Congo Red–magnesium oxalate agar medium. J Clin Microbiol 27:213, 1988.

90. Kandolo K, Wauters G: Pyrazinamidase activity of *Yersinia entercolitica* and related organisms. J Clin Microbiol 21:980–982, 1985.

91. Calderon RL, Dufour AP: Media for detection of *Legionella* spp. in environmental water samples. In Thornsberry C, Belows A, Feeley LC, Jakubowski W (eds): *Legionella:* Proceeding of the 2nd International Symposium, Washington, DC: 1984, pp 290–296.

92. Gilbert J: Detection of *Legionella* spp in water samples by electron capture gas chromatography. In Thornsberry C, Belows A, Feeley JC, Jakubowski W (eds): *Legionella:* Proceeding of the 2nd International Symposium, Washington, DC: 1984, pp 296–298.

93. Martin WT, Barbaree JM, Feeley JC: Detection and quantitation of *Legionella pneumophila* by immune autoradiography. In Thornsberry C, Belows A, Feeley LC, Jakubowski W (eds): *Legionella:* Proceeding of the 2nd International Symposium, Washington, DC: 1984, pp 299–300.

94. Starnbach MN, Falkow S, Tompkins LS: Species-specific detection of *Legionella pneumophila* in water by DNA amplification and hybridization. J Clin Microbiol 27:1257–1261, 1989.

95. Jakubowski W: Detection of *Giardia* in drinking water: state-of-art. In Erlandsen SL, Meyer EA (eds): Giardia and Giardiasis. New York: Plenum Press, 1984, pp 263–285.

96. Sauch JF: Use of immunofluorescence and phase-contrast microscopy for detection and identification of *Giardia* cysts in water samples. Appl Environ Microbiol 50:1434–1438, 1985.

97. Musial CE, Arrowood MJ, Sterling CR, Gerba CP: Detection of *Cryptosporidium* in water using polypropylene cartridge filters. Appl Environ Microbiol 53:687–692, 1987.

98. Ongerth JE, Stibbs HH: Identification of *Cryptosporidium* oocysts in river water. Appl Environ Microbiol 53:672–676, 1987.

99. Sterling CR, Arrowood MJ: Detection of *Cryptosporidium* sp. infections using a direct immunofluorescent assay. Pediatr Infect Dis 5:5139–5142, 1986.

100. Garcia LS, Brewer TC, Bruckner DA: Fluorescent detection of *Cryptosporidium* oocysts in human fecal specimens by using monoclonal antibodies. J Clin Microbiol 25:119–121, 1987.

101. Rose JB: Occurrence and significance of *Cryptosporidium* in water. J Am Water Works Assoc 80:53–58, 1988.

102. Rose JB, Kayed D, Madore MS, Gerba CP, Arrowood MJ, Sterling CR: Methods for the recovery of *Giardia* and *Cryptosporidium* from environmental waters and their comparative occurrence. In Wallis P, Hammond B (eds): Advances in *Giardia* Research. Calgary, Canada: University of Calgary Press, 1988, pp 205–209.

103. Sobsey MD: Methods for detecting enteric viruses in water and waste water. In Berg G, Bodily HL, Lennette EH, Melnick JL, Metcalf TG (eds): Viruses in Water. Washington DC: American Public Health Association, 1976, pp 89–127.

104. Wallis C, Melnick JL, Gerba CP: Concentration of viruses from water by membrane chromatography. Ann Rev Microbiol 33:413–437, 1979.

105. Goyal SM, Gerba CP: Concentration of virus from water by membrane filters. In Gerba CP, Goyal SM (eds): Methods in Environmental Virology. New York: Marcel Dekker, 1982, pp 59–149.

106. Rose JB, Gerba CP: A review of viruses in drinking water. Curr Practices Environ Engr 2:119–141, 1986.

107. Schmidt NJ, Ho HH, Riggs JL, Lennette EH: Comparative sensitivity of various cell culture systems for isolation of viruses from wastewater and fecal samples. Appl Environ Microbiol 36:480–486, 1978.

108. Goddard MR, Sellwood J: Isolation of enteric viruses from sewage. In Butler M, Medlen AR, Morris R (eds): Viruses and Disinfection of Water and Waste Water. Guilford, University of Surrey, 1982, p 16.

109. Steinmann JC: Detection of rotavirus in sewage. Appl Environ Microbiol 41:1043–1045, 1981.

110. Hejkal TW, Keswick B, LaBelle RL, Gerba CP, Sanchez Y, Dreesman G, Hafkin B, Melnick JL: Viruses in a community water supply associated with an outbreak of gastroenteritis and infectious hepatitis. J Am Water Works Assoc 74:318–321, 1982,

111. Kulski JK, Norval M: Nucleic acid probes in diagnosis of viral diseases of man and brief review. Arch Virol 83:3–15, 1985.

112. Jiang X, Estes MK, Metcalf TG, Melnick JL: Detection of hepatitis A virus in seeded estuarine samples by hybridization with cDNA probes. Appl Environ Microbiol 52:711–717, 1986.

113. Margolin AB, Hewlett MJ, Gerba CP: Use of a cDNA dot-blot hybridization technique for detection of enteroviruses in water. In Water Quality Technology Conference Proceedings. Denver, CO, American Water Works Association, 1986, pp 1987–195.

114. Richardson KJ, Margolin AB, Gerba CP: A novel method for liberating viral nucleic acid for assay of water samples with cDNA probes. J Virol Methods 22:13–21, 1988.

115. Dimitrov DH, Graham, DY, Estes MK: Detection of rotavirus by nucleic acid hybridization with cloned DNA of simian rotavirus SA11 genes. J Infect Dis 152:293–300, 1985.

116. Rao VC, Melnick JL: Environmental Microbiology. Washington DC: American Society for Microbiology, 1986.

117. Environmental Protection agency: Drinking water; National Primary Drinking Water Regulations; Filtration Disinfection, Turbidity, *Giardia lamblia*, Viruses, *Legionella* and Hetero-

trophic bacteria (Part II); Total coliforms including fecal coliforms and *E. coli* (Part III)—final rule. Federal Register 54:27486–27568, June 29, 1990 (40 CFR parts 141 and 142).

118. du Molin GC, Sherman IH, Hoaglin DC, Stottmeier KD: *Mycobacterium avium* complex, and emerging pathogen in Massachusetts. J Clin Microbiol 22:9–12, 1985.

119. Goslee S, Wolinsky E: Water as a source of potentially pathogenic mycobacteria. Am Rev Respir Dis 113:287–292, 1976.

120. Chamberlin CE, Mitchell R: A decay model for enteric bacteria in natural waters. In Mitchell R (ed): Water Pollution Microbiology, Vol 2. New York: John Wiley and Sons, 1978 pp 325–348.

CONTROL OF ENTERIC PATHOGENS IN DEVELOPING COUNTRIES

S. CAIRNCROSS

London School of Hygiene and Tropical Medicine, University of London, London, WC1E 7HT England.

1. HISTORICAL INTRODUCTION

Most historians would agree that it was Robert Koch who first used environmental microbiology to plan the control of enteric microbial infections. His discovery of the ''comma bacillus'' causing cholera was enriched by his analysis of the environmental conditions favorable to its growth and hence its transmission (1). The clarity of his perception led him to set out strategies for cholera control that are perfectly valid today, more than a century later (2). Moreover, he was soon well-placed to ensure that his control strategy was implemented. Within a year of his discovery, he had been made a full Professor of Hygiene at the University of Berlin and a member of the Prussian Privy Council and of the Cholera Commission for the German Empire. Certainly, there were grounds for having high hopes of the rapid eradication of cholera and indeed of the other enteric infections whose pathogens were being identified one by one in the closing decades of the century.

Koch's hopes were thwarted, however, in both the short and the long term. In the short term, he faced a long battle against the miasmatists, led by Von Pettenkofer, who believed that cholera and typhoid bacilli had to undergo a period of maturation in the soil to produce a poison that was then transmitted

Environmental Microbiology, pages 157–189, © *1992 Wiley-Liss, Inc.*

through the air. Twenty years after his discovery of *Vibrio cholerae,* Koch was still having to defend his views in court against opposition from miasmatists at the 1904 trial of the directors of the Gelsenkirchen water company, who were accused of causing a typhoid epidemic by their negligence (3).

Two aspects of his struggle are noteworthy, as they have relevance to recent thinking. The first is the contrast between the two positions with regard to their implications for control strategy and their relationship to political tendencies current at the time. Von Pettenkofer's views stressed the importance of local conditions in determining a particular town's susceptibility to a cholera epidemic and the need for interdisciplinary collaboration between specialists in a range of scientific fields, such as meteorology and statistics, for control of the disease. A pathogen in the environment would only cause an epidemic if local conditions, particularly with regard to the soil and the climate, were favorable. His localism implied that every part of the country had its own individual characteristics and needed to be treated accordingly by those who were familiar with them. He also underlined the responsibility of the individual, insisting that ''we do not solve the problem by providing the poor with the most necessary food, housing and clothing unless we at the same time educate them in painstaking cleanliness'' (4).

Koch, by contrast, returned medical scientists to center stage and pioneered the central role of the microbiologist in setting standards. He was also able to lay down simple and universally applicable measures—quarantine, isolation, and disinfection—which could be applied by the central authority of the State.

The second noteworthy aspect of the debate is its inconclusiveness. Koch's microbiological discoveries did not introduce the hypothesis of feco-oral transmission of enteric infections, which had been stated with extraordinary clarity by John Snow in 1855 (5); nor did they prove the hypothesis conclusively. Admittedly, one weakness was that, because of his inability to infect an animal with *V. cholerae,* Koch was unable to fulfill his own famous postulates for proving that it was the agent of the disease. More generally, most microbiological evidence for a particular transmission route—and indeed epidemiological evidence of the sort collected by Snow—is fundamentally circumstantial.

A microorganism caught under a microscope or in an incubation medium is held static, removed from the environment in which it may (or may not) be in the dynamic process of transmission; it cannot normally be caught *in flagrante delicto,* in the process of starting a case of infection. Even where it is found in circumstances, for example, in drinking water, that strongly incriminate one transmission route, its presence there does not prove that this is the only or even the most important route. Nor, in a nonzoonotic infection such as typhoid or cholera, is it ethically possible to prove that the microorganism found is still capable of causing infection.

Thus the battle of words between the ''contagionists'' and the ''miasmatists'' raged throughout most of the nineteenth century. Writing in 1896 about the cholera debates of the Hamburg Doctors' Club in 1831, the club's historian wrote that "the opposing parties fought with roughly the same means as they do today, and just as they do today, everyone stuck at the end of the debate to his original points of view" (4).

In the long term, the enteric diseases are with us still. Diarrheal diseases still kill an estimated 5 million people each year, most of them children under age 5 years (6). Admittedly, diarrhea mortality, and the morbidity caused by the bacterial infections that preoccupied nineteenth century sanitarians, have largely been eliminated from the industrialized nations of the world; but this has been achieved largely by the attainment of living standards unimaginable to most people at that time (7). In the developing countries, where levels of income, hygiene, and sanitary infrastructure provision are nearer to those in Europe in the nineteenth century than today, typhoid, cholera, and other bacterial diarrheas are still a major public health problem.

In this chapter, I hope to show how these historical themes are echoed by recent thinking with regard to environmental measures for the control of enteric infections, particularly in developing countries. It has become increasingly clear that microbiological standards are not enough on their own and cannot be laid down centrally. There are signs of the advent of a new localism and a consciousness that the presence of pathogens in the environment does not necessarily cause infection. I shall focus on two questions where environmental microbiology has played an important role, the reuse of municipal waste water for irrigation and the health impact of improved domestic water supplies.

2. REUSE OF WASTE WATER

2.1 The Past

The reuse of waste water is probably as old a practice as the use of sewers to collect it. The open sewers of many ancient cities, such as Damascus, were used to water extensive fields and orchards, and the latrines of many a medieval monastery fertilized a fish pond. Edwin Chadwick, one of the great public health campaigners of nineteenth century London, held that the poor would benefit at least as much from the improvement in their diet that "sewage farming" could bestow as from the hygienic disposal of their wastes.

Many European and North American cities adopted crop irrigation as their means of waste water disposal at the same time as they installed waterborne sanitation. The first records of the practice date from as early as 1865 in Britain. During the 1870s, the United States, France, Germany, and India followed the example, with Australia starting in 1893 and Mexico in 1904. In spite of Chadwick's early enthusiasm, the main objective was to prevent river pollution rather than to enhance crop production. However, as cities grew and more and more of their population came to be served by sewer systems, the land area required for sewage farming became too great.

Moreover, the prime motive for reuse had diminished in importance. From the early 1900s, with the introduction of the Royal Commission standards for treated waste water, British sanitary engineers (and soon afterwards those of other countries) knew that under certain circumstances river discharge was permissible. The development of modern waste water treatment processes such as bio-filtration and activated sludge during the first two decades of this century made those standards achievable.

The practice of sewage farming disappeared almost completely in many countries soon after the First World War, although the sewage farms at Werribee (Melbourne, Australia), Mexico City, and various locations in India were notable exceptions to this trend and are still in operation today.

Incidentally, it is noteworthy that the Royal Commission standards were not expressed in terms of pathogens or even fecal indicator bacteria—the statistical foundations for the enumeration of coliforms by the multiple tube method were not laid until 1915 (8)—but in terms of biochemical oxygen demand (BOD) and suspended solids content. Accordingly, the new treatment processes were largely designed and evaluated on the basis of their ability to meet these standards rather than to remove pathogens. However, since a negligible and diminishing number of people in the developed world drank untreated river water, and public concern about water pollution was largely based on esthetic and ecological considerations, this was not a completely irrational approach.

If official, municipal reuse of waste water was discontinued, the economic benefits of the practice to the farmer were no less strong, particularly in those areas, such as parts of the United States, where the availability of water resources was a major constraint to agricultural production. Increasing concern led the California State Health Department to set a stringent quality standard for effluent irrigation of vegetables and salad crops some 50 years ago. This time the standard was in terms of indicator bacteria—a maximum of 2 coliforms per 100 ml of effluent. The rationale behind such a demanding requirement was the idea of achieving "zero risk," based on reports of pathogen detection and survival in wastewater and soil, and the idea that the mere presence of pathogens in the environment is evidence of a serious public health risk. Those involved in setting this standard were undeterred by the fact that it was not really achievable in practice, even in a developed country; indeed, this was probably an advantage to them as it effectively outlawed the practice of unrestricted waste water irrigation altogether (9). Nevertheless, the California standard spread rapidly throughout the world as the most commonly accepted guideline for waste water reuse, because no other credible standard or source of evidence on this subject was available.

A more pragmatic stance was adopted by the World Health Organization (10), when for unrestricted irrigation it recommended adoption of what was then considered to be the highest standard achievable under field conditions—not more than 100 coliform organisms per 100 ml of treated waste water in 80% of samples. In addition to conventional waste water treatment, to achieve this would require heavy and carefully controlled chlorination. This is an expensive procedure, and difficult to maintain reliably in most developing countries. For example, disinfection of the wastewater from the city of Amman, Jordan, cost some U.S. $1,000 a day in foreign exchange (11).

2.2 Post-War Developments

The last two decades have seen a considerable increase in the use of waste water for irrigation, particularly in the more arid regions of both developed and developing countries. Bartone and Arlosoroff (12) were able to document waste water irrigation schemes in 16 countries, including the United States. In many of these countries, tens of thousands of hectares are involved. Because of the high cost of meeting the WHO standard, schemes seeking to do this have generally been limited to the industrialized world and the oil-rich Middle East. In many other developing countries, the practice has not been permitted by the authorities, who judged, correctly, that the prevailing standards were not achievable in practice under local conditions. However, this has not prevented peasant farmers and others from yielding to the strong economic incentives to practise illegal or clandestine reuse, usually of waste water or nightsoil that has received no treatment at all. In some cases, as in parts of Peru, they have even been known to break open sewer pipes in order to divert the flow onto their fields (13).

The first official recognition that the "zero risk" approach in the context of waste water reuse was unjustifiably stringent was perhaps the acceptance by a WHO Working Group that "economically and practically a "no-risk" level cannot be obtained, although it may be technologically possible" (14). At the same time, theoretical understanding and epidemiological evidence were beginning to point to the view that "no-risk" levels were not only unfeasible, but also unnecessary.

2.3. Theoretical Considerations

Theoretically, the public health importance of the transmission of a given pathogen through waste water reuse depends not only on the presence of that pathogen in the population contributing the waste water, but is also conditional on a range of intervening factors (Fig. 7.1). The "no-risk" approach or, in other words, the application of the old quality standards for treated waste water, acts only on the first group of these, by preventing the pathogen from reaching the

EXCRETED LOAD

• latency
• multiplication
• persistence
• treatment survival

INFECTIVE DOSE APPLIED TO LAND/WATER

• persistence
• intermediate host
• type of use practice
• type of human exposure

INFECTIVE DOSE REACHES HUMAN HOST

• human behaviour
• pattern of human immunity

RISKS OF INFECTION AND DISEASE

• alternative routes of transmission

PUBLIC HEALTH IMPORTANCE OF EXCRETA AND WASTEWATER USE

Fig. 7.1. Factors influencing the sequence of events between the presence of a pathogen in excreta and the occurrence of measurable disease attributable to waste water reuse. (Reproduced from Blum and Feachem [15], with permission of the publisher.)

fields. A second group can prevent an infective dose of the pathogen, even if it should reach the fields, from coming into contact with a human host. These are

1. The survival time of the pathogen in soil, on crops, or in water
2. The presence (for certain helminthic infections) of the required intermediate host
3. The mode, timing, and frequency of waste water application
4. The type of crop to which the waste water is applied
5. The nature of the exposure of the human host to the contaminated soil, crops, or water

If the dose does reach a human host, it still may not cause infection, depending on behavioral factors, particularly regarding measures taken by exposed individuals to protect their health, and also on their immune status. Finally, even where the pathogen is being transmitted through waste water reuse practices, the public health significance of this fact will depend on the degree to which transmission is also occurring by other routes. These factors are briefly discussed in turn.

1. Persistence. The extensive literature on the survival times of fecal pathogens in soil and on crop surfaces has recently been extensively reviewed (16,17). There are wide variations in reported survival times, which reflect both strain variation and differing climatic factors as well as different analytical techniques. Nevertheless, the available evidence shows that almost all excreted pathogens can survive in soil for long enough to pose potential risks to farm workers. The survival times of intestinal helminth eggs are particularly long; in particular, *Ascaris* eggs can survive in soil for as long as 1 year. Survival times on crops are much shorter, but in some cases they can be long enough to pose potential risks to crop handlers and to consumers, especially if they exceed the growing season of the crop. Some of these survival times are shown in Figure 7.2.

2. Intermediate host. Transmission of *Taenia saginata* and *Taenia solium* can only result from waste water irrigation if cows or pigs, respectively, have access to the irrigated land while the eggs are still viable.

3. Mode and frequency of application. Certain irrigation methods, such as sprinkler irrigation, clearly expose people to infection in ways that cannot result from others, such as trickle irrigation in which the water is applied to the roots of each plant under the cover of a protective plastic sheet. The interval between the last application and harvesting also affects the likely degree of crop contamination.

4. Type of crop and type of exposure. Crops grown for human consumption pose potential risks to those who consume them, as well as to those who produce and handle them. The greatest risk to the consumer is associated with food eaten raw, especially root crops (such as radishes) or vegetables growing close to the soil (for instance, lettuces). For fodder crops, farm workers and those who consume the resulting meat or milk may be at risk. In the case of industrial products, such as sugar beet, cotton, or timber, only farm workers and product handlers are potentially subjected to risk. Where sprinkler irrigation is practised, those living near the irrigated fields, who are at potential risk from pathogens present in wind-dispersed aerosol droplets, form an additional exposure group.

5. Human behavior. Adequate standards of personal and food hygiene and, in the case of occupational exposure, the wearing of protective clothing and footwear, can protect against infection, even in conditions that would otherwise present an extremely high risk of infection. Regular chemotherapy for intestinal helminths is another intervening behavioral factor.

6. Host immunity. The role of immunity is most noticeable in the case of viral infections, where infection at an early age is very common, even in communities with high standards of hygiene, so that most adults are immune to the disease and frequently also to infection. In the case of typhoid, immunization programs can affect the risk of transmission.

7. Alternative transmission routes. The factors outlined above determine the potential health risks associated with waste water reuse. The relative *importance*

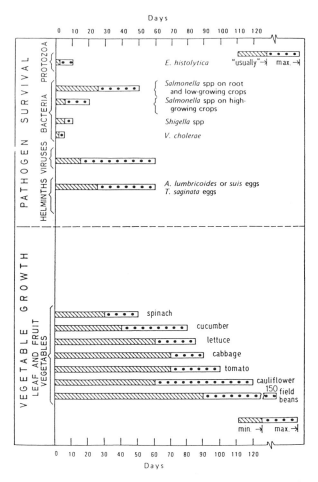

Fig. 7.2. Pathogen survival on crops compared with vegetable growth periods in warm climates. (Reproduced from Strauss [17], with permission of the publisher.)

of such risks depends on the existence of any other routes by which the excreted pathogens reach those at risk. If there are many such alternative routes, waste water reuse may not pose a significant additional risk. To illustrate this, consider the residents of a wealthy, modern city and those of a poor, traditional village who both consume vegetables fertilized with the villagers' excreta. Let us suppose that the standards of personal and domestic hygiene are very high in the city, but very low in the village. Then the only (or almost the only) exposure of the city inhabitants to excreted pathogens is via the vegetables. For the villagers, however, this transmission route will be only one of many, and not necessarily the most important, since the high level of fecal contamination of their immediate environment is likely to give rise to much more direct exposure and consequent infection and disease. Thus prevention of consumption of the vegetables in the city would be an effective control strategy, but similar measures in the village would probably have little effect, if any, on the disease transmission rate.

With these theoretical considerations in mind, Shuval *et al.* (18) deduced *a priori* that the pathogens most effectively transmitted by waste water irrigation would be the intestinal helminths. They are the most persistent, they have a low infectious dose, there is little or no immunity to them, and long-term exposure and continuing reinfections can lead to a build-up of heavy worm loads. They concluded that viruses would present the least risk, because of the tendency for immunity to be developed in childhood through other transmission routes, and that enteric bacteria and protozoa would occupy an intermediate position.

This theoretical approach is characterized by a change in the method of assessment of health risks, from microbiological to epidemiological criteria. In the past, a "potential risk" was said to occur when pathogenic microorganisms were found in waste water or on crops, even if no cases of disease caused by these microorganisms were detected. This is in contrast to an epidemiological definition of risk, which is the probability that an individual will develop a given disease during a specified period. With an epidemiological approach, as advocated by Feachem and Blum (19) decisions are based on "actual risks" rather than potential risks.

Clearly, a "potential risk" can fail to become an "actual risk" because of any of the intervening factors discussed above. The effect of alternative transmission routes can also be taken into account, using the concept of "attributable risk" or "excess risk," which is a measure of the amount of disease associated with a particular transmission route within a population—in this case, the amount associated with waste water irrigation. This can only be measured by comparing two populations, one exposed to waste water, the other not exposed (the control population). The difference in disease between the exposed and control populations, and not simply the amount of disease in the exposed population, is a measure of the risk attributable to waste water reuse. Epidemiological studies are thus the only type of study that can measure actual risks associated with the practice.

2.4. Epidemiological Evidence

In view of both the increasing concern that existing standards might be excessively strict and the growing perception that rational standards should be

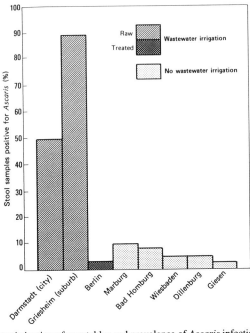

Fig. 7.3. Wastewater irrigation of vegetables and prevalence of *Ascaris* infection in West Berlin and a number of cities in West Germany in 1949. Note: The waste water used in West Berlin was conventionally treated (*i.e.*, with primary sedimentation, biofiltration, and secondary sedimentation). (Reproduced from Shuval et al. [20], with permission of the publisher.)

based on epidemiological evidence, the World Bank commissioned a detailed review of all of the available literature on the subject. The main findings (20) are summarized below.

1. The use of untreated waste water for irrigation exposes farm workers to a serious excess risk of infection with intestinal helminths, particularly *Ascaris* and hookworm. Where salad crops or vegetables generally eaten uncooked are irrigated in this way, the risk also applies to the general population that consumes them. This finding, which bears out the theoretical conclusion above, is supported by strong evidence from various countries. Figure 7.3 presents an interesting example.

2. Cholera can be transmitted to populations consuming vegetable and salad crops irrigated with untreated waste water and also apparently to farm workers. There is evidence to suggest that this is also true for typhoid, particularly where

other normal routes of typhoid transmission are blocked because of fairly good community hygiene.

3. There is no clear evidence of adverse health effects from bacterial and viral diseases caused by transmission in waste water aerosols. Morbidity and serological studies among waste water irrigation workers and treatment plant workers suggest that, though they may suffer from some relatively benign gastrointestinal diseases and develop antibodies to some viral pathogens, particularly in the first months of their contact with waste water, they show few signs of serious health effects. In the case of viral pathogens, it appears that the mild exposure to infection caused by aerosols and by other transmission routes is sufficient to develop immunity without ill effect; as regards the bacteria, it is likely that the low doses received in this way are insufficient to cause serious morbidity.

4. There is evidence for a reduction in negative health effects resulting from effective pathogen removal by waste water treatment. In Figure 7.3, for example, note that the treatment of Berlin's waste water appears to have protected the residents of that city from *Ascaris* infection, even though the form of treatment in use at the time (conventional biological oxidation and sedimentation) does not guarantee full removal of *Ascaris* eggs.

A noteworthy epidemiological study, not included in the review, was conducted by the municipal authorities in Colorado Springs, U.S.A. (21). It compared the incidence of gastroenteritis among inhabitants who had visited parks irrigated with waste water treated to various levels of quality. The study was of sufficient power to detect a difference in incidence of only four cases out of 100, but statistically significant increases in gastrointestinal symptoms were only found to be associated with fecal coliform and total coliform densities above 500 per 100 ml and 3,000 per 100 ml, respectively.

2.5. The New Guidelines

Between 1985 and 1987, various international agencies including the World Bank and the World Health Organization convened two meetings of experts (22,23). These meetings prepared the ground for the meeting that produced the new WHO guidelines for waste water use in agriculture (24), and that was certainly notable and probably unique in deciding to relax an existing WHO bacteriological water quality guideline by nearly three orders of magnitude. Where the old guideline for unrestricted irrigation (10) was expressed in total coliforms, the new one was in terms of fecal coliforms; where the old one stipulated a maximum value, the new one was a geometric mean; and the guideline value was increased from 100 to 1,000 per 100 ml of treated wastewater. Each of these three changes is roughly equivalent to relaxation by a factor of 10.

It was the consensus of all three meetings of experts that a microbial standard of a mean of 1,000 fecal coliforms per 100 ml for unrestricted irrigation was both epidemiologically sound and technologically feasible. It was considered to be in line with the actual quality of river water used for unrestricted irrigation in Europe and North America with no known ill effects. In this context, it is noteworthy that in many such rivers the majority of the fecal bacteria present have originated from treated domestic waste water; it was mentioned above that conventional waste water treatment processes are designed for (and are efficient at) the removal of BOD and suspended solids—not fecal bacteria. The participants also noted that many countries found levels of 1,000 fecal coliforms per 100 ml acceptable for bathing water quality and considered that it was not rational to require a stricter standard for waste water irrigation than those applied to general irrigation and bathing in most of the industrialized countries.

On the other hand, the epidemiological evidence regarding intestinal helminths led to a stricter approach to these pathogens in the new guidelines. A guideline value of 1 helminth egg or less per liter was proposed, which represents a requirement to remove some 99.9% by appropriate treatment processes. It was understood that the strict helminth standard recommended would serve as an indicator for all of the large easily settleable pathogens including the protozoa, such as *Entamoeba* and *Giardia*. The full microbiological quality guidelines are shown in Table 7.1.

While the helminth egg guideline was intended as a design goal for waste water treatment systems and not as a standard requiring routine testing of effluent quality, it is clear that its implementation in practice will require readily available and reliable techniques for the detection of helminth eggs at low concentrations in waste water. Methods for the detection of these pathogens in the environment have hitherto been research tools rather than instruments of policy, and the currently available techniques are laborious, time-consuming, and thus far validated only by one or two groups of researchers (25–29). Major difficulties include the problem of centrifuging the large sample volumes required and of separating the helminth eggs from the quantites of extraneous material suspended in waste water.

The other noteworthy feature of the new guidelines was their recognition that treatment of the waste water was not the only feasible effective measure for the reduction of the health risks of waste water reuse. Treatment may avoid "potential" risks by pathogen removal, but other measures affecting the intervening factors shown in Figure 7.1 can be equally effective in preventing an actual, attributable risk. Moreover, there are some situations, particularly where economic constraints are felt, where full treatment is not feasible or even desirable. It is then necessary, as well as possible, to consider ways of protecting human health other than waste water treatment.

TABLE 7.1. Recommended Microbiological Quality Guidelines for Waste Water Use in Agriculture[a]

Category	Reuse conditions	Exposed group	Intestinal nematodes[b] (arithmetic mean No. of eggs per liter)	Fecal coliforms (geometric mean No. per 100 ml[c])	Waste water treatment expected to achieve the required microbiological quality
A	Irrigation of crops likely to be eaten uncooked; sports fields; public parks[d]	Workers, consumers, public	≤1	≤1,000[d]	A series of stabilization ponds designed to achieve the microbiological quality indicated or equivalent treatment
B	Irrigation of cereal crops, industrial crops, fodder crops, pasture, and trees[e]	Workers	≤1	No standard recommended	Retention in stabilization ponds for 8–10 days or equivalent helminth and fecal coliform removal
C	Localized irrigation of crops in category B if exposure of workers and the public does not occur	None	Not applicable	Not applicable	Pretreatment as required by the irrigation technology, but not less than primary sedimentation

[a]In specific cases, local epidemiological, sociocultural, and environmental factors should be taken into account, and the guidelines modified accordingly. Reproduced from WHO (24), with permission of the publisher.

[b]*Ascaris* and *Trichuris* species and hookworms.

[c]During the irrigation period.

[d]A more stringent guideline (≤200 fecal coliforms per 100 ml) is appropriate for public lawns, such as hotel lawns, with which the public may come into direct contact.

[e]In the case of fruit trees, irrigation should cease 2 weeks before fruit is picked, and no fruit should be picked off the ground. Sprinkler irrigation should not be used.

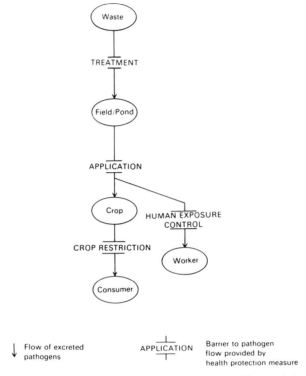

Fig. 7.4. Effect of health protection measures in interrupting potential transmission routes of excreted pathogens. (Reproduced from Mara and Cairncross [30], with permission of the publisher.)

The options for health protection are grouped under four headings by Mara and Cairncross (30). With regard to waste water irrigation, these are

1. Waste water treatment
2. Restriction of the crops grown
3. Choice of irrigation methods
4. Control of human exposure to infection

The points at which these four types of health protection measure can interrupt the potential transmission routes of excreted pathogens are shown in Figure 7.4. While full treatment stops pathogens from reaching the field, the other measures act later in the pathway, preventing them from reaching the persons concerned, the crop consumers, and the agricultural workers.

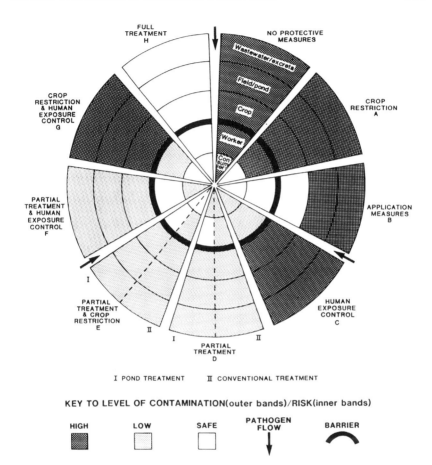

Fig. 7.5. Generalized model to show the level of risk to human health associated with different combinations of control measures for the reuse of waste water in agriculture. (Reproduced from Mara and Cairncross [30], with permission of the publisher.)

Some of these health protection measures may not be considered fully effi-cient in preventing transmission of pathogenic microorganisms, and some do not act on both the consumers and the workers. Accordingly, a generalized model was developed (Fig. 7.5) to show how the various options can be combined to achieve the same degree of protection as full treatment of the waste water (31). The model is intended as an aid to thinking and to decision making, allowing a flexibility of response to different situations. Measures for health protection can be "targeted" toward the specific exposed groups in the population, and the most appropriate mix of such measures can be selected for local circumstances, taking economic, cultural, and technical factors into account.

In this new approach we can see a reappearance, hopefully on a much more informed and rational basis, of the "conditionality" and "localism" of Von Pettenkofer. Pathogens in the environment are not a public health problem unless various other conditions are satisfied. As with Von Pettenkofer's theories, the understanding and control of those conditions require contributions from a variety of professions, in this case, agronomists, engineers, social scientists, health educators and others. And, as asserted by Von Pettenkofer, disease control depends not so much on the rigorous application of centrally determined hygienic standards as on a flexible response that takes account of local conditions. Perhaps it is appropriate for a product of a decade of individualism—the 1980s—that the guidelines should include precautions that are the responsibility of the individual, and not only measures to be applied by the state.

3. HEALTH IMPACT OF WATER SUPPLIES

3.1 Past Controversies

Of the diseases on which improvements in domestic water supply can have an impact, the enteric infections are the group for which that impact has the greatest public health importance. They can be transmitted in drinking water, but also by various other feco-oral routes. These other routes are facilitated by conditions of poor domestic hygiene and are therefore related to the availability of water for hygienic purposes rather than its quality. The policy question of whether, and in what circumstances, water supply improvements can help to control enteric diseases is to some extent a restatement of a scientific question: "To what extent in practice are these infections transmitted by the waterborne route?" Perhaps the importance of the question for policy has contributed over the years to the difficulty of answering it scientifically. For the debate is a very old one, and a search for its origins takes us back to our old friends Snow, Koch, and Von Pettenkofer and to their discussions of cholera and typhoid.

A full account of these early debates on the principal transmission routes of cholera has been provided by Feachem (32), who has shown how Snow was careful to set waterborne transmission in the context of other feco-oral routes. Snow (5) wrote:

> Nothing has been found to favour the extension of cholera more than want of personal cleanliness, whether arising from habit or scarcity of water, although the circumstances till lately remained unexplained. The bed linen always becomes wetted by the cholera evacuations, and as these are devoid of the usual colour and odour, the hands of persons waiting on the patient become soiled without their knowing it; and unless these persons are scrupulously cleanly in their habits, and wash their hands

before taking food, they must accidentally swallow some of the excretion, and leave some on the food they handle or prepare, which has to be eaten by the rest of the family, who, amongst the working classes, often have to take their meals in the sick room. Hence the thousands of instances in which, amongst this class of the population, a case of cholera in one member of the family is followed by other cases.

Clearly, Snow's experience of visiting and treating large numbers of cases had given him an intimate knowledge of their living conditions and of the manifestations of the disease. That intimacy has been less in evidence among microbiologists and other researchers who have studied its transmission in developing countries in more recent times, perhaps because, as Snow pointed out, transmission of cholera occurs largely among the poor, with the notable exception of waterborne transmission:

> I know of no instance in which it has spread through all classes of the community, except where the general supply of water has been contaminated with the contents of the drains and sewers. (5)

Von Pettenkofer, for all his insistence that waterborne transmission was impossible, was at least convinced of the value of making water readily available; he believed that every dwelling should be supplied, because this meant people were more likely to wash frequently than if they had to fetch the water from a distance. He was not deterred from these views when the (unfiltered) water supply built at his insistence for the city of Munich caused a massive typhoid epidemic soon after it was commissioned (4).

Meanwhile, back in London, the debate was warming up, as opposing sides took up positions that many would recognize today. The occasion was the 1866 cholera epidemic in the East End, which killed nearly 4,000 people. As the hue and cry against the water companies concentrated on the accusation that they had allowed waterborne transmission to occur, the engineer of the River Lea Trust wrote that "overcrowding, deficiency of drainage, and inferior articles of food are more likely to have promoted cholera than impurity or deficiency of water" (33).

Koch then entered a debate that focused increasingly on the degree to which water suppliers were responsible for causing epidemics of typhoid and cholera. He took an admirably undogmatic position, conceding even to the localism of his miasmatist opponents (1). Nevertheless, his perception was inevitably colored by the circumstances of his discovery. In his race to beat the French team in Alexandria, he had gone on to Calcutta to look for his bacillus; it is hard to imagine a more water-rich environment, or one in which surface water is more evidently fecally polluted. He even believed that the bacillus could survive and multiply in an aqueous environment (1). In the circumstances, it is remarkable that Koch resisted the pressure to embrace the increasingly prevalent theory of exclusively waterborne transmission. Even under pressure at the Gelsenkirchen trial already mentioned, he insisted that he was "absolutely not a drinking-water fanatic and definitely not committed to any particular theory" (32).

Others were less judicious, however, in their desire to pin the responsibility on the water engineers. Such debates have their modern counterparts, of course, although nowadays we tend to be more concerned about carcinogens and other substances in our drinking water than about pathogenic microorganisms. When scientific debates become matters of public controversy, there is a tendency for the opposing positions to become oversimplified. So it was that the arguments of the contagionists (Snow, Koch, and their followers) were reduced by their supporters to the slogan of waterborne transmission. When their side ultimately won, this simplistic paradigm became the conventional wisdom and held considerable sway for nearly a century.

A Century of Myopia

The waterborne model was applied not only to the classical infections of cholera and typhoid. Few could resist the temptation to apply it to endemic diarrheal disease in general, even though microbiologists were unable, until very recently, to isolate causative organisms for most of them, and epidemiologists had yet to define their relationship to environmental conditions (34). Why was this model so attractive and so enduring?

One reason was that most of the epidemiological accounts in which a specific feco-oral transmission route was identified were accounts of epidemics. Many of these were indeed caused by waterborne transmission and involved a wide range of viral, bacterial, and protozoal enteric pathogens (35,36).

The epidemiology of endemic diarrheas, by contrast, has been less studied in the West, no doubt because their importance as a cause of child mortality declined progressively during the first half of this century. Concern about possible contributory factors, such as the contamination of milk and neglect by wet-nurses, prominent throughout the nineteenth century (37,38), faded from the literature.

Incidentally, it is not certain that the decline in endemic diarrheal disease came in the late nineteenth century, when water treatment was introduced in many Western cities. Although these improvements did have an immediate impact on typhoid, the fall in diarrhea death rates was a more protracted process (39,40). Moreover, diarrhea mortality may have fallen before the incidence of *morbidity* because of diarrhea. There is evidence to suggest that in the industrialized nations, the change from the Third World pattern of a summer peak caused mainly by bacterial pathogens to the winter peak of viral diarrheas found today occurred between the First and Second World wars (41), when the population at large came to benefit from in-house water and sanitation and became increasingly conscious of hygiene.

With regard to the developing countries, where endemic diarrheas are still an important cause of mortality, the concept of waterborne transmission continues to dominate the thinking of many water engineers and policymakers. Two examples, taken from particularly enlightened sources, illustrate the pervasiveness of this view.

The first is WHO's principal text, for more than two decades, on rural water supplies (42), which cites evidence for the health benefits they bestow. The authors comment on a Brazilian study, which showed a similar risk of death from diarrhea in families using treated water from public faucets and in those using outside unprotected wells, but a much smaller one for those with in-house piped water. Ignoring the well-known fact that people with house connections use two to three times more water than those carrying it from outside wells or taps, and their own mention on the previous page of the importance of water availability in the control of diarrhea, they conclude, "This would indicate that the treated water was recontaminated during transport to the houses," (42). The nonsequitur was not noticed when the monograph was reviewed by a number of the most eminent sanitarians of the day.

The second example is Bradley's original classification of water-related disease, in which he coined the phrase "water-washed transmission" to include the nonwaterborne routes, but still did not feel brave enough to take cholera and typhoid from their traditional position as "waterborne diseases" (43).

To some extent, the obsession with waterborne transmission in the Third World is a result of historical factors, and the influence of Western ideas, but there are also local reasons. One is no doubt that, as John Snow pointed out, a waterborne epidemic can attack everyone in the community, rich and poor alike. By contrast, transmission caused by shortcomings in domestic hygiene, such as a lack of readily available water, is not a serious problem to the middle classes and the elite. A contaminated public water supply is likely to lead to a sudden surge of cases of disease and thus to a public outcry. Person-to-person transmission, on the other hand, is largely limited to the poor, and its impact is not sudden and dramatic, but sporadic and inconspicuous.

Another reason is that the responsibility for water quality lies clearly with those who supply the water. When waterborne transmission of disease is suspected, the first to be questioned—and even, sometimes, arrested—are the operators of the water supply. It is not so clear who is responsible when a poor family does not have enough water for its needs, and it is easier to blame the victims. They should have installed a house connection, one might argue, or used more soap.

If epidemiologists and engineers have contributed to a century of myopia, environmental microbiologists have played a supporting role. Their accounts of fecal indicator bacteria in untreated water supplies and in household storage vessels have contributed to the impression that the presence of any such bacteria in drinking water indicates the likelihood of waterborne disease. Most water engineers in developing countries have received more than one laboratory report describing a source of water as ''unfit for human consumption'' because of the presence of a couple of bacteria, often *total* coliforms whose presence does not necessary imply any fecal contamination at all.

Unfortunately, such judgements had the blessing of the World Health Organization, which published drinking water quality standards stipulating the complete absence of fecal coliforms, even from untreated drinking water (44). When these were being redrafted, it was pointed out that they were too stringent to be feasible and that, in some cases, overzealous health officials had closed down village wells because they contained 50 fecal coliforms per 100 ml, thus obliging the villagers to use polluted surface water containing 10,000 per 100 ml (45). In the new version, the title was changed from ''standards'' to ''guidelines,'' but the zero fecal coliform requirement was not changed (46). It is easier to administer microbiological standards than behavioral factors such as hygiene, and easier to sleep soundly after setting a stringent standard, however impractical, than one that allows for the risks of everyday reality.

3.3. A New Consensus

With the development of international aid in the post-war decades, donor agencies invested increasing sums in water supply programs in developing countries. Whatever their real motives, their ostensible rationale for this investment was its health impact. To evaluate their programs, they were willing to pay for epidemiological studies to measure that impact. Mainly in this way, a sizeable literature has built up. By the time Esrey *et al.* (47) came to review it, they were able to select 67 health impact studies of reasonable quality from 28 different countries.

The results of a number of these and subsequent studies have surprised their authors by failing to show any difference in diarrhea incidence between households whose drinking water contained large numbers of fecal bacteria and others who drank water of good microbiological quality (*e.g.*, 48–53). Although some ingenious hypotheses have been put forward to suggest that such evidence is still compatible with waterborne transmission (34,54), they begin to look increasingly implausible. Apart from the internal weaknesses of such models (55), there is the steadily accumulating evidence that health benefits *do* arise when water availability is increased (56–58), or hygiene practices improved (59).

A new consensus seems to be emerging that, whatever their pattern of transmission in some epidemics, most of the *endemic* transmission of enteric infections among poor communities in developing countries is not primarily waterborne, but occurs by other routes such as the contamination of hands, food, clothes, and other fomites—routes that are susceptible to control by improved water availability or, conceivably, by hygiene education. (To some extent, cholera may prove to be an exception to this, for reasons explained below.) This has several important consequences.

First, it means that the presence of fecal bacteria in water does not necessarily imply a risk of enteric disease transmission. As in the case of wastewater reuse, there are a number of intervening factors that determine whether the "potential risk" posed by such bacteria corresponds to an "actual risk" of public health importance. It is to be hoped that the new version of the WHO drinking water quality guidelines, currently undergoing redrafting, will make this clear and will be flexible enough to take it into account.

This does not mean, of course, that the microbiological quality of drinking water is irrelevant to health. No one would dispute that it is desirable that drinking water should be of the highest possible quality, especially when one system serves large numbers of people, in order to prevent any possibility of a waterborne epidemic. Even if a system is to serve a small community, any reasonable measures to protect the water from pollution are worth taking. The point is that perfect microbiological quality is not necessarily attainable in an untreated water supply and that the presence of fecal bacteria is not necessarily a sufficient reason to condemn it; it still may bestow health benefits if it makes water more available to the consumers.

A second consequence is that we need to know much more than we do about the particular hygiene practices that need to be promoted if enteric disease transmission is to be controlled and about how they can best be promoted (60). In recent years, the first part of the question has become the subject of intensive research.

3.4. Hygiene Behavior

Recent studies have provided evidence to suggest that several different hygienic (or unhygienic) forms of behavior are associated with a lower (or higher) than average incidence of diarrhea among those who practise them. Examples are the hygienic disposal of children's stools (61,62) and the unhygienic use of the corner of a mother's sari to wipe faces, plates, children's bottoms, and other items (63). However, these have generally been observational studies and thus share an important weakness. A family that behaves more hygienically in a particular respect than its neighbors may do so for a number of reasons; its members may be wealthier or better educated than average, or they may have a greater awareness of hygiene that leads them to behave hygienically in other ways. If a study finds that they have less diarrhea, it is quite possible that they are protected from the disease by their wealth, their education, or other forms of hygienic behavior not recorded in the study, and not by the specific practice observed.

For the epidemiologist, the acid test of a causal link between hygiene behavior and enteric disease is an intervention study. This would show that an intervention, such as hygiene education, has succeeded in changing a specific practice in a population and also reducing the incidence of diarrheal disease. Unfortunately, relatively few such studies have been carried out thus far, though the need for more of them has been stressed by the WHO Diarrheal Diseases Control Programme (64).

The hygiene-related behavior that has been most studied in this way is hand washing, particularly with soap. During the years up to 1980, considerable microbiological evidence accumulated to suggest that this is likely to be efficacious in interrupting the transmission of enteric pathogens (65–74; see Table 7.2). In 1982, the publication of two epidemiological studies from developing countries confirmed that this was so; one was from Guatemala and one from Bangladesh (75,76).

The results of the Guatemalan study have been reanalyzed by Feachem (59) to show that a general hygiene education program among 106 mothers reduced the incidence of diarrhea in their children under 6 years by 14% over the following year when compared with the children of mothers who did not participate. During the 4 months of the peak diarrhea season, the reduction was 32%.

The Bangladesh study involved a more intense intervention, as the study families were provided with free soap and water pitchers, as well as a lecture on the benefits of hand washing. They had been selected at random from among families of outpatients with *Shigella* dysentry and were followed up for 10 days. The author concluded that his intervention had lowered the incidence of secondary cases among these families by 84%. Assuming that the intervention would not affect the incidence of primary cases (*i.e.,* the rate of transmission between families rather than within them), Feachem (59) has estimated that this is equivalent to an overall reduction of 35% in the attack rate of shigellosis. The attack rate of non-*Shigella* diarrhea was also monitored during the 10 day surveillance period and found to be 37% lower in the soap and water group than in the control group.

Research teams in two neighboring countries, India and Burma, then sought to reproduce the Bangladesh results in a community setting. One team (77) worked in two Calcutta slums. Giving out free soap every fortnight and making educational home visits every 3 months for 1 year, they found a reduction of 41% in the incidence of dysentery among residents of the intervention community over 5 years of age. Among children under 5 years, the relative reduction was 32%, but the sample was too small for a significant result. Their intervention had no detectable impact on watery diarrhea, however.

TABLE 7.2. Hand-Washing and Enteric Pathogen Transmission: Occurrence and Survival of Enteric Pathogens on the Hands

Study	Findings	Reference
1. The transmission of *Klebsiella* spp. in an intensive care unit in London, England	*Klebsiella* were commonly passed from patients (especially patients' hands) to nurses' hands during simple and "clean" nursing procedures. Over 90% of *Klebsiella* on dry hands could survive for at least 2.5 hours.	65
2. Occurrence and survival of *Shigella sonnei* on the hands of children in Southampton, England, during shigellosis outbreaks	In four studies, 0%–49% of children had *S. sonnei* on their hands following a visit to the toilet for urination. *S. sonnei* on the hands survived for at least 3 hours. *S. sonnei* in stools passed through a double thickness of several brands of toilet paper onto the hands.	66
3. Contamination by gram-negative bacteria of the hands of nurses at an intensive care nursery in Florida, U.S.A.	151 hand cultures were made from 13 nurses; 86% of cultures and 100% of nurses were positive for gram-negative bacteria. *Klebsiella pneumoniae* and *Escherichia coli* accounted for 55% of isolates. Evidence was obtained that some gram-negative bacteria, including *E. coli,* could multiply and persist on the hands of nurses.	67
4. Occurrence and survival of enteric bacteria on fingertips studied at the Central Public Health Laboratory in London, England	*E. coli* were not isolated from the fingertips of 100 laboratory staff but were isolated from the fingertips of 12% of butchers in a meat factory. *E. coli* inoculated onto fingertips decreased by 99% or more after 1 hour. With an initial inoculum of 530 per fingertip, *Salmonella anatum* were still detectable after 3 hours. *S. anatum* were frequently isolated from corned beef and cooked ham that had been touched for 5 seconds by contaminated fingertips.	68
5. A known number of coliforms were placed on the hands of the author	After 3 hours, the number of coliforms was "virtually unchanged."	69
6. Bacterial contamination of the hands of staff on the general surgical and medical wards at a hospital in New York, U.S.A.	Coliforms were found in 23% of hand rinses from physicians, 55% from nurses, 67% from nurses' aides, and 67% from other staff; 18% of cultures revealed $>10^3$ coliforms per two hands. Eighty-eight percent of coliforms isolated were in the *Klebsiella-Aerobacter* group, and the remainder were *Escherichia coli,* 92% of coliforms isolated were resistant to one or more antibiotics.	70

continued

TABLE 7.2. Hand-Washing and Enteric Pathogen Transmission: Occurrence and Survival of Enteric Pathogens on the Hands (continued)

Study	Findings	Reference
7. Contamination of the hands of the attendants of 147 children under 5 years hospitalized with acute diarrhea in Dhaka, Bangladesh; 70 of the children had rotavirus diarrhea	Rotavirus antigen was detected in the hand rinses from 79% of the attendants of children with rotavirus diarrhea and from 20% of the attendants of children with nonrotavirus diarrhea.	71
8. The effect of baby handling on hand bacteria of nurses in a hospital nursery in New York, U.S.A.	Changing soiled diapers increased the coliforms on the hands by 10^1- to 10^5-fold.	72

Reproduced from Feachem (59), with permission of the publisher.

The other group (78) was in Rangoon. In their 4 month study of children under 5 years, they also gave out free soap. They found a reduction in diarrhea incidence of 40%, but their sample size was not large enough to detect a reduction in dysentery rates.

The results of these studies may not appear wholly consistent, and the reductions in incidence achieved may not at first seem very large. However, compared with studies of most interventions to control diarrheal diseases, they are very positive results indeed. In view of the magnitude of the problem in developing countries, with a median of 2.2 episodes of diarrhea per year and an estimated diarrhea mortality rate of 14 per 1,000 among children under 5 years (6), a reduction of 30% would be a major advance in public health.

A more fundamental problem is that the interventions studied, with free distribution of soap, are not replicable in most developing countries. The microbiological studies cited above show the value of soap (see Table 7.3), and my reanalysis of Khan's results suggests that promotion of hand washing without soap had no appreciable effect. It remains to be seen whether the promotion of soap use for hand washing, possibly with a partial subsidy for bar soap, can be as effective as a full subsidy. Certainly, most poor households use soap for some purposes, such as laundry and bathing, but in many countries it is only used for hand washing in a minority. In the Calcutta slums, for instance, only 16% were found to use it after defecation and less than 5% before handling food (77).

TABLE 7.3. Hand–Washing and Enteric Pathogen Transmission: Cleansing the Hands by Washing With Water and Soap

Study	Findings	Reference
1. Experiments to determine whether ordinary toilet soap, without antibacterial additives, could act as a vehicle for the dissemination of bacteria. Bacteria used were *E. coli, S. aureus,* two gram-positive micrococci, and *Serratia marcescens.*	Bacteria inoculated onto the surface of soap bars declined in number by at least 5 log_{10} units in 15 minutes. Washing massively contaminated hands transferred bacteria to the soap bar, but these bacteria did not subsequently transfer to the hands of the next user. Bars of soap under ordinary heavy use did not accumulate appreciable bacterial populations, although bars kept in nondraining trays became somewhat more contaminated than those that were allowed to drain.	73
2. Effect of hand-washing with water (rubbing hands together under 45°C running tap water for 20 seconds) and soap and water (rinsing in warm water for 5 seconds, washing with soap for 15 seconds, rinsing for 5 seconds) on removal of inoculated *Klebsiella* from hands of staff in an intensive care unit.	Washing with water alone removed <98% of *Klebsiella*. Washing with plain soap removed >98% of *Klebsiella* in 50% of experiments. Washing with medicated soap removed >98% of *Klebsiella* in 77% of experiments.	65
3 Hands were contaminated with *S. aureus* or *Pseudomonas aeruginosa* and then washed for 30 seconds with soap and running water. Bacterial counts were compared with those on unwashed (control) hands that had been similarly contaminated.	Washing with soap and water reduced the geometric mean counts of *S. aureus* by 99.7%, and of *P. aeruginosa* by 99.8%. Some washing procedures using disinfectants did not remove more bacteria than did washing with soap.	74
4. Hands were contaminated with *S. aureus* and then rinsed for 30 seconds in distilled water. Bacterial counts were compared with those on unrinsed (control) hands that had been similarly contaminated.	Rinsing with distilled water reduced the geometric mean counts of *S. aureus* by 89%. Rinsing with hypochlorite solution did not remove significantly more *S. aureus* than did distilled water.	74

continued

TABLE 7.3. Hand–Washing and Enteric Pathogen Transmission: Cleansing the Hands by Washing With Water and Soap (continued)

Study	Findings	Reference
5. Effect of hand washing (with soap and running water for 15 seconds and then drying on a paper towel) on removing inoculated *S. anatum* from fingertips.	Proportion of experiments in which *S. anatum* could be isolated from the fingertips after hand washing depended on the initial inoculum and was 100% for 10^6 *S. anatum*/fingertip, 30% for 10^3–10^4 *S. anatum*/fingertip, and 0% for $<10^3$ *S. anatum*/fingertip.	68
6. A variety of experiments on the removal of resident and transient skin flora from hands by scrubbing with soap and water.	Bacterial removal was not affected by water temperature (24–56°C), type of soap, drying on a sterile towel, or bacteriological water quality. Inoculated bacteria were reduced by 50% after washing with soap and warm water for 30 seconds.	69
7. Effect of rapid hand washing with soap and water, or water alone, on removing naturally acquired coliforms from the hands of nurses in a hospital nursery.	Hand washing with soap and water removed 67%–100% of coliforms (median, 96%). Hand washing with water alone removed 93%–100% of coliforms (median, 98%). The use of disinfectants in rapid hand washing did not improve coliform removal.	72

Reproduced from Feachem (59) with permission of the publisher.

The good agreement between the results of microbiological and epidemiological studies, both pointing to hands as an important vehicle of enteric pathogens, and to hand washing as a measure to interrupt their transmission, has recently suggested to some workers that the detection of fecal indicator bacteria on people's hands can be used as a measure of the hygiene standards of a community and hence as a means of evaluating the impact of water supply or hygiene education programs. For example, Pinfold et al. (79), working in Lesotho, found a significantly lower level of hand contamination among households with in-house piped water than among those using public standpipes, while Henry and Rahim (80) in Bangladesh found that the density of fecal bacteria on children's hands, but not in their drinking water, was significantly correlated with the diarrhea incidence.

3.5. Food Hygiene

Another possible alternative to the waterborne route for the transmission of enteric pathogens is the contamination of foods, particularly those consumed by small children such as weaning foods and powdered milk. Barrell and Rowland (81,82) have shown the high degree of fecal contamination introduced into such foods during preparation, and the rapid bacterial growth occurring when they are stored, as often occurs, at ambient temperature for 8 hours. The seasonal pattern of diarrheal diseases in many countries, in which diarrheas caused by bacterial pathogens peak in the warm season whereas those caused by viruses peak in the cold, is suggestive of transmission routes that permit pathogen regrowth in the environment, and the foodborne route is the most obvious of these.

Esrey and Feachem (83) have pointed out how little is known about foodborne diarrhea in developing countries and estimated that it accounts for between 15% and 70% of all diarrhea episodes. There are certainly plenty of reports of microbiological contamination of foodstuffs. The offal meat products purchased by low-income groups are particularly likely to be contaminated with the zoonotic pathogens *Salmonella* and *Campylobacter* (84,85). And in conditions of poor hygiene, there are many opportunities for the transfer of fecal pathogens to foodstuffs and from one food item to another during the marketing and preparation of food.

The epidemiological evidence is somewhat contradictory, though (83). Foods containing fecal indicator bacteria, and even pathogens, have sometimes, but not always, been associated with increased diarrhea rates. To the diarrhea epidemiologist, this is not very surprising, as the impact of such contamination may be swamped by other transmission routes or by other diarrhea pathogens. Moreover, the presence of fecal indicator bacteria is not necessarily correlated with the presence of fecal pathogens. Finally, pathogens ingested do not necessarily cause infection, particularly if the host has acquired immunity through being continuously exposed to the same pathogen.

Moreover, we do not know which specific practices in the preparation, handling, and storage of food are the main risk factors for foodborne diarrhea in young children or what feasible interventions might be taken to improve them. It is likely that there is no general answer, as food preparation practices vary between different cultural settings.

3.6. Cholera

No mention of Koch's work in the light of new concepts would be complete without some reference to the current revolution in our understanding of cholera transmission. As mentioned above, the tendency for many decades was to bracket cholera with other "waterborne diseases," such as typhoid. Perhaps the first modern indication that the environmental microbiology of cholera, and hence no doubt its transmission dynamics, would prove to be rather different, came with the discovery (of Isaacson and Smit [86]) that it could multiply in human sweat—or at least in the sweat of people acclimatized to tropical conditions.

More was to come 2 years later, when the existence of aquatic reservoirs of *V. cholerae* was proposed as an explanation for small outbreaks of the disease in Queensland, Australia (87), Louisiana, U.S.A. (88), and Sardinia, Italy (89). *V. cholerae* began to be found in a variety of saline estuarine environments (90,91), and laboratory studies showed that the organism could survive and even multiply in warm waters containing no nutrients but having a salinity of 0.25%–3.0% and a pH of about 8.0 (92,93).

The possibility of a hitherto undiscovered aquatic reservoir for cholera would help to explain the unanswered question of what happens in Bangladesh in the noncholera season, when no infected human carriers of the vibrio can be found. A particularly attractive hypothesis is that this might be in some microhabitat, such as the surfaces of plants or zooplankton, where the necessary physicochemical conditions for survival are maintained and the prevalence of which might increase in a bloom at particular seasons of the year, explaining the sudden simultaneous initiation of the scattered outbreaks that occur each year (94). Recent research (95,96) although not conclusive, has provided further support for this hypothesis, which is remarkably close to Koch's original inspired speculation.

3.7. Conclusion

As the old certainty of waterborne transmission has faded, then, we are faced with a new conditionality and a modern localism. Pathogens in the environment, in water, and even on food do not necessarily cause infection. The appropriate strategies for controlling the transmission of enteric infections are not as simple and universal as the provision of wholesome drinking water, but are complex and site specific, because they involve changes in human behavior. At least we can claim to have advanced beyond the simplistic and doctrinaire positions often adopted in the past to an understanding surprisingly close to the astute perceptions of Snow and Koch and certainly more worthy of those who stand, as we do, "upon their shoulders."

REFERENCES

1. Koch R: An address on cholera and its bacillus. Br Med J 2:403–407, 453–459, 1884.
2. Conference in Berlin for the discussion of cholera. Br Med J 1:1075–1076, 1885.

3. Howard-Jones N: Gelsenkirchen typhoid epidemic in 1901, Robert Koch, and the dead hand of Max Von Pettenkofer. Br Med J 1:103–105, 1973.

4. Evans RJ: Death in Hamburg: Society and Politics in the Cholera Years, 1830–1910. Oxford, UK: Clarendon Press, 1987.

5. Snow J: Snow on Cholera. New York: Hafner, 1965.

6. Snyder JD, Merson MH: The magnitude of the global problem of acute diarrhoeal disease: A review of active surveillance data. Bull WHO 60:605–613, 1982.

7. McKeown T: The Modern Rise of Population. London: Arnold, 1976.

8. Waite WM: A critical appraisal of the coliform test. J Inst Water Eng Sci 39:341–357, 1985.

9. Shuval HI: Rationale for the Engelberg guidelines. IRCWD News 24/25:18–19, 1988.

10. WHO: Reuse of Effluents: Methods of Wastewater Treatment and Health Safeguards. Tech Rep Ser No. 517. Geneva: World Health Organization, 1973.

11. Cairncross S: Re-use of treated wastewater in Jordan. Alexandria, Egypt: WHO Regional Office for the Eastern Mediterranean, 1987.

12. Bartone CR, Arlosoroff S: Irrigation reuse of pond effluents in developing countries. Water Sci Technol 19:289–297, 1987.

13. Strauss M, Blumenthal UJ: Health Aspects of Human Waste Use in Agriculture and Aquaculture: Utilization Practices and Health Perspectives. IRCWD Rep No 6. Duebendorf, Switzerland: International Reference Centre for Waste Disposal, 1990.

14. WHO: The Risk to Health of Microbes in Sewage Sludge Applied to Land. EURO Rep Stud No. 54. Copenhagen: World Health Organization Regional Office for Europe, 1981.

15. Blum D, Feachem RG: Health Aspects of Nightsoil and Sludge Use in Agriculture and Aquaculture. Part III: An Epidemiological Perspective. Report No. 05/85. Duebendorf, Switzerland: International Reference Centre for Waste Disposal, 1985.

16. Feachem RG, Bradley DJ, Garelick H, Mara DD: Sanitation and Disease: Health Aspects of Excreta and Wastewater Management. Chichester, UK: John Wiley and Sons, 1983.

17. Strauss M: Survival of excreted pathogens in excreta and faecal sludges. IRCWD News 23:4–9, 1985.

18. Shuval HI, Yekutiel P, Fattal B: An epidemiological model of the potential health risk associated with various pathogens in wastewater irrigation. Water Sci Technol 18:191–198, 1986.

19. Feachem RG, Blum D: Health aspects of wastewater use. In: Proceedings of the International Symposium on Re-Use of Sewage Effluent. London: Institution of Civil Engineers, pp 237–247, 1984.

20. Shuval HI, Adin A, Fattal B, Rawitz E, Yekutiel P: Wastewater Irrigation in Developing Countries: Health Effects and Technical Solutions. World Bank Tech Paper No. 51. Washington, DC: The World Bank, 1986.

21. Durand RE, Schebach GH, Michael GY, Grimes MM: Epidemiological Investigation of Community Health Effects of Landscape Irrigation Using Reclaimed Wastewater. The Colorado Springs study. Report to the Colorado Department of Health, Colorado Springs: Colorado Springs Wastewater Division, 1986.

22. IRCWD: Health aspects of wastewater and excreta use in agriculture and aquaculture: The Engelberg Report. IRCWD News 23:11–19, 1985.

23. IRCWD: Human wastes: Health aspects of their use in agriculture and aquaculture. IRCWD News 24/25:1, 1988.

24. WHO: Health Guidelines for the Use of Wastewater in Agriculture and Aquaculture. Tech Rep Ser No. 778. Geneva: World Health Organization, 1989.

25. Ockert G: Ehrfahrungen mit einer modifizierten Austriebs-methode zum Nachweis von Wurmeiern. Abwasser Wasserwirtsch Wassertech 16:198–201, 1966.

26. Ockert G: Erfahrungen mit einer modifizierten Austriebs-methode zum Nachweis von Wurmeiern in Abwasser. II. Das Erfassungsvermogen des Verfahrens. Z Ges Hyg Grenzgeb 15: 515–521, 1969.

27. Mara DD, Silva SA: Removal of intestinal nematode eggs in tropical waste stabilisation ponds. J Trop Med Hyg 89:71–74, 1986.

28. Teichmann A: Zur Methodik des quantitativen Nachweises von Helminthenstadien in kommunalen Abwassern. Angewandte Parasitol 27:145–150, 1986.

29. Stien JL, Schwartzbrod J: Viability determination of *Ascaris* eggs recovered from wastewater. Environ Technol Lett 9:401–406, 1988.

30. Mara DD, Cairncross S: Guidelines for the Safe Use of Wastewater and Excreta in Agriculture and Aquaculture. Geneva: World Health Organization, 1989.

31. Blumenthal UJ, Strauss M, Mara DD, Cairncross S: Generalised model of the effect of different control measures in reducing health risks from waste reuse. Water Sci Technol 21:567–577, 1989.

32. Feachem RG: Environmental aspects of cholera epidemiology. III. Transmission and control. Trop Dis Bull 79:1–47, 1982.

33. Luckin W: The final catastrophe—cholera in London, 1866. Med History 21:32–42, 1977.

34. Briscoe J: The role of water supply in improving health in poor countries (with special reference to Bangla Desh). Am J Clin Nutr 31:2100–2113, 1978.

35. Galbraith NS, Barrett NJ, Stanwell-Smith R: Water and disease after Croydon: A review of water-borne and water-associated disease in the UK. J Inst Water Environ Manage 1:1, 7–21, 1987.

36. Lippy EC, Waltrip SC: Waterborne disease outbreaks, 1946–1980: A thirty-five year perspective. J AWWA 76(2):60, 1984.

37. Beaver MW: Population, infant mortality and milk. Popul Stud 27:243–254, 1973.

38. Woods R, Woodward J (eds): Urban Disease and Mortality in Nineteenth-Century England. New York: St Martin's Press, 1984.

39. Condran GE, Cheney RA: Mortality trends in Philadelphia: Age- and cause-specific death rates 1870–1930. Demography 19:97–123, 1982.

40. Preston SH, Van de Walle E: Urban French mortality in the nineteenth century. Popul Stud 32(2):275–297, 1978.

41. Pickles W: Epidemiology in Country Practice. London: Royal College of General Practitioners, 1984.

42. Wagner EG, Lanoix JN: Water Supply for Rural Areas and Small Communities. WHO Monogr Ser No. 42. Geneva: World Health Organization, 1959.

43. White GS, Bradley DJ, White AU: Drawers of Water: Domestic Water Use in East Africa. Chicago: University of Chicago Press, 1972.

44. WHO: International Standards for Drinking-Water, 3rd ed. Geneva: World Health Organisation, 1971.

45. Feachem RG: Bacterial standards for drinking water quality in developing countries. Lancet 2:255–256, 1980.

46. WHO: Guidelines for Drinking-Water Quality. Geneva: World Health Organisation, 1984.

47. Esrey SA, Feachem RG, Hughes JM: Interventions for the control of diarrhoeal diseases among young children: Improving water supplies and excreta disposal facilities. Bull WHO 63:757–772.

48. Levine RJ, Khan MR, D'Souza S, Nalin DR: Failure of sanitary wells to protect against cholera and other diarrhoeas in Bangladesh. Lancet 2:86, 1976.

49. Feachem RG, Burns E, Cairncross S, Cronin A, Cross P, Curtis D, Khan MK, Lamb D, Southall H: Water, Health and Development: An Interdisciplinary Evaluation. London: Tri-Med Books, 1978.

50. Kirchhoff LV, McClelland KE, Pinho MC, Araujo JG, de Sousa MA, Guerrant RL: Feasibility and efficacy of in-home water chlorination in rural North-eastern Brazil. J Hyg Camb 94: 173–180, 1982.

51. Lindskog P: Why Poor Children Stay Sick: Water, Sanitation, Hygiene and Child Health in Rural Malawi. Linkoping, Sweden: Linkoping University, 1976.

52. Young BA, Briscoe J: A case–control study of the effect of environmental sanitation on diarrhoea morbidity in Malawi. J Epidemiol Commun Health 42:83–88, 1988.

53. Esrey SA, Habicht J-P, Latham MC, Sisser DG, Casella G: Drinking water source, diarrhoeal morbidity, and child growth in villages with both traditional and improved water supplies in rural Lesotho, Southern Africa. Am J Public Health 78:11, 1451–1455, 1988.

54. Briscoe J: Intervention studies and the definition of dominant transmission routes. Am J Epidemiol 120:449–455, 1984.

55. Cairncross S: Ingested dose and diarrhoea transmission routes. Am J Epidemiol 125:921–925, 1987.

56. Pickering H, Hayes RJ, Ng'andu N, Smith PG: Social and environmental factors associated with the risk of child mortality in a peri-urban community in the Gambia. Trans R Soc Trop Med Hyg 80:311–316, 1986.

57. Victora CG, Smith PG, Vaughan JP, Nobre LC, Lombardi C, Teixeira AMB, Fuchs SC, Moreira LB, Gigante LP, Barros FC: Water supply, sanitation and housing in relation to the risk of infant mortality from diarrhoea. Int J Epidemiol 17:651–654, 1988.

58. Aziz KMA, Hoque BA, Huttly SRA, Minnatullah KM, Hassan Z, Patwary MK, Rahaman MM, Cairncross S: Water Supply, Sanitation and Hygiene Education: Report of a Health Impact Study in Mirzapur, Bangladesh. Water and Sanitation Rep Ser No. 1. Washington, DC: The World Bank, 1990.

59. Feachem RG: Interventions for the control of diarrhoeal diseases in young children: Promotion of personal and domestic hygiene. Bull WHO 62:467–476, 1984.

60. Cairncross S: Water supply and sanitation: An agenda for research. J Trop Med Hyg 92: 301–314, 1989.

61. Alam N, Wojtyniak B, Henry FJ, Rahaman MM: Mothers' personal and domestic hygiene and diarrhoea incidence in young children in rural Bangladesh. Int J Epidemiol 18:242–247, 1989.

62. Mertens TE, Fernando MA, Cousens SN, Kirkwood BR, Marshall TFDC, Feachem RG: Childhood diarrhoea in Sri Lanka: A case–control study of the impact of improved water sources. Trop Med Parasitol 41:98–104, 1990.

63. Stanton BF, Clements JD: Soiled saris: A vector of disease transmission? Trans R Soc Trop Med Hyg 80:485–488, 1986.

64. WHO: Biomedical and Epidemiological Research Priorities of Global Scientific Working Groups. WHO/CDD/RES/86.8. Geneva: World Health Organisation, 1986.

65. Casewell M, Phillips I: Hands as route of transmission for *Klebsiella* species. Br Med J 2: 1315–1317, 1977.

66. Hutchinson RI: Some observations on the method of spread of Sonne dysentery. Monthly Bull Ministry Health Public Health Lab Serv 15:110–118, 1956.

67. Knittle MA, *et al.:* Role of hand contamination of personnel in the epidemiology of gram-negative nosocomial infections. J Pediatr 86:433–437, 1975.

68. Pether JVS, Gilbert RJ: The survival of salmonellas on finger-tips and transfer of the organisms to foods. J Hyg 69:673–681, 1971.

69. Price PB: The bacteriology of normal skin: A new quantitative test applied to a study of the bacterial flora and the disinfectant action of mechanical cleansing. J Infect Dis 63:301–318, 1938.

70. Salzman TC *et al.:* Hand contamination of personnel as a mechanism of cross-infection in nosocomial infections with antibiotic-resistant *Escherichia coli* and *Klebsiella-Aerobacter.* Antimicrob Agents Chemother 7:97–100, 1967.

71. Samadi AR, *et al.:* Detection of rotavirus in hand-washings of attendants of children with diarrhoea. Br Med J 286:188, 1983.

72. Sprunt K, *et al.:* Antibacterial effectiveness of routine hand washing. Pediatrics 52:264–271, 1973.

73. Bannan EA, Judge LF: Bacteriological studies relating to handwashing. I. The inability of soap bars to transmit bacteria. Am J Public Health 55:915–922, 1965.

74. Lowbury EJL, *et al.:* Disinfection of hands: Removal of transient organisms. Br Med J 2: 230–233, 1964.

75. Torun B: Environmental and educational interventions against diarrhoea in Guatemala. In Chen LC, Scrimshaw NS (eds.): Diarrhea and Malnutrition: Interactions, Mechanisms and Interventions. New York: Plenum Press, pp 235–266, 1982.

76. Khan MU: Interruption of shigellosis by hand-washing. Trans R Soc Trop Med Hyg 76:164–168, 1982.

77. Sircar BK, Sengupta PG, Modal SK, Gupta DN, Saha NC, Ghosh S, Deb BC, Pal SC: Effect of handwashing on the incidence of diarrhoea in a Calcutta slum. J Diarrhoeal Dis Res 5: 112–114, 1987.

78. Han AM, Hlaing T: Prevention of diarrhoea and dysentery by hand washing. Trans R Soc Trop Med Hyg 83:128–131, 1989.

79. Pinfold JV, Horan NJ, Mara DD: The faecal coliform fingertip count: A potential method for evaluating the effectiveness of low cost water supply and sanitation initiatives. J Trop Med Hyg 91:67–70, 1988.

80. Henry FJ, Rahim Z: Transmission of diarrhoea in two crowded areas with different sanitary facilities in Dhaka, Bangladesh. J Trop Med Hyg 93:121–126, 1990.

81. Barrell RAE, Rowland MGM: Commercial milk products and indigenous weaning foods in a rural West African environment: A bacteriological perspective. J Hyg Camb 84:191–202, 1979.

82. Barrell RAE, Rowland MGM: Infant foods as a potential source of diarrhoeal illness in rural West Africa. Trans R Soc Trop Med Hyg 73:85–90, 1979.

83. Esrey SA, Feachem RG: Interventions for the control of diarrhoeal diseases in young children: Food hygiene. WHO/CDD/89.30. Geneva: World Health Organisation, 1989.

84. Richardson NJ, Burnett GM, Koornhof HJ: A bacteriological assessment of meat, offal and other possible sources of human enteric infections in a Bantu township. J Hyg Camb 66: 365–375, 1968.

85. Rasrinaul L, Suthienkul O, Echeverria PD, Taylor DN, Seriwatana J, Bangtrakulnonth A, Lexomboon U: Foods as a source of enteropathogens causing childhood diarrhoea in Thailand. Am J Trop Med Hyg 39:97–102, 1988.

86. Isaacson M, Smit P: The survival and transmission of *V. cholerae* in an artificial tropical environment. Prog Water Technol 11:89–96, 1978.

87. Rogers RC, Cuffe RGCJ, Cossins YM, Murphy DM, Bourke ATC: The Queensland cholera incident of 1977. 2. The epidemiological investigation. Bull WHO 58:665–669, 1980.

88. Blake PA, Allegra DT, Snyder JD, Barrett TJ, McFarland L, Caraway CT, Feeley JC, Craig JP, Lee JV, Puhr ND, Feldman RA: Cholera—a possible endemic focus in the United States. N Engl J Med 302:305–309, 1980.

89. Salmaso S, Greco D, Bonfiglio B, Castellani-Pastoris M, De Felip G, Bracciotti A, Sitzia G, Congiu A, Piu G, Angioni G, Barra L, Zampieri A, Baine WB: Recurrence of pelcypod-associated cholera in Sardinia. Lancet 2:1123–1127, 1980.

90. Bashford DJ, Donovan T, Furniss A, Lee J: *Vibrio cholerae* in Kent. Lancet 1:436–437, 1979.

91. Colwell RR, Seidler R, Kaper J, *et al.*: Occurrence of *Vibrio cholerae* serotype 01 in Maryland and Louisiana estuaries. Appl Environ Microbiol 41:555–558, 1981.

92. Singleton FL, Attwell RW, Jangi S, Colwell RR: Effects of temperature and salinity on *Vibrio cholerae* growth. Appl Environ Microbiol 44:1047–1058, 1982.

93. Miller CJ, Drasar BS, Feachem RG: Response of toxigenic *Vibrio cholerae* 01 to physico-chemical stresses in aquatic environments. J Hyg Camb 93:475–496.

94. Miller CJ, Drasar BS, Feachem RG: Cholera epidemiology in developed and developing countries: new thoughts on transmission, seasonality and control. Lancet *1*:261–263, 1985.

95. Islam MS, Drasar BS, Bradley DJ: Attachment of *Vibrio cholerae* 01 to various freshwater plants and survival with a filamentous green alga. *Rhizoclonium fontanum.* J Trop Med Hyg 92:396–401, 1989.

96. Islam MS, Drasar BS, Bradley DJ: Long-term persistence of toxigenic *Vibrio cholerae* 01 in the mucilaginous sheath of a blue-green alga, *Anabaena variabilis.* J Trop Med Hyg 93:133–139, 1990.

MICROBIAL PROCESSES IN COASTAL POLLUTION

DOUGLAS G. CAPONE

Chesapeake Biological Laboratory, University of Maryland, Solomons, Maryland 20688

JAMES E. BAUER

Department of Oceanography, Florida State University, Tallahassee, Florida 32306

1. INTRODUCTION

Estuaries and coastal areas are among the most biologically productive (1) and economically important areas of the seas (2). The microbiota are intrinsically involved in the productivity of these systems through their participation in the flow of energy and cycling of materials (3). Coastal areas are also sites of discharge and accumulaton of a range of environmental contaminants (Fig. 8.1) (4,5).

The interactions of contaminants and the resident microbiota are often overlooked or given only cursory consideration in many environmental toxicological studies. However, environmental contaminants may affect microbiological activities. Biotic effects can range from stimulation to inhibition, depending on the form and concentration of the contaminant. If pollutants are either acutely or chronically toxic to part or all of the microbial population, alterations in the normal pathways of carbon, energy flow, and productivity within a system may be expected. The microbiota can affect the fate of many contaminants by contributing to their removal through uptake, degradation, (for organic contaminants) (Fig. 8.2) or transformation (for metals) (Fig. 8.3). However, excessive toxification can impair the capacity of a particular system to detoxify itself,

Environmental Microbiology, pages 191–237, © 1992 Wiley-Liss, Inc.

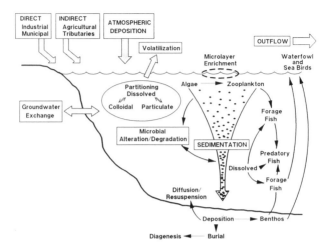

Fig. 8.1. Generalized scheme of contaminant input, dynamics, and removal in coastal waters.

particularly in sediments where contaminants accumulate to relatively high levels. For toxics not susceptible to microbial degradation, or for which toxicity is aggravated by microbial transformations, effects can persist indefinitely with concomitant alterations in overall community structure and function.

At present, however, information on the types of interactions between the natural microbes of coastal areas and environmental contaminants is relatively sparse, making it difficult to generalize or to predict the fate and effects of contaminant compounds in affected areas. In this chapter we discuss the nature and range of some of the interactions that can occur between the microbiota and environmental contaminants and the implications of such interactions.

1.1. Habitats and Populations

Distinct populations of microorganisms inhabit the water column and sediments of coastal and estuarine areas. For example, viable bacteria in shallow, surficial sediments range in numbers between 10^9 and 10^{10} cells per cc of sediment (6,7). In contrast, the densities of bacteria in overlying waters are typically about 10^6 to 10^7 cells per cc (8,9), or about 100 to 10,000 times less per unit volume. Physiological differences between sediment and water column bacteria are also striking. Our understanding of the important ecological roles of each population has accelerated dramatically during the recent past. In addition to differences in microbial population densities and physiologies, the spectrum and

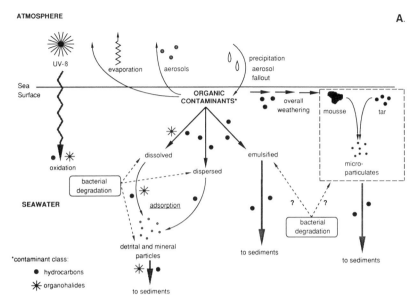

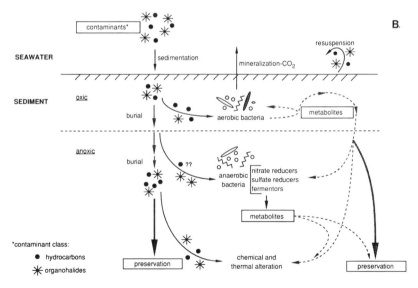

Fig. 8.2. Organic contaminant dynamics in natural waters (**A**) and in sediments (**B**), with emphasis on interactions with the microbiota.

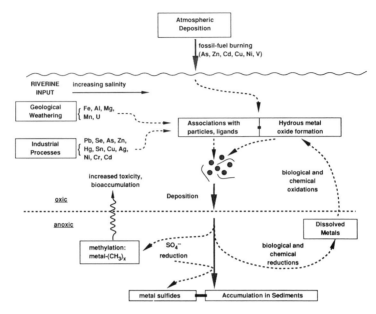

Fig. 8.3. Generalized scheme of metal contaminant dynamics in coastal waters, with emphasis on interations with the microbiota.

concentrations of contaminants are quite different in sediments compared with the water column (5,10–12).

1.1.1 Water Column Bacteria

The recognition of the "microbial loop" over the last two decades as a major pathway of carbon and energy flow has focused attention on the ecological importance of marine bacterioplankton (13–15) (Fig. 8.4). Recent research suggests that the productivity of heterotrophic bacterioplankton is on the order of 10% to 60% of the productivity of phytoplankton (9,16). Heterotrophic bacterioplankton derive their energy and carbon primarily from components of the dissolved organic carbon (DOC) pool (17), which is presumably resupplied by direct algal excretion or from cellular lysis. For labile components of the DOC pool, microbially mediated turnover can be extremely rapid. For instance, some amino acid pools in coastal waters can have residence times of minutes to hours (18).

The energy captured by heterotrophic bacterioplankton becomes available to higher trophic levels after transfer through one or more levels of microzooplankton. Because of energy losses associated with multiple trophic transfers, it is still

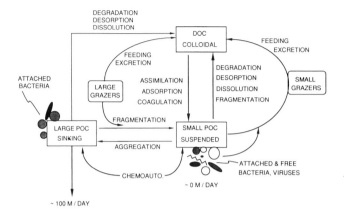

Fig. 8.4. The "microbial loop" of coastal waters, with emphasis on contaminant interactions.

unclear as to whether energy flow through the microbial loop is a significant source for higher trophic levels (*e.g.*, ref. 19). In any event, the flow of carbon and energy through bacterioplankton is quantitatively important and of direct relevance to contaminant dynamics.

Coastal bacterioplankton participate directly in the cycling of nitrogen and other nutrient elements. For nitrogen (N) and phosphorus (P), water column bacteria had been thought to be responsible for much of the regeneration of inorganic N and P from dissolved organic N (DON) and P (DOP) (20). However, several recent observations (21–24), coupled with consideration of bacterial growth rates, growth efficiencies, and the C to N content of bacteria relative to dissolved organic matter (DOM) (25), have prompted reassessment of this generalization. A developing consensus suggests that bacterioplankton may compete with phytoplankton for inorganic nitrogen resources in many instances.

1.1.2. Sediment Microbiota

Sedimentary microorganisms occupy an equally central role in regulating the turnover and degradation of organic material, both natural and synthetic, that is deposited in coastal systems. These microorganisms are also key participants in the major biogeochemical cycles of C, N, P, and S (3). In shallow waters, they affect the overlying water column by driving the efflux of ammonium, nitrite, and nitrate (26,27). Sediment microbes may also be an important food source for deposit-feeding infauna (28,29). Thus accumulation of biologically active contaminants may influence diverse microbial roles in the sediments.

Organic carbon catabolism is probably the most important ecological function of the benthic microbiota (3). The oxidation of a broad range of natural and

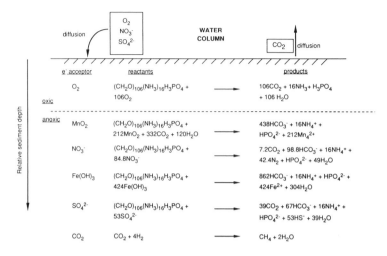

Fig. 8.5. Potential respiratory electron acceptors in marine sediments.

exotic organic compounds is mediated by a variety of physiologically distinct sediment bacteria. The quantity and quality of organic material input to a sedimentary environment, the abundance of electron acceptors, and the activities of bioturbating infauna all interact to determine the relative importance of different and sometimes competing processes in heterotrophic carbon cycling in marine sediments (30,31).

At the sediment surface, organic material is first catabolized by aerobic bacteria, many of which are individually capable of complete oxidation of a wide range of organic compounds. Oxygen provides the highest free energy yield of potential electron acceptors (32). The extent of O_2 penetration into a sediment depends on its rate of downward diffusion (33,34) and on its downward advection by bioturbation (35). Where the organic load is moderate, or in coarsegrained sediments, O_2 may diffuse to several centimeters depth, and aerobic respiration may account for the bulk of carbon oxidation (36).

In areas of high organic input, O_2 may be utilized so rapidly as to be undetectable in surface sediments (33). In the absence of O_2, other, less energetically favorable electron acceptors are used (Fig. 8.5) (32,37). The substrate specificity of anaerobes is more restricted than that of aerobes, and complete oxidation of complex substrates requires several steps and the combined activities of a variety of distinct bacterial types (Fig. 8.6), including various fermentative, nitrate reducing, sulfate reducing, and methanogenic bacteria. In most anaerobic coastal and estuarine sediments, sulfate reduction predominates and can account for up to 50% of the total C oxidation in coastal sediments (38).

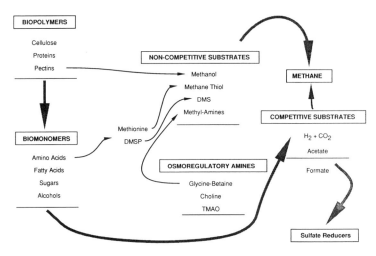

Fig. 8.6. The anaerobic food chain of aquatic sediments (adapted from ref. 307).

1.1.3. Sediment–Water Column Interactions

One feature of coastal and estuarine areas, compared to off-shelf open ocean waters, is a relatively close coupling of processes between the water column and sediments. A shallow water column ensures that a substantial proportion of pelagic primary productivity reaches the sediments (39). Particle flux to the sediments, both organic and inorganic, mediates the transfer of many particle-reactive contaminants (40). In shallow areas, wave action and turbulence may reach the bottom, causing periodic resuspension of sediments and their associated matter into the water column. Benthic metabolism may drive the net efflux of regenerated inorganic nutrients across the sediment water interface (26,27).

1.1.4. Bacterial Attachment/Substrate Availability

The state of attachment of bacterial populations to solid surfaces with respect to their access to sorbed or dissolved inorganic and organic substrates is an area of fundamental importance in microbial ecology (41,42). It is also of direct relevance when considering the interactions of the microbiota and contaminants from both a toxicological and degradative point of view.

Populations of bacteria in the water column are often characterized as either attached or unattached to suspended particulate matter (Fig. 8.4) (43). Recent research indicates that free-living (unattached) bacteria in the water column may be metabolically more important with respect to the utilization of DOC (43–45).

For more complex polymeric organics (46) and organics sorbed to particulates, however, the relative importance of attached and free-living populations is less clear. In a recent review of the interplay of bacterial adhesion to surfaces and substrate utilization, van Loosdrecht *et al.* (47) noted that while surfaces could affect microbial substrate utilization rates either positively or negatively, there is no conclusive evidence that adhesion directly influences bacterial metabolism because of structural modification. Hence, for particle-active contaminants (see below), both attached and free-living populations in the water column may play a role in contaminant uptake, transformation, and transfer to the sediments.

In the sediments, bacterial attachment may be the norm, with numbers of free-living ''interstitial'' bacteria small compared with those sorbed to the sediment matrix (48). This itself has important implications with regard to the relative availability of sorbed contaminants to bacterial populations of sediments (see Secton 2.1.2.).

1.2. Pollutant Types

Most coastal areas are affected to some degree by past and ongoing inputs of contaminants. Anthropogenically generated pollutants, derived from various domestic, industrial, and agricultural activities (12), arrive in the coastal environment through direct point-source discharge, sludge dumping, runoff, atmospheric deposition, and groundwater flow (10).

Several major classes of anthropogenic contaminants that are relevant to microbial processes are found in the coastal marine environment (Fig. 8.7). These include excess inorganic nutrients, heavy metals, and bulk organics, as well as specific organic contaminants such as petroleum hydrocarbons and pesticides. Pathogenic microorganisms constitute yet another form of anthropogenic contamination that may interact with the natural biota (49). From an anthropocentric point of view, the natural microbiota may themselves also be viewed, at times, as pollutants.

1.2.1. Inorganic Nutrients

Inorganic nutrients, particularly forms of N, are often thought to be an important factor limiting primary productivity in coastal waters (50,51). Recent reports have concluded that phosphorus may be the key limiting nutrient in some environments (*e.g.,* 52–54). In any event, nutrient loading is pervasive in many coastal areas, deriving from both point (*e.g.,* sewage) and diffuse (*e.g.,* agricultural) sources (51,55). Nutrient saturation may prevail in some cases (*e.g.,* ref. 56). In general, inorganic inputs are not directly toxic to the indigenous biota. However, they can promote excessive algal growth. Concomitant with such growth is stimulated bacterial heterotrophy, which may contribute to the development of hypoxia or anoxia in the aftermath of algal blooms, even off-shore (57).

The microbiota are extensively involved in various aspects of the natural cycling of N and P. Other microbial transformations of nitrogen may be "N limited" (*e.g.*, nitrification). Hence, N loading can have profound effects on the relative importance of various N cycle reactions (*e.g.*, refs. 58,59). Besides absolute loading, the relative input of N and P (*i.e.*, the N/P ratio) is an important parameter. Departures from natural input ratios can elicit biotic responses such as shifts in species composition as well as the relative importance of particular pathways of limiting nutrients (60).

1.2.2. Heavy Metals

Heavy metals derive from various industrial and domestic discharges (61). In coastal waters, metals are often associated with particulate phases in the water column, or are enriched in surface sediments (11,62). The particular form of a metal has direct influence on its environmental mobility (63) and toxicity (64).

While many metals are essential trace nutrients and may at times be limiting (*e.g.*, Fe), high concentrations can disrupt cellular function and metabolism. A number of bacterial activities can affect metal speciation and availability (Figs. 8.3, 8.5).

1.2.3. Bulk Organics

Large quantities of organic matter are discharged in sewage effluents, as well as from industrial waste sources (*e.g.*, pulp mills, seafood processing) (55). As with algal blooms, organic loading can stimulate excessive heterotrophy and lead to hypoxia or anoxia (*e.g.*, ref. 64).

1.2.4. Organic Contaminants

A wide range of organic contaminants enters the coastal seas (Fig. 8.7). These include some relatively long-lived petroleum-derived aliphatic or aromatic hydrocarbons, chlorinated pesticides, preservatives, and insulators (4,12,65). An increasing portion of the contaminant flux includes the more labile substitutes for persistent herbicides and pesticides, including various organophosphates, carbamates, and triazenes.

Hydrophobic organic pollutants partition rapidly onto particles (66–68). In the water column, particle assimilation can account for substantial flux of these contaminants into the biota (69). High sedimentation rates in coastal areas results in enrichment of such compounds in sediments (12). Marine sediments can become long-term sinks for organic contaminants if they are not degraded. The

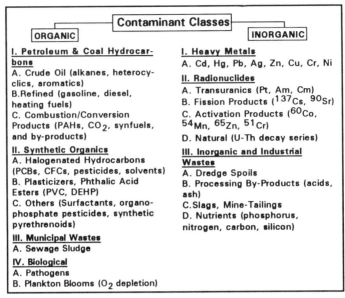

Fig. 8.7. Representative classes of contaminant (adapted from ref. 308).

sediments subsequently become sources of contamination to the water column upon re-release through sediment resuspension (70,71).

Chlorinated hydrocarbons such as DDT and PCBs have been long-standing concerns in coastal marine waters. Despite the severe restriction and reduction in their use, significant residues still remain in many coastal sediments (72). Recognition of their potential to be reductively degraded in anaerobic environments has infused new interest in their environmental dynamics (see Section 3.2.2.). Petroleum hydrocarbon contamination, in contrast, is of continually increasing concern in coastal environments.

1.2.5. Pathogenic Microorganisms

The presence in coastal waters of viable populations of the microbial pathogens of humans has been demonstrated over the last several decades (e.g., refs. 73,74). While many may be recently introduced in sewage effluents, some appear to be part of the resident microbial flora. The interaction of these populations with the microbiota at large is as yet undefined.

1.2.6. Microbial Pollutants

The natural microbiota of coastal waters may also be viewed as a source of contaminants. Biofouling of economically valuable surfaces (e.g., ships' bottoms, heat exchanger tubes) by growth of bacteria and protists is perceived as a nuisance. Excess growth of microbial populations at the expense of organic or inorganic inputs and the associated oxygen demand contribute to hypoxia/anoxia. Sulfide produced by sulfate respirers also augments O_2 demand and is directly toxic to metazoans. Furthermore, sulfate respiration promotes corrosion of metal surfaces.

With the current interest in global warming, the role of the coastal ocean in the production and/or consumption of radiatively important trace gases (RITG) has gained attention. Present estimates suggest that coastal areas are sources of methane (CH_4) (75), nitrous oxide (N_2O) (76), and dimethyl sulfide (DMS) (77), all derived through microbial processes. Thus the coastal microbiota may have an important role in global climate change.

1.3. Interactions

The ways in which contaminants can influence natural marine microbial populations are as varied as the specific pollutants themselves and the environments across which they occur. Contaminants can affect a variety of metabolic functions of marine microorganisms (Fig. 8.8) and can be stimulatory, inhibitory, or neutral. Thus the rate of a specific biochemical process or geochemical transformation mediated by bacteria may be increased, decreased, or unaltered relative to the same process or transformation in the absence of the contaminant.

Of course, given the broad spectrum of contaminants and the complex assemblages of bacteria, there is no *a priori* reason to suspect that responses to environmental contaminants are homogenous or harmonious within any micro-

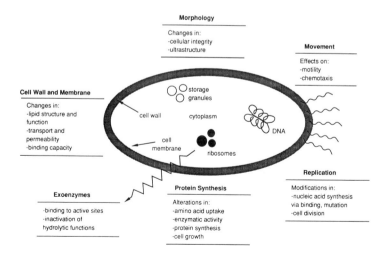

Fig. 8.8. Potential sites of action of toxic contaminants on microorganisms.

bial population. While the observation of changes in the rate of a single process in the presence of a contaminant is often assumed to reflect cause and effect between the pollutant and the process (or even the entire microbial population), a direct causal mechanism should not be assumed unless it can be unequivocally shown (*e.g.*, at the biochemical or molecular level). Because of possible physicochemical interactions between a pollutant and its environment, potential multiple cellular-level sites of action, and the antagonistic, synergistic, or neutralizing effects of multiple pollutants or of naturally occurring organics, the observed effects may reflect only an integrated or indirect response.

1.3.1. Stimulation

A variety of factors may contribute to the stimulation of bacterial populations. Heterotrophic growth stimulation can result from the addition of bulk organics and may be traced to a subset of organic substrates that are readily metabolized. Stimulation of populations or subpopulations can occur indirectly through the abatement of competitive interactions or through the reduction in predation as a result of restricted toxicity of contaminants (*e.g.*, refs. 78,79).

Alternately, macro- or micronutrients that stimulate algal growth and excretion or that are directly limiting to bacterial populations can also stimulate population growth. Numerous inorganic compounds can serve as electron donors or acceptors in respiratory pathways (30–32). Addition of these compounds to the environment can result in a surge in growth of subpopulations. Certain microorganisms may have unusual nutrient requirements (*e.g.*, Mo and high Fe in N_2-fixing bacteria), the shortage of which would otherwise limit their preponderance in a given environment (*e.g.*, ref. 80).

1.3.2. Inhibition

Many of the contaminants discharged into coastal water can have direct toxicological effects on the biota. These effects can be either broad, affecting common metabolic pathways or functions, or narrow, affecting specific physiological groups. Narrow spectrum effects can be expected to alter competitive or mutualistic relationships.

1.3.3. Microbial Transformations

Microbial transformations of contaminants linked to metabolism can change the form, phase, or redox state of the contaminants (81). The result of microbial metabolism of a contaminant, with respect to its environmental impact, may be neutral. Conversely, metabolism may either alleviate or aggrevate the toxicity of particular compounds. The effect of microbial metabolism on contaminants may be either direct or indirect. Complete catabolism of an aliphatic hydrocarbon to

CO_2 and H_2O would represent a direct transformation, whereas sequestration or chelation of a metal by microbial metabolites (*e.g.*, via organic complexation or HS^- precipitation) would be indirect.

During both anabolic and catabolic metabolism, inorganic elements are interconverted with organic forms with corresponding phase and redox transformations of particular elements. Thus the microbiota modify the form of an element in the environment and thereby contribute to its dynamics. Fortuitous transformation, not linked to metabolism, may also occur. Cometabolism of organic compounds (82) or redox conversion of non-nutritional metals (81) are both known to take place.

2. POLLUTANT EFFECTS ON NATURAL MICROBIAL POPULATIONS

Many of the compounds referred to in Figure 8.7. have demonstrated effects on various microbial populations. However, it is not our objective to provide a comprehensive review of the effects of all of these contaminants on natural populations of aquatic bacteria. Rather, we restrict our discussion primarily to petroleum and organochlorine contaminants and to their effects on the microbiota of marine ecosystems, an interaction often overlooked in toxicological studies.

2.1. Petroleum Hydrocarbons

Hydrocarbon contaminants have probably received the most attention among pollutants occurring in coastal marine systems. Most of the 6.1 million metric tons of petroleum hydrocarbons estimated to be discharged to the oceans annually passes through the coastal zone, either directly as the environment of deposition or indirectly as the environment of transit (83). Hydrocarbons in the coastal zone derive from both chronic (river and urban run-off) and acute (accidental releases from their production and transport) inputs. Hydrocarbons generally enter coastal environments as complex petroleum mixtures (Fig. 8.9).

Among the petroleum hydrocarbons, the polynuclear aromatic hydrocarbons (PAHs), a series of compounds composed of fused aromatic rings, have received particular attention, as they are potent mutagens and carcinogens (84). PAHs are produced through petroleum refining, coal coking, and the incomplete combustion of fossil fuels (85–87). Several studies have specifically documented their increased flux to coastal environments as a result of atmospheric deposition and runoff (4,65).

Physicochemical factors affect the impact of petroleum hydrocarbons on water column and sediment bacteria, food webs, and biogeochemical processes. As

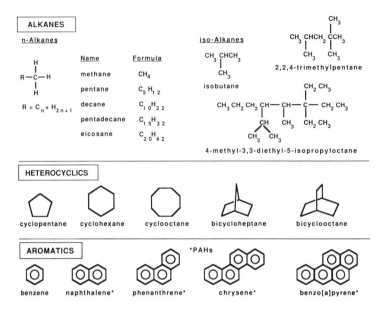

Fig. 8.9. Representative classes of petroleum hydrocarbons.

for many organic pollutants, the bulk of petroleum hydrocarbons initially introduced into the water column rapidly becomes associated with hydrophobic organic matter and suspended particulates (5,66), with most nonvolatiles being deposited to sediments (Fig. 8.1). The component of petroleum that is not dissolved, sorbed, or volatilized, *i.e.*, the emulsion or "mousse," will usually be subjected to a similar fate in the sediment but through direct sinking of the mass rather than through adsorption to particles.

Before any physiological response, either toxicological or degradative, can be observed, contact between the bacterium and the hydrocarbon molecule(s) (either dissolved or in the colloidal/particulate or emulsified form) must presumably occur. In the case of hydrocarbons, mechanisms such as direct attachment of cells to oil globules (88), solubilization of globules to dissolved hydrocarbon components by extracellular enzymes in solution (89), uptake of hydrocarbons directly from the dissolved phase (90), and the inclusion of hydrocarbons into the bacterial cytoplasm (91) have been postulated. However, data are scant on the actual mechanisms of hydrocarbon transport or on the form in which hydrocarbons (*i.e.*, parent molecules, intermediate metabolities, or low-molecular-weight metabolites from initial extracellular metabolism) are transported across the cell membrane.

2.1.1. Effects on Water Column Bacteria

Studies of the effects of petroleum hydrocarbons on water column bacteria have primarily examined the alteration of growth and heterotrophic activity. These studies have focused on either petroleum mixtures or individual aliphatic or aromatic constituents of petroleum.

The limited information available suggests that the growth and metabolism of natural, unacclimated (through prior exposure) bacterioplankton populations are sensitive to even relatively low-level exposures to hydrocarbons. Several studies (92,93) reported the inhibition of heterotrophic utilization and uptake of ^{14}C-glucose and other model substrates by natural water column populations (cf., ref. 94). Processed oils were more toxic than crude oils, and the bacterial populations showed little capacity for acclimation to the oils even after considerable periods of exposure (30 days) to low concentrations (10 μg/liter) (92). Proteolytic, cellulolytic, chitinolytic, and lipolytic enzymatic activities (95) and amino acid turnover (96) have also been reported to be adversely affected by crude oils. Differences among these studies may be attributed to the specific type of oil used as well as to the geographically divergent microbial populations.

Studies focusing on the specific soluble components of petroleum complement research examining the effect of oil and have helped to provide inferences as to the identity of those components responsible for apparent responses to mixtures. Specific dissolved aromatic hydrocarbons decreased the growth of cultures of marine bacteria, with more soluble compounds having greater effect (97). Furthermore, hydroxylated intermediate metabolites having greater solubility decreased growth more than the parent compound. Thus complex mixtures of hydrocarbons of varying toxicity can develop as specific components of oil mixtures are sequentially degraded. The inhibition of growth in a coastal isolate was a function of increased hydrocarbon concentration in the soluble phase following weathering of crude and refined oils (98). Likewise, Henson and Hayasaka (99) found that only the water-soluble portion of crude oils was inhibitory to glutamate uptake by water column microorganisms from coastal waters.

The apparent effects of hydrocarbons on bacteria can also vary with time. Hudak et al. (100) demonstrated acute inhibition of thymidine incorporation in coastal bacterioplankton populations upon exposure to naphthalene and phenanthrene. In longer term (>12 hours) exposures at high levels, naphthalene (10 ppm) stimulated thymidine incorporation; similarly, while phenanthrene affected greater long-term inhibition of thymidine incorporation at lower concentrations (10–500 ppb), cell yields were stimulated after 7 days at 100 and 500 ppb.

Marine bacterial chemotaxis, but not motility or viability, was affected by both crude oils and aromatic hydrocarbons (101,102). This effect was reversed when the hydrocarbons were removed, indicating that, in addition to heterotrophy and growth, a bacterium's ability to locate nutrient- or organic-rich microenvironments may be compromised by hydrocarbon exposure. No effect of oil on bacterial attachment and growth was noted for strains isolated by Holloway *et al.* (103).

The relatively few studies of microbial species' evenness and diversity generally indicate both positive and negative shifts in diversity under the influence of hydrocarbons (104–107). Such changes may be both a result and a cause of concurrent changes in pelagic trophic structure. Hydrocarbon exposure often results in an increase in the numbers of hydrocarbon-utilizing bacteria as well as in total "culturable" and microscopically enumerable bacteria; the magnitude of the response depends on both the specific environment and the amount and type of hydrocarbons present (108). Shifts in community diversity and evenness likely reflect both toxic responses to hydrocarbons as well as competitive interactions and selection within the microbial community.

Substances used to aid in the dispersal of spilled oil can also affect water column microbial activities. The dispersant Corexit inhibited the heterotrophic metabolism of glucose (109). The magnitude of its inhibitory effects was dependent on both its concentration and the presence of inorganic N and P supplementation (110). Corexit (either alone or with crude oil) was found to stimulate bacterial production and increase biomass, presumably through organic enrichment and elimination of bactivorous predators (111).

2.1.2. Effects on Sediment Bacteria

Several factors distinguish the response of the sediment microbiota to organic pollutants compared with water column populations. First, organics in the sediments are primarily present in an adsorbed rather than dissolved form. Thus, while hydrocarbons and other organic pollutants may occur in greater absolute concentrations in sediments (*i.e.*, per unit volume) their effective concentration and bioavailability may be limited. That is, physicochemical characteristics of a given pollutant that determine its toxicity in aqueous solution may not be directly applicable to sediment environments. Second, the microbiota of sediments (at least surface sediments) generally occur in abundances 2 to 3 orders of magnitude greater than those in the water column (see Section 1.1.2.). Thus the microbiota of sediments may have tolerance to higher concentrations of organic pollutants than the water column.

In the event of massive inputs of petroleum, the sediment may be cut off from the overlying water column, eliminating the flux of molecular oxygen from bottom waters into sediments. Negligible biological degradation of nearly all hydrocarbons should result (see Section 3.) along with a shift in the balance of microbially mediated oxidation-reduction reactions within the sediment (112) (Fig. 8.5). In such cases, physical and chemical weathering of the oil must occur before it can be dispersed and mixed into sediments to allow degradation to proceed.

As suggested above, the available evidence indicates that sediment bacteria may be less overtly affected by petroleum contamination than their water column counterparts: indeed, at times their overall metabolism may be stimulated. Microbial heterotrophy decreased and the percentage of substrate respired increased in Arctic sediments exposed to crude oil (50 ppt) (113), suggesting that both cell transport and biosynthesis were adversely affected. However, the inhibition in sediments was lower than the corresponding inhibition in overlying waters (93). Neither sulfate reduction nor methanogenesis was significantly altered in sediments containing weathered oil from the Amoco Cadiz spill off France (114). However, control sediments in this study contained significant concentrations of anthropogenic hydrocarbons. Total sediment metabolism and sulfate reduction were elevated in sediments of a natural hydrocarbon seep (115), possibly because of utilization of low-molecular-weight hydrocarbon gases or volatile fatty acids produced from metabolism of higher molecular weight hydrocarbons. Similar observations of elevated sulfate reduction and methanogenesis in oil-amended sediments were made by Griffiths *et al.* (116).

Apart from their potential toxicity, petroleum and crude oils are carbon-rich substrates for sediment bacteria (117). However, while hydrocarbons can serve as sources of energy, they are N and P poor and thus may be less immediately important as a source of cell C without additional N or P sources. C-rich/N-poor substrates often select for the development of N_2-fixing populations (118). However, N_2 fixation was not found to be elevated in hydrocarbon-rich sediments (116,119,120). Griffiths *et al.* (116,120) observed a net decrease in denitrification and phosphatase activity in contaminated sediments. Whether this was a toxic or an adaptive (*i.e.,* to retain N or P within the system) response of the microbial community is unknown. While no acute effects on either sediment N_2 fixation or denitrification to two types of crude oil were observed, one of the oils tested had significant inhibitory effects on both processes after chronic exposure (121). Thus both overall heterotrophic activity and specific respiratory pathways of biogeochemical significance (*e.g.,* denitrification, sulfate reduction) may be altered by complex hydrocarbon mixtures. The net effect is likely a result of hydrocarbon toxicity and organic enrichment and stimulation.

Individual components of hydrocarbon mixtures have been found to affect a broad range of microbial activities in coastal marine sediments. The PAHs naphthalene and anthracene inhibited glucose metabolism and thymidine incorporation in salt marsh sediments at high (10–1,000 ppm) concentrations (122). Anoxic sediments were more strongly inhibited by PAH treatment and less able to acclimate than oxic sediments, presumably as a result of greater PAH persistence in anoxic sediments (see Section 3.). In anaerobic salt marsh sediments, high concentrations (1,000 ppm) of naphthalene inhibited sulfate reduction and methanogenesis, but benzene and phenanthrene did not (123). Phenanthrene (1,000 ppm) significantly reduced nitrogenase activity and denitrification by salt marsh sediments, whereas naphthalene strongly inhibited denitrification but had no significant effect on nitrogenase activity (124).

While oils and individual hydrocarbons are most often the focus of microbiological impact studies, petroleum exploration and drilling activities in coastal regions also account for the release of other compounds and mixtures that may affect microbial activity. Foremost among these compounds are drilling muds, which contain variable amounts of petroleum-based fluids as well as clays and heavy metals. Enhanced microbial heterotrophic activity is observed in areas where these muds are discarded from drilling platforms, and activities are greater for lower-toxicity (*i.e.*, kerosene-based) than higher-toxicity (*i.e.*, diesel-based) muds (125). Water-based muds were found to alter the sediment microbial community considerably less than hydrocarbon-based muds and for a much shorter period after their application (126). Hydrocarbon-based muds affected the sediment microbiota in much the same way as petroleum mixtures and oils, *i.e.*, by decreasing sediment oxidation-reduction potential, increasing anaerobic metabolism, and rendering hydrocarbon degradation negligible (see Section 3.).

2.2. Effects of Organochlorine Contaminants

In contrast to hydrocarbons, organochlorine contaminants are generally introduced to coastal areas as individual compounds (*e.g.*, pesticides) or groups of compounds of similar molecular structure (*e.g.*, polychlorinated biphenyls [PCBs]) (Fig. 8.10) through agricultural application or industrial discharge. The important routes for their introduction into coastal waters are atmospheric transport, urban run-off and wastewater discharges, riverine discharge, and sewage sludge dumping (Fig. 8.1) (127). Many of the organochlorine contaminants have ubiquitous worldwide distributions. Two of the most common groups of organochlorine pesticides found in coastal environments are DDT (2,2 bis[p-chlorophenyl]-1,1,1-trichloroethane) and its analogs (*e.g.*, Methoxychlor) and the cyclodienes (Chlordane, Heptochlor, Aldrin, Dieldrin, Mirex, Endosulfan) (128). The other major class of organochlorine pollutant, the PCBs, is used primarily in industrial dielectric, heat transfer, and lubricating fluids. The PCBs consist of 210 possible isomers of varying numbers and arrangements of Cl atoms attached to the biphenyl molecule (129). The physicochemical characteristics as well as the toxicity and persistence of PCBs is directly related to the pattern of chlorination. Most PCBs are not introduced as individual molecules but rather as commercially available mixtures such as the Arochlors. Other important individual organochlorine pesticides include PCP (pentachlorphenol), 2,4-D-(2,4-dichlorophenoxyacetic acid), and Kepone. Because of their limited

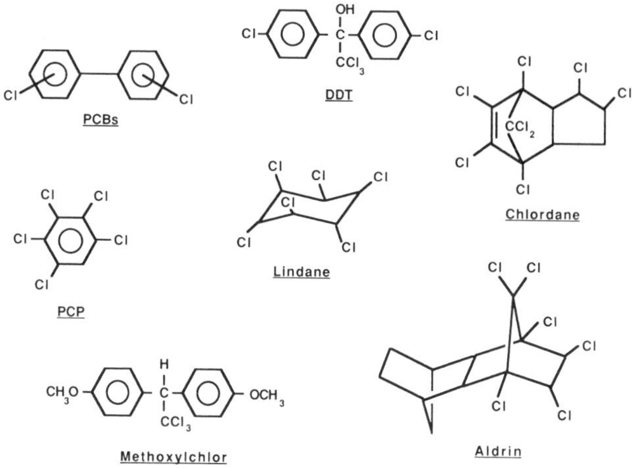

Fig. 8.10. Representative organochlorine contaminants.

solubilities (130) and hydrophobic nature, pesticides and PCBs will also tend to partition onto particulate and sedimentary phases (131).

The cellular-level changes induced by organochlorine compounds that affect microbial physiology and metabolic activity are relatively well understood for certain bacteria. Much of the acute toxicity of organochlorines may be attributed to their hydrophobic (*i.e.*, lipophilic) nature, which allows them to interact readily with the cell membrane and its components (*e.g.*, phospholipids) (132), thus interfering with membrane transport. Long-term effects may arise from either direct or indirect interference of the compound with enzymatic activity, protein synthesis, and cell replication (133,134).

In spite of the broad documentation of their environmental distributions, information regarding the effects of organochlorine pollutants on growth, metabolism, and geochemically mediated activities in aquatic microorganisms is very sparse. Relatively high concentrations of PCBs inhibited growth of isolates of estuarine heterotrophs (135). Growth of 2 of 17 open ocean isolates was consistently inhibited at relatively low concentrations (10 ppb) of Arochlor 1254 (136). A corresponding inhibition of the respiration of glucose and amino acids was not observed, suggesting a selective interference with cellular replication but not metabolic pathways. Blakemore (137) went on to determine that growth inhibition could be attributed to a decrease in the rate of nucleic acid synthesis, resulting in lower DNA and RNA contents per cell in Arochlor-treated cultures. The lower rates of nucleic acid synthesis were further hypothesized to be a result of inhibited adenine transport across the cell membrane. This study is noteworthy in demonstrating a specific molecular-level mechanism behind an observed effect for a marine bacterial–pollutant interaction.

Turnover of amino acids by and viable cell counts of estuarine microbes were not affected by low (ppb) concentrations of Kepone, although higher concentrations inhibited both parameters (138). High concentrations of Kepone, which had initially inhibitory affects on bacterioplankton incorporation of thymidine, significantly stimulated incorporation after 1 day (139). In the same study, PCP inhibited rates of thymidine incorporation for up to 3 days, after which it recovered and, in some cases, exceeded control rates of incorporation. In salt marsh sediments, high concentrations (1–1,000 ppm) of DDT had no significant effect on either thymidine incorporation or glucose uptake by sediment bacteria, but PCP significantly inhibited both processes at all concentrations (122). The inhibitory effects of PCP were much more severe and sustained in anoxic than oxic sediments. The growth of mixed cultures of estuarine bacteria was inhibited by 1 ppm of Kepone, but it was less inhibited by PCP at higher concentrations (3 ppm) (140), contrary to most other findings of the comparative toxicity of these two pesticides.

In anoxic salt marsh sediments, Arochlor 1254 (1,000 ppm) stimulated methanogenesis, whereas Chlordane, Lindane, DDT, Heptachlor, Kepone, and PCP (all at 1,000 ppm) were inhibitory (123). Of the same suite of compounds, only DDT and PCP (≥ 100 ppm) inhibited sulfate reduction. While 1,000 ppm levels of Heptachlor, Arochlor 1254, DDT, Endrin, Chlordane, and Lindane had little effect on denitrification in salt marsh sediments, PCP and Toxaphene strongly inhibited this process (124). Toxaphene stimulated nitrogenase activity, PCP and Lindane had little effect, while Heptachlor, DDT, Endrin, and Chlordane all inhibited nitrogenase activity in the same sediments (124). Thus, in some cases (*e.g.*, Toxaphene), N dynamics could be severely altered by the concurrent inhibition of N removal and stimulation of N input. The relatively high concentrations generally required in these sediment studies to demonstrate a response suggests that the toxic effects of the contaminants are mitigated because of adsorption to the sediment matrix when compared with the much lower concentrations needed to elicit a response in the marine water column.

3. MITIGATION OF CONTAMINANT EFFECTS BY THE NATURAL MICROBIOTA

The effects of contaminants reaching coastal waters can be mitigated by the resident microbiota through their metabolism or transformation in such a manner that the deleterious impacts of such contaminants are reduced or eliminated. This potential exists for inorganic nutrients, metals, and organic contaminants.

3.1 Metals

Microbially mediated transformations of metals to less toxic forms may result from direct microbial uptake and incorporation (their metabolism) or through indirect complexation or precipitation. Metals may also be converted to potentially more toxic forms through methylation (see Gilmour, this volume).

3.1.1. Metal Precipitation

Metals such as Mn and Fe have active biogeochemical cycles in coastal waters, serving alternately as electron donors and acceptors (Figs. 8.3, 8.5). Recent studies have confirmed that Fe(III) can serve as a biological electron acceptor for some anaerobic heterotrophs, and this process is coupled with energetic metabolism and growth (141) (Fig. 8.5). Similarly, Fe(II)-oxidizing autotrophic bacteria are often localized above zones of Fe reduction. Similar reactions are thought to occur with Mn (142). Even U is not immune to microbial transformation, as recently shown by Lovley et al. (143). Abiological reduction of some metals by sulfide appears to occur (144,145).

The relative solubilities of metals are strongly influenced by their oxidation state. Reduced species are often more soluble (63), resulting in increased mobility and loss from the environment in which biological or chemical reduction occurs. Mobility through increased solubility implies transfer to other zones where uptake may occur or where toxic effects may be manifested. Oxidation of reduced forms result in decreased mobility and precipitation of the oxyhydroxides that form (63).

Anaerobic bacterial metabolism can also indirectly affect metal solubility and availability. The predominant mode of anaerobic respiration in most marine sediments is sulfate reduction (30,38), and an important end product of this activity is sulfide. At the pH and redox of most sediments, precipitation of metal sulfides is rapid and the resulting metal sulfides are highly insoluble (146,147). Several studies in coastal sediments have concluded that the fluxes of many metals are predominantly controlled by sulfide formation and, therefore, sulfate reduction (146,148). The importance of sulfide in controlling toxic metal availability depends to a large extent on the relative capacity of Fe quantitatively to sequester sulfide (148).

3.1.2. Metal Sorption and Complexation

Marine bacteria can remove toxic metals through uptake, sorption to their surfaces, or complexation with organic compounds they produce (149). As has been shown for soil bacteria (150), the sorption of metals such as Cu by bacteria can be far more significant than uptake (151).

Chelation of toxic metals by organic molecules can reduce their toxicity (*e.g.*, ref. 152). Some marine bacteria produce specific extracellular proteins that can bind metals and allow resumption of growth after initial metal-induced cessation (153,154). Should this be common among the heterotrophic bacteria, it may indicate a major role for these organisms in controlling metal toxicity in marine systems.

3.2. Degradation of Organic Contaminants

Biological transformation and degradation of organic contaminants has been widely studied in aquatic and terrestrial environments (155). We herein restrict ourselves to an overview of organic contaminant degradation relevant to coastal marine environments.

The fate of organic contaminants that enter and remain dissolved in coastal waters will depend on a variety of factors (Fig. 8.2a). Besides direct microbiological attack, abiotic factors that come into play include physicochemical adsorption, volatilization, dilution, and, for some organic contaminants in the euphotic zone, photodegradation. These processes all affect qualitative and quantitative changes in the distribution of a given contaminant after its discharge to coastal waters.

High rates of sedimentation in estuaries and ocean margins promote the rapid deposition of hydrophobic organic contaminants to the sediments where they may be subjected to a completely different suite of deterministic factors (Fig. 8.2b). In many cases, the rate of sediment burial below the oxic–anoxic interface may control whether an organic contaminant will be degraded or preserved. Where turbulent regimes dominate, contaminants may be re-introduced into the water column after desorption from resuspended contaminated sediments (60,70).

Oxygen availability is a key factor affecting the mode, rate, and extent of organic contaminant transformation. In the water column where oxygen is usually abundant and the potential for complete mineralization of specific organic contaminants may be high, the short residence time before removal of hydrophobic organics, whether as parent or intermediate compounds, may minimize the extent of degradation. Conversely, in sediments where contaminants accumulate, rates of transformation may be limited by oxygen.

Complete mineralization of an organic contaminant to CO_2 is probably the exception rather than the rule in marine sediments. More often, contaminants are partially transformed to yield mixtures of intermediate metabolites and mineralized end products. Intermediates may have a greater inherent toxicity than the parent compound, thus complicating simple predictions of degradation and residence time based on parent compound characteristics (see below).

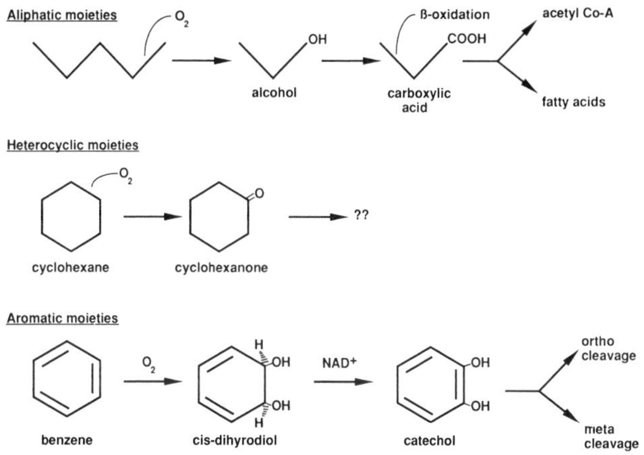

Fig. 8.11. Generalized pathways of degradation of petroleum hydrocarbon classes.

3.2.1. Degradation of Hydrocarbons

Hydrocarbon degradation has been investigated in both complex mixtures as well as with individual model constituents. While considering the material as it generally occurs in nature, studies of the degradation of petroleum and crude oil as mixtures are difficult to interpret because of the vast array of compounds present. Furthermore, effects may reflect synergistic or antagonistic interactions. As might be expected, a wide range of degradation rates of such mixtures has been reported. These have been related to inherent compositional differences of the petroleum mixtures examined, experimental conditions, and the relative importance of such phenomena as cooxidation. The sulfur content, relative abundances of aromatic and aliphatic hydrocarbons, and state of weathering have all been cited as factors affecting the rates of degradation of the various constituents of oil in marine and estuarine systems (156–159). The generally accepted pattern of susceptibility of hydrocarbon component groups to microbial degradation is n-alkanes > branched alkanes > low-molecular-weight aromatics > polycyclics, but system-specific exceptions to this pattern have been found (159).

The degradation of individual petroleum hydrocarbon constituents has also been examined in coastal environments (Fig. 8.11). Under oxic conditions, aliphatic hydrocarbons generally undergo attack at one or both ends of the molecule to yield, sequentially, alcohol, aldehyde, and carboxylic acid functional groups (160,161) (Fig. 8.11). Subsequent oxidations at one (β-oxidation) or both (ω-oxidation) ends of the molecule (162) result in correspondingly smaller fatty acids. Aromatic hydrocarbons are oxidized by bacterial dioxygenases using molecular oxygen to dihydroxy intermediates and, following ring cleavage, to pyruvic and muconic acids (163,164) (Fig. 8.11). The alicylic hydrocarbons are among those most resistant to microbial degradation; they also require oxygen incorporation into the ring before ring cleavage can occur to yield the carboxylic acid intermediates (165).

Limited degradation of specific hydrocarbon compounds can occur in the absence of oxygen, especially under denitrifying conditions (166). Certain monoaromatic compounds with oxygen-containing side chains may be degraded without oxygen (167). Unsaturated hydrocarbons with terminal double bonds can be degraded by methanogenic cultures (168). However, these mechanisms for hydrocarbon degradation appear to be of minor importance in marine environments. Anoxic degradation of petroleum hydrocarbons occurs at negligible rates relative to oxic degradation (158,169–173). In organically enriched sediments, an oxygenated zone may be extremely limited or entirely absent because of increased rates of microbial O_2 consumption and sulfide production. Sediments receiving large, acute inputs of hydrocarbon pollutants in the form of petroleum may thus be less capable of eliminating these compounds as rapidly as sediments that receive low-level, chronic inputs of lower concentrations which allow the sediments to remain partially oxic. Physicochemical (wind-induced mixing) or biological (bioturbation) processes that enhance the flux of oxygen into sediments (174) may stimulate rates of microbial hydrocarbon degradation (175,176).

3.2.2. Degradation of Organochlorine Compounds

The role of oxygen in the degradation of many halogenated compounds may be more complex than for hydrocarbons. While degradation of the carbon skeleton may be more rapid under oxic conditions, the initial removal of halogen atoms (dehalogenation), a key element for the further breakdown of many organohalide molecules, proceeds more vigorously under anoxic conditions (177). Nonetheless, some organohalides (*e.g.*, PCBs and halobenzenes) may be dechlorinated and degraded under oxic conditions (178). Other factors affecting the susceptibility of halogenated organics to degradation include the specific halogen in the compound (179), the number and position of halogen atoms (180,181), aryl ring substituents (182), and the availability of oxygen (183).

In aerobic environments, organochlorine substances may undergo dechlorination, resulting in a hydroxylated daughter product (Fig. 8.12) (184,185). Once sufficient removal of chlorine atoms has occurred, the remainder of the molecule can be metabolized by mechanisms similar, if not identical, to those of aerobic aliphatic and aromatic hydrocarbon degradation (see above).

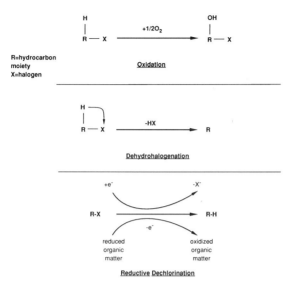

Fig. 8.12. General degradation reactions of organochlorine compounds.

Aerobic dehalogenation of PCBs can occur through the action of substrate-specific dehalidohydrolases, which directly displace the halide by –OH, or oxygenases, which indirectly displace the halide by incorporation of molecular oxygen into the aromatic or biphenyl ring prior to ring cleavage (164,178). Some PCBs are degraded aerobically in aquatic environments, with the extent of chlorination and solubility being factors regulating degradation (186). *Pseudomonas* and *Vibrio* spp. were the predominant genera found in seawater that were capable of growth on and mineralization of PCB mixtures, but in estuarine waters up to seven distinct genera were found (187). The chlorinated benzoates, major metabolites of PCB metabolism, can also be degraded under aerobic conditions either directly or through cooxidative processes (188,189).

Several other pesticide and organochlorine contaminants can be attacked under aerobic conditions in marine environments. These include methyl parathion and malathion (190–193), parathion (194), p-nitrophenol (193), and PCP (195,196). DDT can undergo degradation to the more persistent DDE in aerobic environments (185).

Many organohalide contaminants are sorbed to particles and are concentrated in coastal sediments (12), where they may experience a wide range of redox conditions and hence diverse means of biochemical alteration. Where surficial sediments contain oxygen, the depth and extent of oxygenation can vary dramatically on both spatial and temporal bases (Fig. 8.13) (see Section 1.1.2.). Those organohalides that do not rapidly undergo aerobic pathways of degrada-

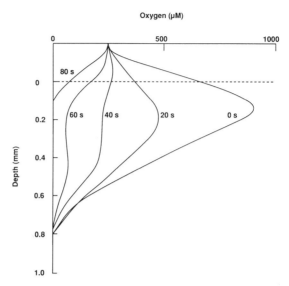

Fig. 8.13. Microelectrode profiles of oxygen in sediments darkened for different lengths of time. (Reproduced from Revsbech *et al.* [217], with permission of the publisher.)

tion in oxic waters or sediments may eventually be subducted to anaerobic strata. The persistence of a contaminant will then depend on the efficacy of anoxic or coupled oxic–anoxic mechanisms of degradation.

Many chlorinated pesticides are initially degraded to nonchlorinated intermediates in anoxic environments by reductive dehalogenation (177). Electrons are transferred from a reduced organic substance (often naturally occurring organic matter) to the chlorinated compound via electron-transferring intermediates; a proton replaces Cl^- on the contaminant molecule. Significant rates of reductive dechlorination may occur when the redox potential of the system is <350 mV (197).

The relatively simple halogenated benzoates are readily degraded to the non-halogenated benzoic acid under anoxic conditions (180). Benzoic acid is one of the few aromatic compounds that can be completely degraded under anoxic conditions (198–200). PCP may also be susceptible to reductive dechlorination, but is only modified to less chlorinated intermediates (201); complete degradation is observed in aerobic systems (195,196). DDT, methoxychlor, and heptachlor degrade more rapidly and completely in anaerobic systems, and the rate of degradation increases with increasing organic matter (*i.e.*, reductant) content (202). DDT degrades much more rapidly under reducing conditions in marine sediments (203). Reductive dechlorination of many other compounds, including chlorobenzenes (204), and chlorinated biphenyls (205) also occurs (206).

Marine sediments are likely places for dehalogenating microbial populations to have developed, particularly given the abundance and types of natural halogenated compounds that occur in the sea (207,208). In fact, the thermodynamic yield for reductive dechlorination is at least as good under sulfate-reducing as under methanogenic conditions. However, the majority of studies of reductive dechlorination have been performed using methanogenic consortia or in methanogenic environments of low salinity and sulfate concentration, since dechlorination was not observed initially under sulfate-reducing conditions (209). More recently, however, evidence has been presented for reductive dechlorination under sulfate-reducing conditions as well. The anaerobic dechlorinating bacterium DCB-1 dechlorinates at sulfate concentrations up to 20 mM. Kohring *et al.* (210) demonstrated co-occurrence of sulfate reduction and reductive dechlorination of chlorophenols in freshwater lake sediments amended with sulfate. Both Bosma *et al.* (204) and Haggblom and Young (208) found chlorobenzene and chlorophenol dechlorination to occur under fully sulfate-reducing conditions in marine sediments. Genthner *et al.* (211) recently presented evidence for anaerobic degradation of chlorobenzoates and chlorophenols in estuarine environments after relatively long lag periods in the absence of sulfate.

Transformation of certain pesticides, particularly the cyclodienes, may occur by means of nonmicrobiologically mediated redox couples. Toxaphene, for example, is dechlorinated by the $Fe(II)/Fe(III)$ redox couple in the absence of microbial activity (212,213), but it is also readily degraded by anaerobes (214). Thus, effective microbial degradation of chlorinated organic pollutants may be supplemented by abiotic reactions.

Temporal variation of oxidation-reduction conditions in marine sediments may subject organic contaminants to a variety of metabolic pathways that may promote or retard their breakdown. While anoxia may be more effective for dechlorination of particular compounds, oxic conditions may result in more rapid cleavage of carbon–carbon bonds contained within the hydrocarbon moiety of these pollutants. Pfaender and Alexander (215) noted that complete degradation of DDT occurred only when an anoxic regime was followed by aerobic conditions. Fogel *et al.* (216) found methoxychlor (a DDT analog) to undergo 70-fold greater degradation under conditions of alternating redox than under either fully oxic or anoxic conditions.

Shallow marine sediments occurring in photic zones have highly dynamic O_2 cycles with surficial sediment conditions varying from oxic to anoxic on a daily basis (Fig. 8.13) (217,218). In many coastal areas, seasonal water column anoxia provides variable redox conditions over a somewhat longer time scale. Bioturbation, irrigation, turbulence, and sediment resuspension can all serve to oxygenate previously anoxic sediments. Thus, a reasonable potential exists in marine systems for degradation under sequentially anoxic and oxic conditions.

3.2.3. Other Environmental Factors

In addition to oxygen, other environmental factors such as temperature, nutrient concentrations, and salinity are important in hydrocarbon degradation (see refs. 158,159). Generally, decreased rates of degradation have been observed with decreasing temperature (219,220), but certain cold-adapted microbial populations may degrade significant amounts of hydrocarbons (221). Temperature may be a crucial factor limiting hydrocarbon degradation in arctic regions (158).

Hydrocarbons represent a significant source of additional C for coastal marine systems, but are relatively poor in N and P. Several studies have concluded that inorganic N and P limit hydrocarbon degradation (117,222,223). However, turbulent or diffusive mixing of N and P to bacterial degraders associated with congealed forms of hydrocarbons may be restricted (158). Recent major bioremediation efforts that promoted oil degradation with inorganic nutrient additions (224,225) seem to confirm that inorganic nutrient limitation can, at times, be a crucial factor limiting natural biodegradation rates (see below).

The relatively few studies that have examined the effects of salinity on hydrocarbon degradation in coastal environments provide, at first glance, a somewhat confusing picture. Bartholomew and Pfaender (226) found that degradation rates of various organic pollutants (including chlorobenzene and 1,2,4-trichlorobenzene) in the water column decreased in an estuarine gradient as conditions became more marine. Similarly, Ward and Brock (169) observed decreasing rates of hydrocarbon degradation with increasing salinity in the water column of a hypersaline lake. In contrast, Shiaris (220) found increasing rates of hydrocarbon degradation with increasing salinity in estuarine sediments. Kerr and Capone (227) noted that estuarine sediments were able to degrade hydrocarbons over a broader range of salinities than those from less saline waters upstream. These results are not surprising. The bacterioplankton of rivers introduced to estuaries are largely freshwater and may be salinity stressed. Further seaward, as water volumes and residence times increase, the proportions of indigenous bacterioplankton adapted to higher salinity increases (228). Similarly, sediment bacteria probably represent a less transient population than those in the water column.

3.2.4. Compound-Dependent Factors

Another suite of factors affecting contaminant degradation is operative at the compound-microorganism level. These factors include contaminant concentration, adaptational response of the microbial community, co-oxidation, and cross-acclimation because of the presence of compound analogues.

Concentration dependence of the rate and extent of pesticide degradation has been observed in marine systems, with lower concentrations eliciting lower rates of degradation (229,230). For hydrocarbons, Tagger *et al.* (231) initially proposed that hydrocarbon degradation may cease below certain threshold concentrations. The limited information available (172,232,233) suggests that the rate and total extent of degradation of aromatic hydrocarbons is also directly dependent on the exposure concentration and solubility.

Several mechanisms operative over different time scales may account for enhanced degradation rates. Relatively short-term (minutes to hours) increases in degradation rates in response to increases in organic contaminant concentrations may simply be a result of increased reaction velocity of extant, subsaturated enzyme systems exhibiting Michaelis-Menten type saturation kinetics. Enhanced degradation rates after somewhat longer (hours to days) time scales following significant lag periods may reflect induction of enzyme systems specific for the transformation of a compound in those organisms with the appropriate catabolic pathways.

Increased degradation at increased substrate concentration over longer exposure periods may also reflect microbial community adaptation. The microbiota can respond through the selection of strains that are tolerant of a particular compound and/or capable of efficiently utilizing and degrading it. Alternately, the transfer and proliferation of genetic factors (*e.g.,* plasmids encoding for specific degrading abilities) among microorganisms that impart biodegradation capability could promote species shifts.

Adaptational responses of coastal marine microbial communities to hydrocarbon exposure have been observed. In several studies, the numbers of hydrocarbon-utilizing heterotrophic bacteria increased dramatically upon exposure of a sample to hydrocarbons (for reviews, see refs. 158,234). Marine and estuarine bacterial isolates capable of degrading either alkanes or PAHs were unable to perform the reciprocal degradation (235), indicating that the simultaneous degradation of different classes of hydrocarbons may require co-occurring guilds of microorganisms.

A lag period often precedes significant contaminant degradation as hydrocarbon-utilizing populations develop. However, prior exposure may result in shortened lags and enhanced rates of hydrocarbon degradation in both water column (100,172,227,236–241) populations in coastal systems. Sediment microbial communities often retain enhanced hydrocarbon-degrading abilities for years following initial exposure (221).

Adaptational responses have also been noted in the degradation of pesticides in certain aquatic environments. Riverine sediment/water systems degraded p-nitrophenol more rapidly following an initial exposure, in contrast to estuarine, saltmarsh, and marine sediment/water system, which did not adapt (230,242).

Plasmids are suspected to play a role in the adaptation of natural microbial communities to pollutants (see above). Considerable information exists on the hydrocarbon-encoding functions of plasmids for simple aromatic hydrocarbons, with less known for alkane-degrading–encoding plasmids (for review, see ref.

243). Plasmids have been found to encode for several compound-specific metabolic pathways, especially for mono- and diaromatic hydrocarbons (244). Catabolic plasmid incidence in isolates, as well as the percentage of isolates with multiple plasmids, can be greater in populations of microorganisms exposed to hydrocarbons and other contaminants than in those that have not been exposed (245,246). However, Leahy *et al.* (247) found no significant differences in plasmid occurrence with proximity to an oil field in the Gulf of Mexico, possibly because of the generally low concentrations of hydrocarbons overall.

Plasmids have also been implicated in the expression of enzymatic activity for the breakdown of pesticides (248,249). However, beyond these circumstantial data that plasmids are important components of the adaptational responses of natural microbial populations, there is little information on the rate at which plasmids are broadcast after exposure to a contaminant and the extent to which specific plasmids are retained within the general population once a particular contaminant falls below a critical concentration.

Co–oxidation and cross-acclimation are two other organism-contaminant factors of potential importance in marine microbial degradation. Both of these mechanisms are of greater probable importance where mixtures (*e.g.*, petroleum) or multiple contaminants are introduced into the coastal environment. Co-oxidation, the coincidental oxidation of co-occurring nongrowth substrates by the same oxygenases that oxidize the growth substrate, yields no energy or cellular carbon from the fortuitous oxidation. While the co-oxidized substrate may or may not undergo complete degradation (*i.e.*, mineralization) by this process, it may be transformed to a more labile form. Various aliphatic, alicyclic, and aromatic hydrocarbon compounds undergo co-oxidation (82,250), as do pesticides such as DDT (251). In arctic coastal waters, hydrocarbon co-oxidation may have been responsible for different hydrocarbon components (alkanes, alicyclics, and aromatics) being degraded at similar rates (252).

Finally, a system exposed to a particular compound may exhibit a degree of adaptation when it is exposed to compounds of the same or similar molecular structure at a later time. This phenomenon has been examined for very few systems. Barnsley (253) initially found that naphthalene and phenanthrene could enhance each others' degradation through prior pre-exposures in cultures. In salt marsh sediments, Bauer and Capone (254) noted that pre-exposure to one PAH enhanced the subsequent degradation of only certain other PAHs. This suggests either that common pathways for the degradation of different PAHs exist or that microbial populations selected for the degradation of a given PAH have the ability (through several oxygenases or metabolic pathways) to degrade a number of PAHs. Similar cross-acclimation has been found for 2,4-dichlorophenoxyacetic acid and 2-methyl-4-chlorophenoxyacetic acid (255). Cultures of a 2,4,5-trichlorophenoxyacetic acid-degrading *Pseudomonas* were found to degrade various chlorophenols readily, probably as a result of 2,4,5-T metabo-

lism for which a degradative pathway was established by initial 2,4,5-T exposure
(256).

3.2.5. Changes in Organic Pollutant Toxicity and Recalcitrance

The degradation of complex organic molecules by bacteria entails the sequential formation and metabolism of numerous intermediate compounds *en route* to
complete mineralization or terminal metabolite formation. The degradation of all
but the simplest compounds results in multiple intermediates produced and sequentially transformed concurrent with the metabolism of the parent compound.
In general, larger compounds probably yield greater numbers of intermediates.
Intermediate compounds may have characteristics of toxicity and recalcitrance
distinct from the parent contaminant. Therefore, as physical, chemical, and
biological processes transform the initial compound(s) with time, different components of the water column or sediment microbiota (and, indeed, other faunal
and floral components of the communities) may encounter compounds of increasing or decreasing toxicity and biodegradability.

The effects of intermediates are probably best studied by amendment of
specific individual intermediates to systems. Unfortunately, there exists little
information on the interactions between metabolic intermediates of contaminants
and the microbiota of marine systems. In the case of petroleum hydrocarbons,
intermediates are often hydroxylated or carboxylated aliphatic or aromatic derivatives of the original compounds. Hydroxylated intermediates often elicit a
greater toxic response, probably because of their increased polarity and hence
solubility. Calder and Lader (97) found that 2,3-dihydroxynaphthalene, an intermediate breakdown product of the PAH naphthalene, was more effective in
decreasing the growth of two species of marine bacteria than was naphthalene
itself. In estuarine microcosms in which naphthalene was used as a model compound, a variety of polar hydroxylated intermediate metabolites were found,
including *cis*-1,2-dihydroxy-1,2-dihydronaphthalene, 1-naphthol, salicylic acid,
and catechol (241). Aliphatic hydrocarbon degradation results in the formation of
numerous fatty acid intermediates (160) that serve as substrates for both aerobic
and anaerobic bacteria (257,258) but that may be toxic to hydrocarbon degraders
(223). When mixed hydrocarbon substrates such as petroleum are considered,
the number of potential intermediates formed may be very large. Further study
of the effects of intermediates arising from model hydrocarbon compounds are
warranted in order to assess the impact of metabolites on marine microbial
communities.

Degradation of organochlorine pesticides may also affect microbial activity.
Time-dependent fluctuations in sediment and soil respiration following application of various pesticides was attributed to the sequential accumulation and
depletion of the metabolities of the parent pesticides (259). As noted above for

hydrocarbons, multiple organochlorine pesticides and contaminants may occur simultaneously (*e.g.,* in areas of heavy agriculture or as PCBs in commercial mixtures). Alterations in microbial activities may thus result from synergistic or antagonistic interactions among contaminants.

3.3. Removal of Inorganic Nitrogen: Denitrification

In many systems, nutrient availability, often as N, limits the rate of primary productivity by the photoautotrophs. This generalization holds true for planktonic microalgae (50,51) as well as for benthic macrophytes (53,54). The underlying sediments of these communities, however, generally contain an abundance of critical nutrients, derived from the microbial degradation of organic material in the sediments. Several researchers have estimated that nearshore sediments could supply much of the calculated N demand of phytoplankton (*e.g.,* refs. 27,51).

For N, several other microbial transformations occur in coastal waters and sediments and may affect exchanges and availability of N. These include nitrification, N_2 fixation, and denitrification (3). High rates of N_2 fixation have been noted in coastal and estuarine sediments (118,260). Denitrification is of particular concern because it represents a loss of biologically available N from the system and may promote N limitation (51,261). Alternately, denitrification may provide a biological relief valve for excess N (261). Very high rates of denitrification have been reported in numerous shallow sediment environments (261, 262,263). Indeed, several studies have reported elevated rates of denitrification (59,264) concurrent with nutrient loading and coastal eutrophication.

4. EMERGENT PROBLEMS, POTENTIAL SOLUTIONS

The coastal marine environment is a dynamic system. Physical processes that operate from very brief (minutes to hours; *e.g.,* wind-driven circulation) to geologic (*e.g.,* sea level change) time scales force natural fluctuations in the biological communities of these systems. Cyclic fluctuations are well characterized, and even predictable, over daily and seasonal scales. Dramatic and sudden increases (blooms), shifts, and die-offs in components of the biological communities have also been noted over the last several decades in various coastal systems and have often been related to anthropogenic perturbations. Such unpredictable phenomena may be expected to continue and to intensify as human populations along the coastline increase.

As detailed in the preceding discussion, the microbiota are intimately involved in many of the more obvious manifestations of change in coastal marine communities, in both a trophodynamic and biogeochemical context. Microbiological processes in coastal areas also contribute to ecosystem changes at larger scales. Where human-induced changes have been adjudged to be negative, the natural microbiota or introduced surrogates may be a useful means for some level of restoration through bioremediation. However, bioremediation at increasingly larger scales provokes serious discussion of the implications and ethics of premeditated environmental intervention and manipulation.

In our final section, and within the context of the coastal environment, we highlight several areas of developing interest and concern that, of necessity, involve a confluence of biogeochemistry, microbial ecology, and applied and public health microbiology and that range in relevance from local/regional to oceanic/global scales.

4.1. Antibiotic Use in Fish Mariculture

Large-scale fish mariculture, particularly of salmonid fish and penaeid shrimp, is a rapidly burgeoning industry in many coastal areas around the world (265,266). For instance, the biomass of farmed salmonids in Norway, the largest producers of net-pen salmon at present, increased from 20,000 tons in 1981 to over 120,000 tons in 1988 (265). There are several environmental impacts associated with intensive fish culture, not least of which is organic enrichment of underlying sediments and general water column eutrophication. However, of particular microbiological interest is the use of antibiotics as a tool of the fish farmer (267). Antibiotics, generally delivered in the fish feed, are used in both therapeutic and, to a lesser extent, prophylactic treatment of the fish stocks, particularly to combat vibriosis (265). A number of drugs have been used throughout the world in this application, including chloramphenicol (266,267), although only oxytetracycline (OTC), Romet-30 (a combination of ormetoprin and sulfadimethoxine), and sulfamerizine are approved for use in fish farming in the United States by the Food and Drug Administration. In 1988, 32.6 metric tons of antibiotics were used in Norway, about 0.36 kg of antibiotics per ton of fish produced (D. Weston, personal communication). In the nascent salmon industry in Puget Sound, Washington, 1.13 tons were used in 1990 in the production of 4,500 tons of fish (0.25 kg per ton of fish) (268,269).

Because of wasted feed and low digestive efficiencies of some antibiotics (*e.g.*, OTC) (270), a substantial fraction of the drug can be released to the environment and is predominantly deposited to the sediments. Studies of widely used OTC indicate that while it may degrade within a few weeks under ideal conditions (271), under conditions of high sedimentation and rapid burial, as may be found in some farms, it persists for much longer periods (272,273). At present, pathways of degradation for OTC are not characterized (271) and in any event appear to be unimportant *in situ* (273). In contrast, other drugs used in fish farming (*e.g.*, Furazolidone) may be rapidly metabolized (274). As might be expected, development of resistant strains of bacteria in the gastrointestinal tract of the farmed fish may occur (*e.g.*, 275).

Residues in the sediments below treated farms can easily exceed minimum inhibitory concentrations for extended periods after antibiotic treatment (272, 273). Among the various potential results of antibiotic use and its accumulation in the sediments are inhibition and disruption of the natural microbiota, as well as the development of antibiotic resistance in sediment populations. Preliminary evidence seems to confirm these possibilities (274,276,277). With the prognosis of expansion in this industry, it is important that the environmental impacts of antibiotic use by the industry be carefully evaluated (278).

As a final footnote, it is interesting to point out that while antibiotics are most often derived from organisms obtained from diverse natural environments, the relevance and dynamics of antibiotics in the environment are only poorly understood in any natural system (279).

4.2. Bioremediation of Oil Spills

The development of effective and appropriate methods of bioremediation has been a long-standing, if not explicit, objective of many of the studies of the microbial ecology and biodegradation potential of contaminated environments. For coastal waters, the bulk of previous bioremediation efforts has been on hydrocarbon contamination (280). A string of major oil spills over the last few years (*i.e. M.V. Bahia Paraiso,* Antarctica; *M.V. Exxon Valdez,* Prince William Sound, Alaska; *M.V. Mega Borg,* Gulf of Mexico; and Persian Gulf war spills) has intensified that focus considerably.

The development of bioremediation as a functional tool has depended most heavily on the basic understanding of the physiology and ecology of native bacterial populations. Recent advances in biochemical and molecular technologies are rapidly finding important applications in bioremediation efforts in various systems, including coastal marine environments (see Olson and Tsai, Bouwer, and Boyle, this volume). Several specific approaches have been proposed to enhance oil biodegradation (280): 1) optimizing environmental conditions in order to maximize the contaminant-degrading potential of native bacteria, *e.g.*, through supplementation of nutritional factors limiting the growth of hydrocarbon degraders; 2) seeding the contaminated environment with strains of bacteria that are tolerant and capable of rapid degradation of a given contaminant and thereby supplementing the natural resident microbiota; and 3) supplementing natural populations with bacteria that have been genomically altered ("engineered") to provide proliferation of a given metabolic pathway.

While genetically engineered hydrocarbon degraders have existed for almost two decades and have received considerable attention and notoriety, they possibly represent the least practical approach at present. The successful application of genetically engineered contaminant-degrading microorganisms has been impeded by the necessity of assessing the potential and perceived impact of such microorganisms on the natural microflora and on the environment as a whole (281,282). The second strategy, seeding of bacteria for hydrocarbon removal following petroleum spillage, has achieved mixed and sometimes disappointing results (280). The relative success of exotic bacteria, whether engineered or not, when added to a given system will depend on many factors including competitive interactions with native bacteria, their rate of growth in the system, and their tolerance of the physicochemical environment (159,283,284).

Of these three strategies, the first, which is the least complicated or controversial, has provided the most promising results. With the development of oleophilic fertilizers that enhance inorganic nutrient delivery in the vicinity of the oil (223), several successful field tests have provided evidence of the efficacy of this approach (280). The most dramatic, positive results come from the recent pilot and field tests in the aftermath of the *Exxon Valdez* spill (225). An existing population of hydrocarbon degraders in the vicinity of the spill was clearly constrained by the availability of inorganic N and P. Appropriate additions of fertilizer were sufficient to increase degradation rates up to threefold. Thus natural biodegradation enhancement may become a part of future oil spill responses.

It is interesting to note here the convergence of two relatively removed areas of current research. Whereas microbial ecologists concerned with heterotrophic utilization of N- and P-poor petroleum hydrocarbons have long appreciated that availability of inorganic nutrients might be an important limiting factor to these heterotrophs (see Section 3.2.3.), it is only relatively recently that the heterotrophic population of the marine water column has been reassessed with respect to its relationship to inorganic nutrients (see Section 1.1.1.). Whereas heterotrophic planktobacteria were considered to be net regenerators of inorganic nutrients, it appears now that they may in fact be limited by and compete for inorganic nutrient resources (22,24).

4.3. Marine Viruses: Ecopathological Agents in the Sea?

While viruses have long been recognized as part of the natural microbial flora of the seas, their ecological role, if any, was unclear because of their apparently low abundances in the sea (see ref. 285). However, with the recent revival of studies of the quantitative ecology of marine viruses, the abundance of viruses appears to be several orders of magnitude greater than previously thought, up to 10^8 ml^{-1} (285,286), and hence a more prominent ecological role is indicated. Indeed, a role for viruses in the lytic infection and regulation of heterotrophic bacteria, cyanobacteria, and eukaryotic algae has been put forth (285,287).

A possibly related phenomenon, first reported slightly before the rediscovery of marine viruses, has been the large-scale mortality of marine mammals. Episodes of beachings of dolphins and whales in the mid 1980s were variously related to ingestion of phytotoxins, specifically, brevetoxin derived from the dinoflagellate *Ptychodiscus breve* through food chain concentration (288). Alternately, exposure to organochlorine contaminants, specifically, PCBs, with immunosuppresant activity (288) or infection by indigenous pathogens such as *Vibrio* (289) (or both) have been offered as alternate explanations.

More recently, however, massive seal deaths in the Baltic Sea were traced to infection by a previously unknown virus of the morbillivirus (*e.g.*, canine distemper virus; CDV) group (290–293). Subsequently, a similar virus was detected in porpoises and Mediterranean striped dolphins (294,295).

It has been suggested that this virus may have been endemic in harp seal populations of the extreme north and was only introduced after an invasion of the North Sea by this species in 1987 (296). Populations of cetaceans and pinnapeds along the Atlantic coast of North America may also be at risk (291). Further investigation as to the natural incidence and spread of this virus is clearly warrented.

Thus one line of research has identified a previously unknown, possibly marine, virus in the etiology of disease(s) of marine mammals. Independently, the viral flora in general has been revealed as a component of the microbial communities of the sea that needs to be brought into better ecological perspective. Our understanding of the ecological role of marine viruses should see unprecedented advances in the coming decade.

4.4. Radiatively Important Trace Gases (RITGs), Climate Change, and the Coastal Microbiota

It is now generally accepted that several trace gases (*e.g.*, CO_2, CH_4, and N_2O) in the atmosphere are increasing steadily in concentration; the consensus suggests that this increase may contribute to excess global greenhouse warming (297). The atmospheric increase in CO_2 is greatest ($>1\%$ per year) of all these gases, and CO_2 may play the major role in global warming. However, CH_4 and N_2O are both far more potent greenhouse gases (on a per molecule basis) and may contribute a substantial fraction to overall increases in IR absorption and warming (297).

Various complex factors contribute to the increases in these RITGs. Microbial processes are involved in the biogeochemical cycles of all three gases and are particularly well recognized for CH_4 (75) and N_2O (76,298). Another trace gas with an active microbial connection, DMS, is also involved in climate dynamics (299).

The oceans are considered to be sources for CH_4, N_2O, and DMS (75,76, 300). However, such projections are generally extrapolations from relatively limited data (relative to the vast areal scale being considered), and there is considerable debate as to the absolute magnitude of these fluxes, as well as to the relative importance of various oceanic subsystems. Current research is attempting to assess the contribution of nearshore and coastal areas to the total oceanic source of these gases, as well as to determine the important factors and feedbacks that control the microbial processes producing or consuming CH_4 (75), N_2O (76), and DMS (77,301,302).

Nonetheless, CO_2 remains the focus of the most intensive interest and concern with respect to greenhouse warming. Proposals have recently been put forth that consider the possibility of artifically fertilizing areas of the ocean with Fe in order to stimulate algal productivity (thought to be Fe limited in these areas) and thereby promote biological uptake of CO_2 (303). Ideally, the biomass accumulated from this enhanced production (and its associated C) would settle and be sequestered as remineralized CO_2 in deep waters for a relatively long period thereby, offsetting atmospheric increases. Recent modeling efforts indicate that any bioremediation effort of this sort would require sustained fertilization to have any substantive effect, and overall the mitigation of atmospheric CO_2 increases are likely to be small (304,305). Furthermore, the enhanced flux of organic C to depth is projected to cause hypoxia or anoxia in waters below the euphotic zone (305).

There are important microbiological implications of the development of wide-scale hypoxia or anoxia (306). Besides fundamental disruption of the trophodynamic structure, the potential exists for enhancement of anaerobic bacterial processes leading to the production of CH_4 and N_2O (306). Thus efforts to sequester one greenhouse gas (CO_2) could result in the net production of other, more potent gases (*i.e.*, N_2O, CH_4).

ACKNOWLEDGMENTS

The authors thank Ron Kiene, Glenn Lopez, Jennifer Slater, Bob Kerr, Junko Kazumi, and Cindy Gilmour for their past efforts, collaboration, and discussion. Recent input from Don Weston, Veronica Miller, and Jed Fuhrman is appreciated. Financial support in our efforts over the years culminating in this review derived from the U.S. National Science Foundation, National Oceanic and Atmospheric Administration, Environmental Protection Agency, and Hudson River Foundation.

REFERENCES

1. Woodwell G, Rich P, Hall C: Carbon in estuaries. In Woodwell G, Pecan E (eds): Carbon in the Biosphere. Natl. Tech. Inf. Serv. CONF-720510, 1973, p 221.

2. McHugh L: Management of estuarine fisheries. In Smith RF, Schwartz AH, Massman WH (eds): A Symposium on Estuarine Fisheries. Spec. Publ. No. 3. Washington, DC: American Fisheries and Society, 1966; p 133.

3. Fenchel T, Blackburn TH: Bacterial and Mineral Cycling. London: Academic Press, 1979, 255 pp.

4. Hites RA, Biemann K: Science 178:158, 1972.

5. Shaw DG: Hydrocarbons in the water column. In Wolfe DA (ed): Fate and Effects of Petroleum Hydrocarbons in Marine Ecosystems and Organisms. New York: Pergamon Press, 1978, p 8.

6. Rublee PA: Bacteria and microbial distribution in estuarine sediments. In Kennedy VS (ed): Estuarine Comparisons. New York: Academic Press, 1982; p 159.

7. McDaniel JA, Capone DG: J Microbiol Methods 3:291, 1985.

8. Daley R, Hobbie J: Limnol Oceanogr 20:875, 1975.

9. Ducklow HW: Production and fate of bacteria in the oceans. Bioscience 33:494–499, 1983.

10. Bedding ND, McIntyre AE, Perry R, Lester JN: Sci Total Environ 25:143, 1982.

11. Carmody DJ, Pearce JB, Yasso WE: Mar Pollut Bull 4:132, 1973.

12. Weaver G: Environ Sci Technol 18:22A, 1984.

13. Pomeroy LR: The ocean's food web, a changing paradigm. Bioscience 24:499–504, 1974.

14. Williams PJL: Bacterial production in the marine food chain: The emperor's new suit of clothes? In Fasham MJR (ed): Flows of Energy and Materials in Marine Ecosystems. New York: Plenum Press, 1984, p 271.

15. Fuhrman JA, Ammerman JW, Azam F: Mar Biol 60:201, 1980.

16. Cole JJ, Findlay SF, Pace ML: Mar Ecol Prog Ser 43:1, 1988.

17. Azam F, Ammerman J: Cycling of organic matter by bacterioplankton in pelagic marine ecosystems. In Fasham MJR (ed): Flows of Energy and Materials in Marine Ecosystems. New York: Plenum Press, 1984, p 345.

18. Fuhrman JA: Close coupling between release and uptake of dissolved free amino acids in sea water studied by an isotope dilution approach. Mar Ecol Prog Ser 37:45–52, 1987.

19. Ducklow HW, Purdie DA, Williams JBL, Davies JM: Bacterioplankton: A sink for carbon in a coastal plankton community. Science 232:865–867, 1986.

20. Harrison WG: Limnol Oceanogr 23:684, 1978.

21. Currie DJ, Kalff J: Limnol Oceanogr 29:311, 1984.

22. Wheeler PA, Kirchman DL: Utilization of inorganic and organic nitrogen by bacteria in marine systems. Limnol Oceanogr 31:998–1009, 1986.

23. DiTullio GR, Laws EA: Limnol Oceanogr 28:177, 1983.

24. Fuhrman JA, Horrigan S, Capone D: Mar Ecol Prog Ser 45:271, 1988.

25. Goldman JC, Caron D, Dennett M: Regulation of gross growth efficiency and ammonium regeneration in bacteria by substarte C:N ratio. Limnol Oceanogr 32:1239–1252, 1987.

26. Nixon SW, Oviatt CA, Hale SS: Nitrogen Regeneration and the Metabolism of Coastal Marine Bottom Communities. In Anderson JM, Mcfaydden A (eds): The Role of Terrestrial and Aquatic Organisms in Decomposition Processes. Oxford: Blackwell, 1976, p 269.

27. Rowe GT, Clifford CH, Smith KL Jr: Nature 255:215, 1975.

28. Harper RM, Fry JC, Lerner MA: Oikos 36:211, 1981.

29. Riemann F, Schrage M: Oecologia 34:75, 1978.

30. Capone DG, Kiene RP: Comparison of microbial dynamics in marine and freshwater sediments: Contrasts in anaerobic carbon metabolism. Limnol Oceanogr 33:725–749, 1988.
31. Reeburgh WS: Annu Rev Earth Planet Sci 11:269, 1983.
32. Thauer RK, Jungerman K, Decker K: Bacteriol Rev 41:100, 1977.
33. Revsbech NP, Sorensen J, Blackburn T: Limnol Oceanogr 25:403, 1980.
34. Novitsky JA, Kepkay PE: Mar Ecol Prog Ser 4:1, 1980.
35. Rhoads DC, Yingst JY, McCall PL: Am Sci 66:577, 1978.
36. Sorensen J, Jorgensen BB, Revsbech N: Microbial Ecol 5:105, 1979.
37. Berner RA: Early Diagenesis—A Theoretical Approach. Princeton, NJ: Princeton University Press, 1980.
38. Jorgensen BB: Nature 296:643, 1982.
39. Suess E: Nature 288:260, 1980.
40. Curl H, et al.: Geochemical processes. In Kimrey LW, Burns RE (eds): Pollutant Transfer by Particles Workshop. Washington, DC: U.S. Dept. Commerce, NOAA, 1982, p 17.
41. Zobell C: J Bacteriol 46:39, 1943.
42. Marshall K (ed): Interfaces in Microbial Ecology. Cambridge, MA: Harvard University Press, 1976.
43. van Es FB, Meyer-Reil LA: Adv Microbiol Ecol 6:111, 1982.
44. Azam F, Fenchel T, Field JG, Gray JS, Meyer-Reil LA, Thingstad F: Mar Ecol Prog Ser 10:257, 1983.
45. Simon M, Alldredge A, Azam F: Mar Ecol Prog Ser 65:205, 1990.
46. Cho BC, Azam F: Major role of bacteria in biogeochemical fluxes in the ocean's interior. Nature 332:441–443, 1988.
47. van Loosdrecht MC, Lyklema J, Norde W, Zehnder AJ: Microbiol Rev 54:75, 1990.
48. Lopez G: personal communication.
49. Colwell RR (ed): *Vibrios* in the Environment. New York: John Wiley, 1984.
50. Ryther JH, Dunstan WM: Science 171:1008, 1971.
51. Nixon SW, Pilson MEQ: Nitrogen in estuarine and coastal marine ecosystems. In Carpenter EJ, Capone DG (eds): Nitrogen in the Marine Environment. New York: Academic Press, 1983, p 565.
52. Smith SV: Limnol Oceanogr 29:1149, 1984.
53. Short FT, Davis MW, Gibson RA, Zimmerman CF: Est Coast Shelf Sci 20:419, 1985.
54. LaPointe B: Bull Mar Sci 44:312, 1989.
55. Laws E: Aquatic Pollution. New York: Wiley-Interscience, 1981, 482 pp.
56. Malone TC: Estuarine Coast Mar Sci 5:157, 1977.
57. Segar D, Berberian G: Limnol Oceanogr Special Symp 2:220, 1976.
58. Suttle CA: Fuhrman J, Capone D: Limnol Oceanogr 35:424, 1990.
59. Jensen M, Andersen T, Sorensen J: Mar Ecol Prog Ser 48:155, 1988.
60. Schindler DW: Science 195:260, 1977.
61. Williams SC, Simpson H, Olsen C, Bopp R: Mar Chem 6:195, 1978.
62. Greig BA, McGrath RA: Mar Pollut Bull 8:188, 1977.
63. Bruland K: Trace elements in seawater. In Riley JP, Chester R (eds): Chemical Oceanography. New York: Academic Press, 1983, p 157.

64. Cloern J, Oremland RS: Estuaries 6:399, 1983.

65. Hoffman EJ, Mills GL, Latimer JS, Quinn JG: Can J Fish Aquat Sci 2:41, 1980.

66. Gjessing ET, Berglind L: Arch Hydrobiol 1:92, 1981.

67. Karickhoff SW, Brown DS, Scott TA: Water Res 13:241, 1979.

68. Chiou CT, Peters LJ, Freed JH: Science 206:831, 1979.

69. Mearns A, *et al.:* Biochemical processes. In Kimrey LW, Burns RE (eds): Pollutant Transfer by Particles Workshop. Washington, DC: U.S. Dept. Commerce, NOAA, 1982, p 33.

70. Horzempa L, DiToro D: Unpublished manuscript. Manhattan College, NY, 1981.

71. Baker JE, Eisenreich SJ, Johnson TC, Halfman BM: Environ Sci Technol 19:854, 1985.

72. Oliver BG, Charlton M, Durham R: Environ Sci Technol 23:200, 1989.

73. dePaola A, *et al.:* Appl Environ Microbiol 56:2299, 1990.

74. Venkatewsaran K, *et al.:* Appl Environ Microbiol 55:1591, 1989.

75. Cicerone R, Oremland RS: Global Biogeochem Cycles 2:299, 1988.

76. Capone DG: Aspects of the marine nitrogen cycle with relevance to the dynamics of nitrous and nitric oxide. In Whitman WB, Rogers JE (eds): Microbial Production and Consumption of Radiatively Important Trace Gases. Washington, DC: American Society for Microbiology, Special Publication, ASM Press, 1991, p 255.

77. Steudler PA, Peterson BJ: Nature 311:455, 1984.

78. Oremland RS, Taylor BF: Geochim Cosmochim Acta 42:209, 1978.

79. Fuhrman J, McManus G: Science 224:1257, 1984.

80. Marino R, Howarth R, Shamess J, Prepas E: Limnol Oceanogr 35:245, 1990.

81. Ehrlich H: Geomicrobiology. New York: Marcel Dekker, 1981, 393 pp.

82. Horvath RS: Bacteriol Rev 36:146, 1972.

83. National Academy of Sciences: Petroleum in the Marine Environment. Washington, DC: NAS, 1975.

84. Nagao M, Sagimura T: Mutagenesis: Microbial systems. In Gelboin HV, T'so P (eds): Poly-cyclic Aromatic Hydrocarbons and Cancer, vol 1. New York: Academic Press, 1978, p 99.

85. Guerin MR: Energy sources of polycyclic aromatic hydrocarbons. In Gelboin HV, T'so P (eds): Polycyclic Aromatic Hydrocarbons and Cancer, vol 1. New York: Academic Press, 1978, p 3.

86. Neff JM: Polycyclic Aromatic Hydrocarbons in the Aquatic Environment: Source, Fates and Biological Effects. London: Applied Science Publishers, 1979, p 262.

87. Tripp BW, Farrington J, Teal J: Mar Poll Bull 12:122, 1981.

88. Podlech PA, Borzani W: Biotech Bioeng 14:43, 1972.

89. Chakrabarty M, Singh H, Baruah J: Biotech Bioeng 17:399, 1975.

90. Yoshida F, Yamane T, Yagi H: Biotech Bioeng 13:215, 1974.

91. Kennedy RS, Finnerty WR: Arch Microbiol 102:88, 1975.

92. Hodson RE, Azam F, Lee RF: Bull Mar Sci 27:119, 1977.

93. Griffiths RP, McNamara TM, Caldwell BA, Morita RY: Appl Environ Microbiol 41:1400, 1981.

94. Alexander SK, Schwartz J: Appl Environ Microbiol 40:341, 1980.

95. Walker J, Austin H, Colwell R: J Gen Appl Microbiol 21:27, 1975.

96. Buckley EN: Effects of Petroleum Hydrocarbons on Metabolic Activity and Community Com-position of Suspended Heterotrophic Salt Marsh Bacteria. Ph.D. thesis, University of North Carolina, Chapel Hill, 1980.

97. Calder JA, Lader JH: Appl Environ Microbiol 32:95, 1976.

98. Griffin LF, Calder JA: Appl Environ Microbiol 33:1092, 1977.

99. Henson JM, Hayasaka SS: Mar Environ Res 6:205, 1982.

100. Hudak JP: McDaniel J, Lee S, Fuhrman JA: Mar Ecol Prog Ser 47:97, 1988.

101. Young LY, Mitchell RW: Appl Microbiol 25:972, 1973.

102. Walsh F, Mitchell R: Inhibition of bacterial chemoreception by hydrocarbons. In Ahearn DG, Meyers SP (eds): Microbial Degradation of Oil Pollutants. LSU-SG-73-01. Baton Rouge, LA: Louisiana State University, 1973, p 275.

103. Holloway SL, Faw G, Sizemore R: Mar Pollut Bull 11:153, 1980.

104. Cobet AB, Guard HE: Effects of a bunker crude oil on the beach bacterial flora. In Proceedings of the Joint Conference on Prevention and Control of Oil Spills. Washington, DC: American Petroleum Institute, 1973, 815 pp.

105. Crow SA, Hood MA, Meyers SP: Microbiological aspects of oil intrusion in southeastern Louisiana. In Bourquin AW, Ahearn DG, Meyers SP (eds): Impact of the Use of Microorganisms on the Aquatic Environment. Corvallis, OR: EPA, 1975, p 221.

106. Hood M, Bishop W, Meyers S, Whelan T: Appl Microbiol 30:892, 1975.

107. Westlake DW, Cook FD: Petroleum biodegradation potential in northern Puget Sound and Strait of Juan de Fuca. EPA-600/7-80-133. Washington, DC: EPA, 1980.

108. Pfaender FK, Buckley EN: Effects of petroleum on microbial communities. In Atlas RM (ed): Petroleum Microbiology. New York: MacMillan, 1984, p 507.

109. Griffiths RP, McNamara TM, Caldwell BA, Morita RY: Mar Environ Res 5:83, 1981.

110. Foght JM, Westlake DW: Can J Microbiol 20:117, 1982.

111. Lee K, Wong CS, Cretney WJ, Whitney FA, Parsons TR, Lalli CM, Wu J: Microb Ecol 11:337, 1985.

112. Griffiths RP, Morita RY: Study of Microbial Activity and Crude Oil-Microbial Interaction in the Water and Sediments of Cook Inlet and the Beaufort Sea. Final Report OCSEAP Contract 03-5-022-68. Juneau, AL: NOAA Project Office, 1980.

113. Griffiths RP, Caldwell BA, Broich WA, Morita RY: Appl Environ Microbiol 42:792, 1981.

114. Winfrey MR, Beck E, Boehm P, Ward D: Mar Environ Res 7:175, 1982.

115. Bauer JE, Montagna PA, Spies RB, Prieto MC, Hardin D: Limnol Oceanogr 33:1493, 1988.

116. Griffiths RP, Caldwell BA, Broich WA, Morita RY: Mar Pollut Bull 13:273, 1982.

117. Floodgate GD: Nutrient limitation. In Bourquin AW, Pritchard PH (eds): Microbial Degradation of Pollutants in Marine Environments. Proceedings of a Workshop. EPA-66019-79-012. Gulf Breeze, FL: Environmental Research Laboratory, 1979, p 107.

118. Capone DG: Benthic nitrogen fixation. In Blackburn TH, Sorensen J (eds): Nitrogen Cycling in Coastal Marine Environments. SCOPE Series. New York: Wiley, 1988, p 85.

119. Knowles R, Wishart C: Environ Pollut 13:133, 1977.

120. Griffiths RP, Caldwell BA, Broich WA, Morita RY: Estuarine Coast Shelf Sci 15:183, 1982.

121. Haines JR, Atlas RM, Griffiths RP, Morita RY: Appl Environ Microbiol 41:412, 1981.

122. Bauer JE, Capone DG: Appl Environ Microbiol 49:828, 1985.

123. Kiene R, Capone DG: Mar Environ Res 13:141, 1984.

124. Slater J, Capone DG: Persistent pollutants and nitrogen fixation and denitrification in coastal marine sediments. In Capuzzo JM, Kester DR (eds): Oceanic Processes in Marine Pollution vol 1. Biological Processes and Wastes in the Ocean. Malabar, FL: Krieger, 1987, p 71.

125. Sanders PF, Tibbetts PJ: Philos Trans R Soc Lond [Biol] 316:567, 1987.

126. Dow FK, Davies J, Raffaelli D: Mar Environ Res 29:103, 1990.

127. Phillips DJ: Use of organisms to quantify PCBs in marine and estuarine environments. In Waid JS (ed): PCBs in the Environment, vol. II, Boca Raton, FL: CRC Press, 1986, p 127.

128. Gupta SK: Insecticides and microbial environments. In Lal R (ed): Insecticide Microbiology. New York: Springer-Verlag, 1984, pp 3–39.

129. Mullin MD, *et al.:* Environ Sci Technol 18:468, 1984.

130. Dexter RN, Pavlou SP: Mar Chem 6:41, 1978.

131. Bedding ND, McIntyre AE, Perry R, Lester JN: Sci Total Environ 26:255, 1983.

132. Rosas SB, Secco MD, Ghittoni NE: Appl Environ Microbiol 40:231, 1980.

133. Handy MK, Gooch JA: Uptake, retention and depuration of PCBs by organisms. In Waid JS (ed): PCBs in the Environment, vol II. Boca Raton, FL: CRC Press, 1984, p 63.

134. Lal R: Factors affecting microbe/insecticide interactions. CRC Rev 10:261, 1984.

135. Bourquin AW, Cassidy S: Appl Microbiol 29:125, 1975.

136. Blakemore RP, Carey AE: Appl Environ Microbiol 35:323, 1978.

137. Blakemore RP: Appl Environ Microbiol 35:329, 1978.

138. Orndorff SA, Colwell RR: Microbiol Ecol 6:357, 1980.

139. Hudak JP, Fuhrman JA: Mar Ecol Prog Ser 47:185, 1988.

140. Mahaffey WR, Pritchard PH, Bourquin AW: Appl Environ Microbiol 43:1419, 1982.

141. Lovley DR, Phillips E: Appl Environ Microbiol 54:1472, 1988.

142. Nealson KH: The bacterial manganese cycle. In Krumbein WE (ed): Microbial Geochemistry. Oxford: Blackwell, 1983, p 191.

143. Lovley DR, Phillips E, Gorby Y, Landa E: Nature 350:413, 1991.

144. Ghiorse WC: In Zehnder AJB (ed): Biology of Anaerobic Microorganisms. New York: Wiley, 1988.

145. Smillie RH, Hunter K, Loutit M: Water Res 15:1351, 1981.

146. Boulegue J, Lord CJ III, Church T: Geochim Cosmochim Acta 46:453, 1982.

147. Davies-Colley R, Nelson P, Williamson K: Mar Chem 16:173, 1985.

148. Hines ME, Spencer MJ, Tugel JB, Lyons WB, Jones GE: In Caldwell DE, Brierly JA, Brierly CL (eds): Planetary Ecology. New York: Van Nostrand Reinhold, 1985, p 222.

149. Gadd GM, Griffiths AJ: Microb Ecol 4:303, 1978.

150. Bollag JM, Duszota M: Arch Environ Contam Toxicol 13:265, 1984.

151. Gordon AS, Millero FJ: Va J Sci 38:194, 1987.

152. Ortner PB, Kreader C, Harvey GR: Nature 301:57, 1983.

153. Schreiber DR, Millero FJ, Gordon AS: Mar Chem 28:275, 1990.

154. Harwood-Sears V, Gordon A: Appl Environ Microbiol 56:1327, 1990.

155. Gibson DT (ed): Microbial Degradation of Organic Compounds. New York: Marcel Dekker, 1984.

156. Walker JD, Petrakis L, Colwell R: Can J Microbiol 22:598, 1976.

157. Walker J, Colwell R, Petrakis L: Can J Microbiol 22:1209, 1976.

158. Atlas RM: Microbial degradation of petroleum hydrocarbons: An environmental perspective. Microbiol Rev 45:180–209, 1981.

159. Leahy JG, Colwell RR: Microbiol Rev 54:305, 1990.

160. Foster JW: Bacterial oxidation of hydrocarbons. In Hiyashi O (ed): Oxygenases. New York: Academic Press, 1962, p 241.

161. Ratledge C: Degradation of aliphatic hydrocarbons. In Watkinson JR (ed): Developments in Biodegradation of Hydrocarbons, vol. 1. London: Applied Science, 1978, p 1.

162. Kester A, Foster J: J Bacteriol 85:589, 1963.

163. Gibson DT: Biodegradation of aromatic petroleum hydrocarbons. In Wolfe DA (ed): Fate and Effects of Petroleum Hydrocarbons in Marine Ecosystems and Organisms. New York: Pergamon, 1977, p 36.

164. Cerniglia CE: Adv Appl Microbiol 30:119, 1984.

165. Perry JJ, Microbial metabolism of cyclic alkanes. In Atlas RM (ed): Petroleum Microbiology. New York: MacMillan, 1984, p 61.

166. Mihelcic JR, Luthy RG: Appl Environ Microbiol 54:1182, 1988.

167. Evans WC, Fuchs G: Annu Rev Microbiol 42:289, 1988.

168. Schink B, FEMS Micbiol Ecol 31:69, 1985.

169. Ward DM, Brock TD, Appl Environ Microbiol 35:353, 1978.

170. Hambrick GA, DeLaune R, Hambrick W: Appl Environ Microbiol 40:365, 1980.

171. Ward DM, Atlas RM, Boehm PD, Calder JA: Ambio 9:277, 1980.

172. Bauer JE, Capone DG: Appl Environ Microbiol 50:81, 1985.

173. Mille G, Mulyono M, El Jamal T, Bertrand J: Estuarine Coast Shelf Sci 27:283, 1988.

174. Krantzberg F, Environ Pollut (Ser A) 39:99, 1985.

175. Gordon DC, Dale J, Veizer P: J Fish Res Bd Can 38:591, 1978.

176. Bauer JE, Kerr R, Bautista M, Decker CJ, Capone DG: Mar Environ Res 25:63, 1988.

177. Kobayashi H, Rittman BE: Environ Sci Technol 16:170A, 1982.

178. Colwell RR, Sayler GS: Microbial degradation of industrial chemicals. In Mitchell R (ed): Water Pollution Microbiology, vol 2. New York: Wiley, 1978, p 111.

179. Suflita JM, Horowitz A, Shelton DR, Tiedje JM: Science 218:1115, 1982.

180. Hankin L, Sawhney BL: Soil Sci 137:401, 1984.

181. Kong HL, Sayler GS: Appl Environ Microbiol 46:666, 1983.

182. Dolfing J, Tiedje J: Appl Environ Microbiol 57:820, 1991.

183. Reineke W, Knackmuss HJ: Annu Rev Microbiol 42:263, 1988.

184. Lal R, Saxena DM: Microbiol Rev 46:95, 1982.

185. Barik S: Metabolism of insecticides by organisms. In Lal R (ed): Insecticide Microbiology. New York: Springer-Verlag, 1984, p 87.

186. Parsons J, Veerkamp W, Hutzinger O: Toxicol Environ Chem 6:326, 1983.

187. Sayler GS, Thomas R, Colwell RR: Estuarine Coast Mar Sci 6:553, 1978.

188. Schmidt E, Hellwig M, Knackmuss H-J: Appl Environ Microbiol 46:1038, 1983.

189. Furukawa K: Modification of PCBs by bacteria and other microorganisms. In Waid JS (ed): PCBs and the Environment, vol 2. Boca Raton, FL: CRC Press, 1984, p 89.

190. Walker WW: J Environ Qual 5:210, 1976.

191. Bourquin AW: Appl Environ Microbiol 33:356, 1977.

192. Bourquin AW, Hood M, Garnas R: Dev Ind Microbiol 18:185, 1977.

193. Pritchard PH, Bourquin AW, Frederickson HI, Maziarz T: System design factors affecting environmental fate in microcosms. In Bourquin AW, Pritchard PH (eds): Microbial Degrada-

tion of Pollutants in Marine Environments. Gulf Breeze, FL: Environmental Research Laboratory, 1979, p 251.

194. Weber K: Water Res 10:237, 1976.

195. De Laune R, Gambrell R, Reddy K: Environ Pollut Ser B 6:297, 1983.

196. Pignatello JJ, Martinson MM, Steirt JG, Carlson RE, Crawford RL: Appl Environ Microbiol 46:1024, 1983.

197. Esaac EG, Matsumura F: Pharm Ther 9:1, 1980.

198. Taylor BF, Campbell WL, Chinoy I: J Bacteriol 102:430, 1976.

199. Young L: Anaerobic degradation of aromatic compounds. In Gibson DT (ed): Microbial Degradation of Organic Compounds. New York: Marcel Dekker, 1984; p 487.

200. Horowitz A, Suflita JM, Tiedje JM: Appl Environ Microbiol 9:1, 1983.

201. Ide A, Niki Y, Sakamoto F, Watanabe I, Watanabe H: Agric Biol Chem 36:1937, 1972.

202. Castro TF, Yoshida T: J Agric Food Chem 19:1168, 1971.

203. Gambrell RP, Reddy CN, Collard V, Green G, Patrick WH: Water Pollut Control Fed 56:174, 1984.

204. Bosma TR, van der Meer JR, Schraa G, Tros ME, Zender AJ: FEMS Microbiol Ecol 53:223, 1988.

205. Quensen JF, Tiedje JM, Boyd SA: Science 242:752, 1988.

206. Sahm H, Brunner M, Schoberth SM: Microb Ecol 12:147, 1986.

207. King GM: Appl Environ Microbiol 54:3079, 1988.

208. Haggbloom M, Young L: Appl Environ Microbiol 56:3255, 1990.

209. Gibson D, Suflita J: Appl Environ Microbiol 52:681, 1986.

210. Kohring G-W, Zhang X, Weigel J: Appl Environ Microbiol 55:2735, 1989.

211. Genthner BR, Price WA III, Pritchard PH: Appl Environ Microbiol 55:1466, 1989.

212. Khalifa S, Holmstead R, Casida J: J Agric Food Chem 24:277, 1976.

213. Williams RR, Ridleman TF: J Agric Food Chem 26:280, 1978.

214. Parr JF, Smith S: Soil Sci 121:842, 1976.

215. Pfaender FK, Alexander M: J Agric Food Chem 20:842, 1972.

216. Fogel S, Lancione RL, Sewall AE: Appl Environ Microbiol 44:113, 1982.

217. Revsbech N, Madsen B, Jorgensen B: Limnol Oceanogr 31:293, 1986.

218. Revsbeck NP, Jorgensen BB: Adv Microbial Ecol 9:293, 1986.

219. Atlas R, Horowitz A, Busdosh M: J Fish Res Board Can 35:585, 1978.

220. Shiaris MP: Appl Environ Microbiol 55:1391, 1989.

221. Colwell RR, Mills AL, Walker JD, Garcia-Tello P, Campos-P V: J Fish Res Board Can 35:573, 1978.

222. Atlas RM, Bartha R: Biotechnol Bioeng 14:309, 1972.

223. Atlas RM, Bartha R: Environ Sci Technol 7:538, 1973.

224. Aldhouse P: Nature 349:447, 1991.

225. Pritchard PH, Costa CF: Environ Sci Technol 25:372, 1991.

226. Bartholomew GW, Pfaender F: Appl Environ Microbiol 45:103, 1983.

227. Kerr RP, Capone DG: Mar Environ Res 26:181, 1988

228. Guerin WF, Jones GE: Estuarine Coast Shelf Sci 29:115, 1989.

229. Boethling R, Alexander M: Appl Environ Microbiol 37:1211, 1979.

230. Spain JC, Van Veld PA: Appl Environ Microbiol 45:428, 1983.

231. Tagger S, Deveze L, LePetit J: Mar Pollut Bull 7:172, 1976.

232. Button DK, Robertson BR: Limnol Oceanogr 31:101, 1986.

233. Robertson BR, Button D: Appl Environ Microbiol 53:2193, 1987.

234. Colwell RR, Walker JD: Crit Rev Microbiol 5:423, 1977.

235. Foght JM, Fedorak P, Westlake D: Can J Microbiol 36:169, 1990.

236. Walker JD, Colwell RR: Appl Environ Microbiol 31:189, 1976.

237. Lee RF, Anderson SW: Bull Mar Sci 27:127, 1977.

238. Caparello DM, LaRock PA: Microbiol Ecol 2:28, 1975.

239. Walker JD, Colwell RR: Appl Environ Microbiol 31:189, 1975.

240. Walker J, Colwell R, Petrakis L: Can J Microbiol 22:1209, 1976.

241. Heitcamp MA, Freeman JP, Cerniglia CE: Appl Environ Microbiol 53:129, 1987.

242. Spain JC, Pritchare PH, Abourquin AW: Appl Environ Microbiol 40:726, 1980.

243. Singer JT, Finnerty WR: Genetics of hydrocarbon-utilizing microorganisms. In Atlas RM (ed): Petroleum Microbiology. New York: MacMillan, 1984, p 299.

244. Williams PA, Worsey WJ: Biochem Soc Trans 466, 1977.

245. Devereux R, Sizemore RK: Mar Pollut Bull 13:198, 1982.

246. Hada HS, Sizemore RK: Appl Environ Microbiol 41:199, 1981.

247. Leahy JG, Somerville CC, Cunningham KA, Adamantiades GA, Byrd JJ, Colwell RR: Appl Environ Microbiol 56:1656, 1990.

248. Pemberton JM, Fisher PR: Nature 278:732, 1977.

249. Serdar CM, Gibson DT, Munnedke DM, Lancaster JH: Appl Environ Microbiol 44:246, 1982.

250. Perry JJ: Microbiol Rev 43:59, 1979.

251. Francis AJ, Spanggord RJ, Ouchi GI, Bohonos N: Appl Environ Microbiol 35:364, 1978.

252. Horowitz A, Atlas RM: Microbial seeding to enhance petroleum hydrocarbon biodegradation in aquatic Arctigc ecosystems. In Oxley TA, Becker G, Allsopp D, (eds): Proceedings of the Fourth International Biodeterioration Symposium. London: Pitman, 1977, p 15.

253. Barnsley EA: J Bacteriol 153:1069, 1982.

254. Bauer JE, Capone DG: Appl Environ Microbiol 54:1649, 1988.

255. Torstensson NT, Stark J, Goransson G: Weed Res 15:159, 1975.

256. Karns JS, Kilbane JH, Duttagupta S, Charkrabarty AM: Appl Environ Microbiol 46:1176, 1983.

257. Gottshalk G: Bacterial Metabolism, 2nd ed. New York: Springer-Verlag, 1984.

258. Christensen D: Limnol Oceanogr 29:189, 1984.

259. Bartha R, Linsilotta R, Pramer D: Appl Microbiol 15:67, 1967.

260. Capone DG: Benthic nitrogen fixation. In Carpenter EJ, Capone DG (eds): Nitrogen in the Marine Environment. New York: Academic Press, 1983, p 105.

261. Seitzinger SP: Limnol Oceanogr 33:702, 1988.

262. Hattori A: Denitrification and dissimilatory nitrate reduction. In Carpenter EJ, Capone DG (eds): Nitrogen in the Marine Environment. New York: Academic Press, 1983, p 191.

263. Sorensen J, Koike I: Nitrate reduction and denitrification in marine sediments. In Blackburn TH, Sorensen J, Roswall T (eds): Nitrogen Cycling in Marine Coastal Environments. SCOPE Series. New York: Wiley, 1988, p 251.

264. Seitzinger S, Nixon S: Limnol Oceanogr 30:1332, 1985.
265. Grave K, Engelstad M, Soli N, Hastein T: Aquaculture 86:347, 1990.
266. Brown JH: World Aquaculture 20:34, 1989.
267. Austin B: Microbiol Sci 4:113, 1985.
268. Weston D: personal communication.
269. Gibson, C: personal communication.
270. Cravedi J-P, Choubert G, Delous G: Aquaculture 60:133, 1987.
271. Samuelsen OB: Aquaculture 83:7, 1988.
272. Jacobsen P, Berglind L: Aquaculture 70:365, 1988.
273. Bjorklund H, Bondestam J, Bylund G: Aquaculture 86:359, 1990.
274. Samuelsen OB, Solheim E, Lunestad BT: Sci Total Environ (in press), 1991.
275. Austin B, Al-Zahrani AMJ: J Fish Biol 33:1, 1988.
276. Samuelsen OB: The fate of antibiotics/chemotherapeutics in marine aquaculture sediments. In Problems of Chemotherapy in Aquaculture: From Theory to Reality. Paris: O.I.E., 1991, p 87.
277. Husevag B, Lunestad BT, Johannessen PJ, Enger O, Samuelsen OB: J Fish Dis (in press), 1991.
278. Lunestad BT: "Fate and effects of antibacterial agents in aquatic environments. In Problems of Chemotherapy in Aquaculture: From Theory to Reality. Paris: O.I.E., 1991, p 97.
279. Williams ST, Vickers JC: Microbiol Ecol 12:43, 1986.
280. Bartha R: Microbiol Ecol 12:155, 1986.
281. Sharples FE: Recomb DNA Tech Bull 6:43, 1983.
282. Simberloff D: Predicting ecological effects of novel organisms: Evidence from higher organisms. In Halvorson HO, Pramer D, Rogul M (eds): Engineered Organisms in the Environment: Scientific Issues. Washington, DC: American Society for Microbiology, 1985, p 152.
283. Atlas RM: Crit Rev Microbiol 5:371, 1977.
284. Scanferlato YS, Lacy GH, Cairns J: Microbiol Ecol 20:11, 1990.
285. Procter LM, Fuhrman JA: Nature 343:60, 1990.
286. Bergh O, Borsheim K, Bratbak G, Heldal M: Nature 340:467, 1989.
287. Suttle CA, Chan AM, Cottrell MT: Nature 347:467, 1990.
288. Brody M: Am Soc Microbiol News 55:595, 1989.
289. Smith HL: Am Soc Microbiol News 56:249, 1990.
290. Osterhaus A, Vedder EJ: Nature 335:20, 1988.
291. Kennedy S, Smyth JA, McCullough SJ, Allan GM, McNeilly F, McQuaid S: Nature 335:404, 1988.
292. Mahy B, Barrett T, Evans S, Anderson EC, Bostock CJ: Nature 336:115, 1988.
293. Cosby SL, et al.: Nature 336:116, 1988.
294. Kennedy S, Smyth JA, Cush PF, McCullough SJ, Allan GM: Nature 336:21, 1988.
295. Domingo M, Ferrer L, Pumarola M, Marco A, Plana J, Kennedy S, McAlisky M, Rima BK: Nature 348:21, 1990.
296. Goodhart CB: Nature 336:21, 1988.
297. Dickenson RE, Cicerone RJ: Future global warming from atmospheric trace gasses. Nature 319:109–115, 1986.
298. Cicerone R: J Geophys Res 94:18, 1989.

299. Lovelock JE, Nature 344:100, 1990.

300. Andreae MO, Raemdomock A: Science 221:744, 1983.

301. Dacey JWH, King GM, Wakeham SG: Nature 330:643, 1987.

302. Kiene RP, Bates TS: Nature 345:702, 1990.

303. Martin JH, Gordon RM, Fitzwater SE: Nature 345:156, 1990.

304. Peng T-H, Borecker WS: Nature 349:227, 1991.

305. Joos F, Sarmiento JL, Siegenthaler U: Nature 349:772, 1991.

306. Fuhrman JA, Capone DG: Limnol Oceanogr (submitted), 1991.

307. Oremland RS: The biogeochemistry of methanogenic bacteria. In Zehnder AJB (ed): Biology of Anaerobic Microorganisms. New York: Wiley, 1988, p 641.

308. Olsen CR, Cutshall NH, Nelsen TA: Pollutant-particle dynamics in geochemical cycles. In Kimrey LW, Burns RE (eds): Pollutant Transfer by Particles Workshop. Washington, DC: U.S. Dept. Commerce, NOAA, 1982, p 145.

9

MOLECULAR APPROACHES TO ENVIRONMENTAL MANAGEMENT

BETTY H. OLSON
YU-LI TSAI

Program in Social Ecology, University of California, Irvine, California 92717

1. INTRODUCTION

Advances in molecular biology have led to increased understanding of gene structure and function and have provided the genetic basis to apply molecular methods to environmental study. Historically, genes were thought to be units of heredity. As our knowledge has increased we have learned that there are three types of genes: 1) structural genes that are transcribed into messenger RNA (mRNA) and that then are translated into polypeptides, 2) genes that are transcribed into ribosomal RNA (rRNA) or transfer RNA (tRNA), and 3) regulatory genes that are not transcribed but serve as recognition sites for the initiation of function of an operon. One of the traits of structural genes that makes them ideal for use in technological approaches is that each gene codes for a specific polypeptide (protein), and thus a unique gene codes specifically for an enzyme or its subunit. This feature of genes allows us to identify and follow specific function potential in the environment.

The term *potential* is used, because genes must be expressed in order to carry out their function. Therefore, the presence of a gene that directs the production of a certain enzyme does not mean that the gene is always actively manufacturing

Environmental Microbiology, pages 239–263, © 1992 Wiley-Liss, Inc.

that protein. One can assess the ability of genes to function in the environment by determining expression. Expression can be determined in several ways. Traditionally it has been assessed biochemically, by looking for certain degradation products (1). Today it can be carried out by using antibody preparations to specific enzymes (2), by measurement of mRNA (3), or by the introduction of marker genes into the operon of interest (4). In the last case the marker genes give a signal that indicates that the system is turned on or being expressed.

DNA is organized into two main units in cells: the chromosome(s) and the extrachromosomal DNA (plasmids). The chromosome(s) contain all the information needed for growth and reproduction. Plasmids are nonessential for basic cell growth and reproduction, often stably maintained and autonomous replicating units. Plasmids have been found in a wide variety of microorganisms. These structures often code for phenotypic traits such as drug resistance, detoxification of metals, and degradation of halogenated hydrocarbons. In addition they often code for initiation of vegetative replication (origin of replication) of the plasmid, efficient partition of plasmids into daughter cells, and a means to ensure horizontal gene transfer.

In cells, genetic traits can exist in single or multiple sets. The number of copies of a genetic determinant that are carried within the cell is referred to as *copy number*. The term can refer to a single gene, to an operon, or to an entire plasmid. The concept of copy number is independent of genetic amplification systems and is simply a numerical representation of the amount of DNA within a cell relating to a specific function. Copy number is always a calculated value and therefore in reality represents an average occurrence in a population of microorganisms.

Plasmid copy number is the number of plasmid copies that are maintained in a population of cells. Thus copy number is actually an average, with some cells containing more copies than others. Plasmid copy number may also be a function of the growth phase of the population. Thus an exponentially growing culture may contain a different average copy number than cells in stationary phase. Copy number is usually controlled within the plasmid by a mechanism of negative feedback: by synthesis of a repressor compound, by production of a protein that dilutes the promoter of replication, or by the instability of the promoter protein or related constituent.

Plasmids exist in low or high copy number. In low-number plasmids, cell division must be accompanied by active plasmid-partitioning for maintenance of the plasmid in the population, whereas in high-number systems this is carried out by random partitioning. In the latter case this is usually sufficient to maintain the plasmid in the population unless the plasmids form multimers (plasmid copies remain attached to one another) in which case daughter cells may not receive the plasmid.

Plasmid copy number is either under stringent replicative control, resulting in low numbers (e.g., 1–2), or under relaxed control, which may result in several hundred copies per cell but is usually about 20. Molecular biologists have found ways to increase copy number artificially by the addition of certain chemicals. Antibiotics are often used. An excellent example is chloramphenicol addition to a late log phase bacterial culture containing the plasmid ColEl, which increases the copy number from 20 to between 1,000 and 3,000 (5).

Each plasmid has a distinct region of DNA that controls replication. This origin of replication is also important in the types of plasmids that can remain in a cell simultaneously. If two different plasmids have the same origin of replication, then they will not remain in a cell simultaneously and are called *incompatible*. If the origins are different, then the plasmids may both be replicated and maintained within a cell and are termed *compatible*. Approximately 20 different incompatibility groups have been identified.

Most of our knowledge of plasmid copy number and the controlling factors in replication is restricted to a small group of plasmids. These have great importance as cloning vectors. In natural populations in the environment there is little information on plasmid copy number. Genetic ecology, by delineating the copy number of environmentally important operons and plasmids in natural populations, will allow us to determine the capacity of an environment to detoxify a waste compound within it. Furthermore, it may be possible to increase low copy number systems through the addition of chemicals that interfere with mechanisms that limit plasmid replication.

Transposons (Tn) allow the genetic elements associated with them to duplicate and move onto additional locations within the cell chromosome(s) or plasmid(s). This is one mechanism by which copy number can be increased within a cell. Two basic arrangements of a transposon are illustrated in Figure 9.1. The first arrangement contains copies of insertion sequence (IS) elements flanking drug resistance genes. The second consists of a sequence flanked by short, inverted repeats (IRs). Between the IRs are the genes encoding transposition functions and those encoding drug resistance. For example, resistance to the antibiotic chloramphenicol is contained in Tn9. Transposons may be associated with other types of genes. For example, mercury is known to stimulate transposition in Tn501, while the factors that stimulate transposition in Tn21 are not known (6). The genes contained within these two transpositional units (Tn501 and Tn21) code for mercury detoxification.

Increased and controlled transposition could be applied to certain degradation or detoxification operons. Increased transposition or selection for subpopulations within a bacterial strain has been demonstrated in pure cultures and microcosms containing soil and sediments contaminated with mercury and naphthalene (7,8). Although transposition can increase the copy number of a genetic determinant

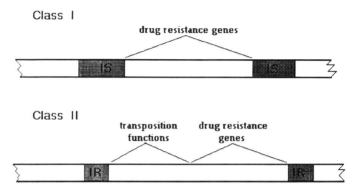

Fig. 9.1. Schematic of class I and class II transposons. In class I transposons, the drug resistance genes are bounded by insertion sequences (ISs). In class II, transpositional units are characterized by inverted repeats (IRs).

intracellularly, it also enables genetic determinants to be transferred horizontally if the determinant is transposed onto conjugative plasmids.

Outside of the laboratory environment little is known regarding the effect of environmental factors on transposition. It has been shown that the compound being detoxified by the operon may induce transposition but this is not always the case (6). Furthermore, many stressor conditions such as temperature and nutrients are believed to play an integral role in stimulating transposition (9,10).

2. EXTRACELLULAR GENETIC TRANSFER

Genetic material can be passed from one organism to another by transformation, conjugation, or transduction. Each of these mechanisms is described in Figure 9.2. Conjugation is the most understood, although our knowledge about activity under natural conditions is still limited. Recent work by the U.S. Environmental Protection Agency has provided valuable and needed insights into all means of DNA transfer among different microorganisms (11,12). The work has focused on the hazards of release of genetically engineered organisms and the probability that introduced DNA would be transferred by any of the mechanisms described below. However, an understanding of genetic ecology can be used to maximize the distribution of genes among members of the bacterial community through any of these mechanisms, in order to improve the environment, and is discussed in a later section of this chapter.

A number of physical and chemical factors affect the various types of gene exchange in the environment. These include temperature, pH, nutrient concen-

Conjugation

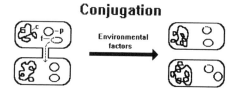

Transformation

Transduction

Fig. 9.2. Mechanisms of extracellular gene exchange. c, chromosome; f, foreign DNA; p, plasmid.

tration, redox potential, cation exchange capacity, as well as types of clays present in soils and sediments (13). Other factors that may be of importance are the geographical origin of the recipients and donors and the ability of seeded bacteria to function in the same manner as natural bacteria in a given environment (14).

2.1. Conjugation

Conjugation is the process of intercellular transfer of DNA and requires cell to cell contact. Conjugation is regulated by specific genes that are usually located on a plasmid. A plasmid may promote transfer if it contains these genes. Plasmids that do not contain such transfer genes are referred to as *nonconjugative,* meaning that they cannot promote their own transfer. Cells containing conjugative plasmids are considered donor cells. Once recipient cells receive a conjugative plasmid they become donors. Nonconjugative plasmids can be transferred if conjugative plasmids are contained in the same cell. Chromosomal DNA can be mobilized especially if the plasmid becomes integrated into the chromosome. Horizontal gene transfer occurs in gram-positive bacteria by a slightly different system. Because many of the currently identified bacteria capable of degrading toxic organics are gram negative, our discussion of conjugation is limited to gram-negative systems. But this limitation is most likely an artifact of the level of knowledge. The genes involved in conjugation can be divided generally into those encoding cell–cell contact, conjugal DNA metabolism, control of conjugation, and specific transfer origins.

Different plasmids are transferred from donors to recipients at different rates. For example, there are high-frequency transfer plasmids in which transfer occurs in 100% of the recipients and very low transfer frequencies where one recipient in 10^7 will receive the plasmid. In consideration of a community's ability to transfer DNA, both of these factors (transfer ability and frequency) must be taken into account. Increasing the amount of DNA in the community through conjugation is a feasible process, but there are limitations. Therefore, each system under study must be separately evaluated to determine if a specific genetic determinant can be transferred via conjugation.

Most environmental studies have been conducted using donor and recipient strains in microcosms or pure cultures. It has been found that bacterial conjugation is stimulated under simulated environmental conditions by optimizing pH, temperature, and nutrient concentration (15–17). It also appears from these results that conjugation is geographically based, with conjugation occurring between 100% of the donors from one location and recipients from the same location versus 60% between donors from one location and recipients from another (14).

Few studies have been conducted *in situ*. Those involving the epilithon (slime layers on rocks in streams) have shown that conjugation readily occurs and that low temperature reduces the frequency of gene exchange but does not eliminate it even at 4°C (18).

2.2. Transformation

Transformation is the incorporation of naked DNA into a cell. It is widely used in the laboratory in the creation of nucleic acid probes, cloning, and the development of genetically engineered organisms. It has been demonstrated in a wide range of microorganisms in pure culture studies. Two factors are important for successful transformation. These are the ability of the host cell to take up the DNA (called *competence*) and the integration into the host cell DNA. Many cells contain endonucleases that destroy foreign DNA. The competence factor is thought to allow the foreign DNA to attach nonspecifically to the cell surface.

The internal control of competence shift, such as starvation and unbalanced growth in natural environments, could favor the induction of competence in transformable bacteria. The competency of certain bacteria in natural environments requires an appropriate concentration of a competence factor that is subject to degradation by proteases (19), leading to increasing frequency of transformation. Additionally, the following environmental factors may affect the binding of DNA to the cells: 1) ionic strength, 2) pH, 3) shear force, and 4) DNase. In natural habitats the enzymatic digestion of transforming DNA plays a significant role in reducing transformation potential (20,21).

2.3. Transduction

Viral DNA is introduced into bacterial cells via transduction. Bacterial viruses are classified as temperate or virulent. In the former the viral DNA may be incorporated into the host chromosome or plasmid or exist in the cytoplasm. It can also enter a lytic phase where bacterial DNA may become associated and transferred with the viral DNA. In the latter the viral DNA utilizes the host cell machinery to replicate itself, resulting in cell lysis and dissemination of the viral particles. Again a variety of physical and chemical factors in the natural environment will affect frequency. Transduction has been shown to occur in microcosms inserted into natural aquatic environment habitats (22).

3. GENETIC MODIFICATION

The use of restriction enzymes has enabled molecular biologists to map DNA and to isolate particular DNA fragments. This technique can give us information about relatedness and polymorphism among microorganisms of the same genus. Following digestion with restriction enzymes, one can join fragments and clone particular DNA segments. These technqiues as a whole are now called *recombinant DNA technology* or *genetic engineering*.

3.1. Molecular Cloning

The two basic stages involved in recombinant DNA techniques are 1) joining a DNA fragment of interest to another DNA molecule that is capable of replicating (vector or cloning vehicle) and 2) introducing the fragment into a bacterium that allows propagation of the joined unit. Plasmids are widely used as cloning vectors because of the following desirable properties: 1) low molecular weight, 2) multiple copies, 3) capability of expressing readily selectable phenotypic traits on host cells, and 4) single sites for many restriction endonucleases in genes with a scorable phenotype.

The low molecular weight makes the cloning vehicles much easier to handle and more resistant to damage by shearing during the isolation procedures. Multiple copies of a vector are useful for gene cloning, since a high yield of cloned DNA is desirable. It is advantageous if insertion of foreign DNA inactivates a gene whose phenotype is readily scorable, and in this way it is possible to distinguish strains with recombinant DNA from those containing self-annealed plasmids. Both naturally occurring and constructed plasmids have been used as cloning vehicles. For example, pSC101, a "natural" plasmid, has been used for cloning to detect expression of *Staphylococcus* plasmid genes in *Escherichia coli* (23) and to clone eukaryotic DNA in *E. coli* (24). Plasmid pBR322 is a typical example of an artificial cloning vector, which is constructed from DNA fragments derived from three different plasmids (25). This plasmid has been widely used as a model system for the study of prokaryotic transcription and translation. Itakura *et al.* (26) used pBR322 as a vector for expression in *E. coli* of a chemically synthesized gene for the hormone somatostatin. A functional gene product from a chemically synthesized gene indicated that genetic engineering had turned yesterday's dream into today's reality.

The success of plasmids in recombinant DNA research led also to the use of temperate phages as vectors, because many phages had already demonstrated their ability to act as transducing agents between donor and recipient bacteria. Lambda is an extensively studied virus of *E. coli*. The DNA of lambda is a linear duplex molecule of about 48.5 kb, which is larger than most plasmid vectors, and its entire DNA sequence has been determined (27). Because wild-type lambda DNA contains several target sites for most of the commonly used restriction endonucleases, it is not suitable as a vector. Derivatives of the wild type have been produced to overcome this problem. Sometimes the goal of gene manipulation is to promote a cloned gene to amplify the synthesis of a desirable gene product. The foreign genes can be inserted into the chromosome of phage lambda derivatives, and they can be expressed very efficiently through lambda promoters (28,29).

Additionally, plasmids have been constructed to contain a fragment of lambda DNA including the *cos* site (30), which is required for recognition by the phage packaging system. These plasmids have been termed *cosmids* and can be used as cloning vectors for large fragments of eukaryotic DNA via the packaging system. Technically, the cosmids are more difficult to use than the lambda phage vectors.

3.2. Molecular Breeding

Recombinant DNA technology has recently been used to introduce exogenous DNA into cultured animal fertilized eggs (31) or plant protoplasts (32) to produce transgenic animals or plants. The production of transgenic mice has now become a routine procedure in a large number of laboratories. Although there are still considerable technical difficulties, gene transfer has been accomplished in farm animals (33,34). The technique may offer opportunities for the researchers to investigate the genetic improvement of traditionally important characteristics such as growth rate and reproductive performance. Clark *et al.* (35) reported that the corresponding protein of a functional human specific RNA was found in the milk of transgenic sheep. The expression of foreign DNA in transgenic mice has been well documented (36–38).

During the past decade there have been rapid developments in transgenic plants using DNA virus or bacterial plasmids as cloning vectors for exogenous DNA. Very intensive research has been conducted on the vector system of the bacterium *Agrobacterium tumerfaciens* and alternatively on transfection of plant protoplasts. The *Agrobacterium* system can be utilized for the transfer of desired genes to plant cells. Traits transmitted to plants through *Agrobacterium* include herbicide resistance (39) and resistance to viruses (40) or insects (41). This indicates that plant molecular biology has enormous biotechnological implications in the creation of plants with useful genetically engineered characteristics. The application of *Agrobacterium* for crop improvement is less limited by the vector system than by the *in vitro* cell culture technology needed to regenerate the transformed cells or transfected protoplasts into complete fertile plants. However, progress in the latter area is rapid and it is believed that in the near future transgenic plants will be obtained from many crops.

4. TRACKING GENES IN THE ENVIRONMENT

Genes present a unique way to assess specific functions in the environment. Although genes share many base pairs in common, the arrangement within each is unique enough to allow us to distinguish among genes on the basis of DNA sequences.

4.1. Nucleic Acid Probes

Nucleic acid probes can be constructed to identify genes that are specific for the information of interest. A basic application of nucleic acid probes is the detection of specific DNA sequences coding for certain traits and the identification of microorganisms. DNA probes have been used to detect *Salmonella* and enterotoxigenic *E. coli* in food (42), to screen water samples for enterotoxigenic *E. coli* (43), to measure the densities of nitrogen-fixing microorganisms in soil bacterial communities (44), to detect mercury resistance genes in gram-negative bacteria (45) and catabolic genotypes in environmental bacterial communities (46). Although RNA probes have been reported to show higher sensitivity in cytological studies (47), they are not commonly used because they are less stable and harder to construct than DNA probes. However, they are frequently used in enterovirus detection.

It is necessary to consider the methods available for labeling nucleic acids in order to detect positive responses easily. The length of a nucleic acid probe can vary from 20 to thousands of base pairs. The short oligonucleotides may achieve higher specificity than the long nucleotides that could cause a false-positive result because of nonspecific hybridization. There are several basic methods for labeling nucleic acids: 1) nick translation (48), 2) primer extension (49), 3) end-labeling (50), and 4) methods based on phage RNA polymerase (51).

Because the density of bacteria in natural habitats is normally low and therefore sequences of interest are also very low in abundance, high probe sensitivity is required. Generally, a radioactive probe with a highly sensitive radionuclide such as ^{32}P will be used to detect the scarce target nucleic acid sequence by autoradiography (52).

A number of nonradioactive labels for probes are also available, but the only one in widespread use is biotin, which can be incorporated into nucleic acid fragments by nick translation or end-labeling with biotinylated nucleotides (biotin-11-dCTP or biotin-11-UTP) as the substrate (53,54). Recently, a nonradioactive enhanced chemiluminescence method (ECL) was developed to detect nucleic acids hybridized on both nylon and nitrocellulose membranes (55). This ECL gene detection system has shortened the labeling time and increased the sensitivity of nonradioactive probes. Although the nonradioactive labels suffer the disadvantage of low sensitivity and high background, they possess obvious advantages, including 1) no hazard associated with radioactivity, 2) longer probe life, 3) rapid detection without autoradiography, and 4) inexpensive reagents.

4.2. Methods for Determining Occurrence and Quantification of DNA and RNA

Of basic importance to determining gene occurrence or genetic potential *in situ* is the ability to extract and purify nucleic acids. The application of nucleic acid extraction methods to environmental samples can obviate the need for cell cultivation, which has the disadvantage of yielding only a very small proportion of the total microbial community. Numerous techniques have been described for direct detection and extraction of DNA from aquatic environments (56–61) and from soil and sediments (44, 62–64). Table 9.1 shows a comparison of four recent methods for DNA extraction. For the extraction of DNA from soil and sediment environments, direct DNA extraction is very efficient (65).

Direct extraction of RNA from a cow's ruminal fluid has been successfully used to study phylogenetic relationships of *Bacteroides* sp. in ruminal microbial communities (66). Furthermore, total RNA extracted from a hot-spring cyanobacterial mat has been used to study species-specific 16S rRNA sequences in natural microbial consortia (67,68). Recently, a rapid method for direct extraction of mRNA from soil has been developed (69) that will allow us to determine *in situ* gene expression. The study of environmental bacterial isolates has shown

TABLE 9.1. Methods of Direct Extraction of DNA From Soil and Sediments[a]

	Ogram et al. (62)	Holben et al. (44)	Tsai and Olson (64)	Rochelle and Olson (81)
Substrates	Sediments	Soil	Soil and sediments	Soil and sediments
Sample size	100 g	50 g	1 g	1 g
Gel embedding	No	No	No	Yes
Cell extraction	No	Yes	No	No
Lysis	SDS	Lysozyme	Lysozyme	Lysozyme
	Bead-beating	Pronase	SDS	SDS
		Sarkosyl	Freeze-thawing	
Phenol-chloroform extraction	Yes	No	Yes	No
Time required	18–20 hours	20–24 hours	6–8 hours	20–24 hours
Recovery efficiency	ND	33%–35%	>90%	>80%

[a]SDS, sodium dodecyl sulfate; ND, not determined.

that the gene expression of metal resistance can be enhanced in the presence of the specific metal in pure cultures and microcosms (3,7).

Because methods for direct extraction of DNA and RNA have been developed, it is possible to determine the occurrence and expression of a gene of interest if the nucleic acid probes are available. The gene dosage or mRNA can also be quantified by radiolabeled nucleic acid hybridization technology and the use of radioanalytical imaging systems (Fig. 9.3). In a practical sense, these newly developed methods not only can detect the presence of a particular microbial gene but also can measure its activity in the natural environment. This will help microbial ecologists to understand the *in situ* cycling of elements as well as biodegradation activities in chemically polluted environments. One potential drawback to using this technique in environmental samples is the question of viability of the microorganisms. There are no good estimates for the survival of free RNA or DNA in the environment. Messenger RNA is so short lived that this could become a problem in extraction. All studies should determine the rate of degradation for accurate quantification. Certain researchers have approached this problem by keeping the organisms intact for as long as possible in the extraction process. The full application of these techniques to public health issues requires that this problem be resolved.

4.3. Nucleic Acid Hybridization

Techniques of nucleic acid hybridization have been widely used to detect specific microorganisms in environmental samples by means of nucleic acid

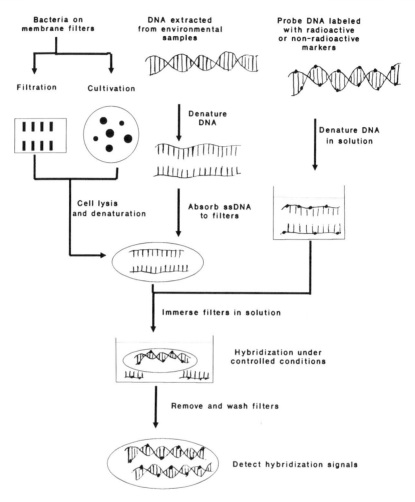

Fig. 9.3. Use of nucleic acid hybridization for microbial detection in natural environments.

probes to target DNA or RNA sequences (Fig. 9.3). This technology is potentially specific and sensitive, can be automated, and is now the basis for a range
of techniques in use in microbial ecology. The rationale behind all the methods
used in nucleic acid hybridization is the double helix with two complementary
DNA strands. Hybridization is currently used to describe the *in vitro* formation
of sequence-specific base-paired duplexes from a combination of nucleic acid
fragments.

The melting temperature (T_m), at which the strands are half dissociated, is used to measure the stability of a DNA–DNA or DNA–RNA hybrid. The rate of reassociation of the duplex is affected by temperature, concentration of salt, base mismatch, fragment length, and complexity of the DNA (52).

High stringency hybridization conditions are often used to detect specific DNA sequences (45) from bacteria in natural habitats. However, recent research indicates that evolutionary divergence occurs in some genes such as *merR* and *merA*. These findings indicate that in natural environments it may be more appropriate to use a lower stringency than 90% in order to accommodate the evolutionary divergence that has occurred. However, caution must be applied in these situations because as stringency is lowered the possibility for cross-reactivity with unrelated genes increases. Two studies have recently indicated the divergence of mercury genes in the environment (14,70). To relax stringency, nucleotide sequence data of the probe being used must be available because T_m is based on the GC ratio (52).

The hybridization reaction can be carried out using two basic formats: 1) with either one of the two nucleic acids immobilized to a filter or 2) with both nucleic acids in solution. Solutions are especially useful in detecting small amounts of target DNA against a large background of DNA from environmental samples (71). The ease of removing nonhybridized probe DNA by washing also explains the popularity of using filter hybridization in detecting gene occurrence in environmental samples (44,62). One use of direct extraction of DNA and T_m is the study of community relatedness. Bacterial DNA extracted from various locations has been allowed to reanneal. Researchers found that the more rapidly reannealing occurred, the more closely related the bacterial DNA (72,73).

The variety of ways that nucleic acid probes are used is shown in Table 9.2. Probes can be universal in origin (identify a class or group of organisms) or specific (identifying a single genus and species). Identification is usually based on ribosomal RNA or on specific genes that are unique to the organism. Each rRNA unit (*i.e.,* 5S, 16S, or 23S) in bacteria has a region of RNA that is unique and another that is conserved. Using homologous probes to the conserved regions allows the study of the relatedness of bacteria within a genus or family such as *Enterbacteraceae*. The region of variation can also be used to separate the level of substrains.

Nucleic acid probes to structural and regulatory genes are commonly used to study function. They allow the determination of potential for a specific activity. Functions include catabolic processes, nitrogen fixation, and detoxification.

In the development of a nucleic acid probe several factors must be considered. These include specificity, the abundance of target sequence, hybridization strategy, and the sensitivity of labeled probe (radioactive or nonradioactive). Once an adequate probe has been developed, it is important to ensure that the other methodologies applied in assessing gene or organism occurrence are properly

TABLE 9.2. **Applications of Nucleic Acid Probes in Microbial Ecology Studies**

Probes	Examples	References
Polynucleotide		
DNA	Detection of *Salmonella* spp. in foods	42
	Determining genotypic adaptation of sediment microbial communities to mercury stress	82
	Detection of an engineered DNA sequence in soil	83
	Detection of *Salmonella* spp. in estuaries	84
	Detection of polychlorinated biphenyl-degrading genotypes in contaminated soils	85
	Detection of heavy metal–resistant bacteria from soils	86
	Identification of *Mycobacterium* sp. from fecal or intestinal tissue samples from animals	87
cDNA	Identification of mRNAs that are differentially or developmentally regulated in different microorganisms	88,89
	Detection of hepatitis A virus in estuarine samples	90
	Detection of plum pox virus in plants	91
	Detection of papaya mosaic potexvirus in plants	92
RNA	Detection of potato spindle tuber viroid	93
	Detection of heat-stable enterotoxigenic *E. coli*	94
	Detection of species-specific microorganisms from natural microbial communities	67
Oligonucleotide		
	Identification of species- and subspecies-specific *Proteus* spp.	95
	Detection of genetically engineered bacteria in environmental samples	96
	Identification of species-specific *Clostridium*	97
	Detection of species-specific *Legionella pneumophila* in water	78
	Identification of *Frankia* strains in root nodules	98
	Detection and identification of enteroviruses	99
	Detection and identification of enterotoxin-producing *Staphylococcus aureus* from clinical isolates	100

controlled. Factors affecting hybridization are cell cultivation, cell lysis, and stringency. Cell cultivation greatly reduces the ability to recover all possible genotypes present. It is estimated that at most cultivation media recover less than 0.1% of the microorganisms present (74). To overcome this problem, a number of techniques have been developed for the direct extraction of DNA as described in the previous section.

4.4. Environmental Studies

Very little is known about gene density in natural environments. Most studies have focused on gene occurrence over a relatively short time period, used only high stringency, and have not dealt with the importance of expression accompanying gene occurrence unless selective media or conditions were used.

It is important to take into consideration both internal and external variations of methodology and sample material in the design and analysis of gene occurrence, especially if polluted environments are being examined. These habitats are often very heterogeneous in the distribution of the pollutant of interest and therefore most likely also the gene of interest.

Furthermore, if colony hybridization is being used, it is important to ensure that cell lysis has occurred. The specificity of the probe under study and the number of colonies need to be assessed in order to make some statistically valid statement regarding occurrence within an environment. Beyond these considerations, those of the isolation medium and its selective nature and their effects on the questions being asked need to be determined.

Other factors become important if direct extraction techniques are being used in a quantitative manner. The lysis efficacy is important. The recovery efficiency of DNA or RNA from environmental samples as well as the variability of recovery efficiency among sample types must be determined. Proteinaceous material in animals and humic and fulvic acids in waters, sediments, and soils may dramatically affect recovery and hybridization efficiency. Interference by environmental chemicals, for example, may block hybridization. The heterogeneous nature of pollutants often leads to low recovery of nucleic acid material. An even more contentious problem is the likelihood of obtaining both low yield of DNA or RNA and variable results among samples taken in close proximity. As these problems increase, more replicate sampling is required to ensure validity of the data.

5. GENETIC ECOLOGY

Genetic ecology is defined as the study of the interrelationships of specific genetic determinants with their environments. Its foundation is in population ecology, molecular biology, and microbiology. It is concerned with the study of structural and regulatory genes and how they can be managed *in situ* to maximize expression (production of enzyme). The application of genetic ecology to environmental problems draws heavily on the field of engineering. The success of the approach is founded in the inherent stability of existing microbial populations in the environment, the conservation of important genetic information within those

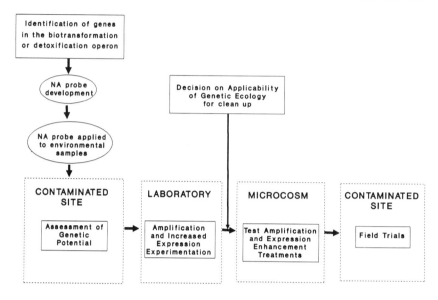

Fig. 9.4. Application of genetic ecology to a contaminated site to determine the appropriate remediation technology.

TABLE 9.3. Polymerase Reaction Applied to Environmental Samples

Organism	Sample type	Primer	References
E. coli	Water	Single set	77
Legionella	Water	Single set	78
Enteroviruses, panspecific	Water	Nested and single sets	101
Rotavirus	Feces	Single set	102
Adenovirus	Feces	Single set	103

populations, and their ability to utilize these characteristics when the environment requires. Thus genetic ecology seeks to identify those physical and chemical properties of the environment that trigger ecological responses and to utilize these properties to direct certain aspects of the chemical activity of a given environment. Stated more simply, one goal of genetic ecology is to identify physicochemical parameters that affect both gene amplification and expression in given environments. It is used to investigate the potential of utilizing genes in microbial populations to direct chemical processes in the environment to decrease pollution.

Another potential benefit from genetic ecology in the remediation of various types of pollutants is the amplification of the structural genes encoding enzymes that are capable of carrying out the biochemical pathway of interest. Transposition may be very important in amplification of genes in natural environments. Through transposition the gene copy number may increase, increasing the production of critical enzymes. The success of the approach is favored by the evidence that many operons that control the degradation or detoxification of pollutants are associated with transposons.

A genetic ecology field approach is shown in Figure 9.4. There is an intimate link between field assessment data and laboratory enhancement feasibility studies, culminating in field treatment schema. Thus far the technology has only progressed to the laboratory phase.

6. OTHER USES OF BIOTECHNOLOGY TO ASSESS AND CONTROL POLLUTION

6.1. Polymerase Chain Reaction (PCR)

The largest potential application of PCR in the environment is in the identification of rare genes, genetically modified DNA, or pathogens. In theory PCR permits extraction and amplification of a single DNA molecule from complex mixtures to give a detectable signal and in many cases produces enough product to carry out further manipulations. It involves the *in vitro* synthesis of nucleic acids by which a particular sequence is replicated (Fig. 9.5). In the polymerase reaction, each cycle doubles the amount of target DNA present; thus there is an exponential increase of target DNA (2^n, where n is the number of cycles). PCR technology is new, with the first DNA sequence having been amplified in 1985 (75). The applications of PCR in the detection of pathogens are summarized in Table 9.3. The application in the medical field is extensive, but fewer uses have been applied to the environment. However, with the advent of direct extraction procedures (Table 9.1), it may well be possible to detect a single organism in an environmental sample or perhaps several copies of a gene.

As with any new technology there are numerous problems that may arise particularly if amplification is proceeding with an extract from soils, sediments, or sludges. Common problems include but are not limited to low yield of desired product, high background caused by mispriming, formation of primer dimers that limit the amount of amplification product produced, and mutations or heterogeneity caused by misincorporation of foreign DNA sequences. It is important to be aware that the polymerase enzymes available can have sequences in common with the target DNA. This can cause false-positive results or limit the ability to detect single sequences.

The use of PCR in public health, especially in aquatic environments, is likely to revolutionize the field within the next decade. It is likely that a water or waste

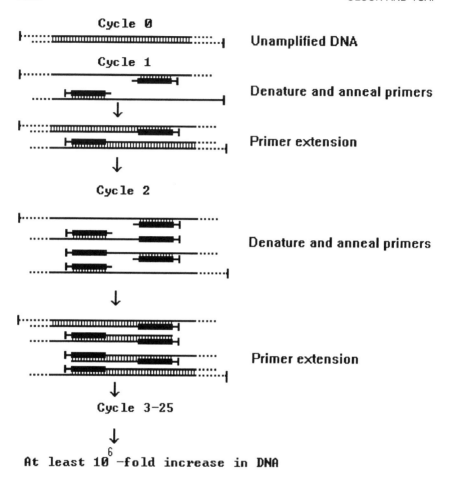

Fig. 9.5. Amplification of a DNA sequence using polymerase and thermocycling.

water utility will be able to screen for a variety of pathogens using molecular techniques. Already nucleic acid probes exist for a wide range of bacteria and viruses (Table 9.2). Fewer probes have been developed for the protozoan pathogens that are important in drinking water, such as *Cryptosporidium, Giardia,* and *Entamoeba histolytica.* Currently, PCR has been used to detect a variety of human enteroviruses (76), pathogenic varieties of *E. coli,* and other environmentally important bacteria such as *Aeromonas* and *Legionella* and is being developed for use with protozoans such as *Giardia* (77–79). The application of PCR to microorganisms of public health concern again raises the question of viability/infectivity. PCR techniques try to keep organisms intact for as long as possible during the extraction procedure to minimize the potential nucleic acid from nonviable cells to be recovered.

Interestingly, in the Bej *et al.* (77) study it was found that the *lacZ* gene that was supposed to be specific for *E. coli* was also present in *Shigella*. This finding points out the importance of careful survey work regarding the cross-reactivity of genetic systems, especially for those genera that are likely to exchange genetic information via conjugation, transformation, or transduction. The application of PCR to microorganisms of public health importance and those strains relevant to detoxification or biodegradation of organic pollutants or metals should proceed cautiously because of very limited knowledge on sequence homology among these groups.

6.2. Biosensors

The concept of biosensors or reporters in the environment suggests the use of bacteria to determine the availability of a compound or the ''health'' status of an environment through the sequestering of bacteria in a probe that is used *in situ* and only allows chemicals to come in contact with the bacteria. The development of this technology is very recent and has not been commercialized. It promises to be important in generating future standards regarding both bioavailability and toxicity of emissions being released into the environment and detection of the ability of bacteria or fungi to utilize pollutants *in situ*.

One newly developed reporter system uses the genes that control luminescence. These genes are placed behind the promoter for a specific operon of interest. The promoter controls the regulation of the operon and will ''turn on'' if the inducer compound or the substrate is being utilized. When the promoter is on, a light signal is transmitted to a photodetector (4). This system is highly sensitive in determining the availability of pollutants *in situ*. It is not quantitative, but rather indicates potential activity. Research conducted in field trials of gasoline bioremediation sites has indicated that bacteria may be starved or injured because of the production of toxic intermediates and thus unable to utilize the substrate even if it is available for bacterial action (80). Reporter systems in this instance might produce results indicating that degradation can occur, when in reality factors other than substrate availability are limiting. The technique can increase our understanding of the availability of pollutants in the environment and the potential for microbial activity.

REFERENCES

1. Mondello FJ: Cloning and expression in *Escherichia coli* of *Pseudomonas* strain LB400 genes encoding polychlorinated biphenyl degradation. J Bacteriol 171:1725, 1989.

2. Coyne MS, Arunakumari A, Averill BA, Tiedje JM: Immunological identification and distribution of dissimulatory heme cd1 and nonheme copper nitrite reductases in denitrifying bacteria. Appl Environ Microbiol 55:2924, 1989.

3. Tsai Y-L, Olson BH: Effects of Hg^{2+}, CH_3-Hg^+, and temperature on the expression of mercury resistance genes in environmental bacteria. Appl Environ Microbiol 56:3266, 1990.

4. King JMH, DiGrazia PM, Applegate B, Burlage R, Sanseverino J, Dunbar P, Larimer F, Sayler GS: Rapid, sensitive bioluminescent reporter technology for naphthalene exposure and biodegradation. Science 249:778, 1990.

5. Hershfield V, Boyer HW, Yanofsky C, Lovett MA, Helinski DR: Plasmid ColE1 as a molecular vehicle for cloning and amplification of DNA. Proc Nat Acad Sci USA, 71:3455, 1974.

6. Brown NL, Ford SJ, Pridmore RD, Fritzinger DC: Nucleotide sequence of a gene from the *Pseudomonas* transposon Tn501 encoding mercuric reductase. Biochemistry 22:4089, 1983.

7. Ogunseitan OA, Olson BH: Selection for multiple copies of the mercuric reductase gene in bacteria exposed to high Hg^{2+}. Environ Sci Technol (in press).

8. Ogunseitan OA, Delgado IL, Tsai Y-L, Olson BH: Effect of 2-hydroxybenzoate on the maintenance of naphthalene-degrading pseudomonades in seeded and unseeded soil. Appl Environ Microbiol 57:xxx, 1991 (in press).

9. Berg DE: Transposable elements in prokaryotes. In Levy SB, Miller RV (eds): Gene Transfer in the Environment. New York: McGraw-Hill Publishing, 1989, p 99.

10. Sherratt D: Tn3 and related transposable elements: Site-specific recombination and transposition. In Berg DE, Howe MM (eds): Mobile DNA. Washington, DC: American Society for Microbiology, 1989, p 163.

11. Levin M, Seidler R, Rogul M (eds): Microbial Ecology: Principles, Applications and Methods. New York: McGraw-Hill, 1991.

12. Levy S, Miller RV (eds): Gene Transfer in the Environment. New York: McGraw-Hill, 1989.

13. Stotzky G: Gene transfer among bacteria in soil. In Levy SB, Miller RV (eds): Gene Transfer in the Environment. New York: McGraw-Hill, 1989, p 165.

14. Rochelle PA, Wetherbee M, Olson BH: Distribution of DNA sequences encoding narrow- and broad-spectrum mercury resistance. Appl Environ Microbiol 57:1581, 1991.

15. Kelly WJ, Reanney DC: Mercury resistance among soil bacteria: Ecology and transferability of genes encoding resistance. Soil Biol Biochem 16:1, 1984.

16. Schofield PR, Gibson AH, Dudman WF, Watson JM: Evidence for genetic exchange and recombination of *Rhizobium* symbiotic plasmids in a soil population. Appl Environ Microbiol 53:2942, 1987.

17. Richaume A, Angle JS, Sadowsky MJ: Influence of soil variables on in situ plasmid transfer from *Escherichia coli* to *Rhizobium fredii*. Appl Environ Microbiol 55:1730, 1989.

18. Bale MJ, Fry JC, Day MJ: Transfer and occurrence of large mercury resistance plasmids in river epilithon. Appl Environ Microbiol 54:972, 1988.

19. Stewart GJ: The mechanism of natural transformation. In Levy SB, Miller RV (eds): Gene Transfer in the Environment. New York: McGraw-Hill, 1989, p 139.

20. Aardema BW, Lorenz MG, Krumbein WE: Protection of sediment-adsorbed transforming DNA against enzymatic inactivation. Appl Environ Microbiol 46:417, 1983.

21. Stewart GJ, Sinigalliano CD: Detection of horizontal gene transfer by natural transformation in native and introduced species of bacteria in marine and synthetic sediments. Appl Environ Microbiol 56:1818, 1990.

22. Saye DJ, Ogunseitan OA, Sayler GS, Miller RV: Transduction of linked chromosomal genes between *Pseudomonas aeruginosa* strains during incubation in situ in a freshwater habitat. Appl Environ Microbiol 56:140, 1990.

23. Chang ACY, Cohen SN: Genome construction between bacterial species *in vitro:* Replication and expression of *Staphylococcus* plasmid genes in *Escherichia coli.* Proc Natl Acad Sci USA 71:1030, 1974.

24. Morrow JF, Cohen SN, Chang ACY, Boyer HW, Goodman HM, Helling RB: Replication and transcription of eukaryotic DNA in *Escherichia coli.* Proc Natl Acad Sci USA 71:1743, 1974.

25. Bolivar F, Rodriguez RL, Greene PJ, Betlach MC, Heyneker HL, Boyer HW, Crosa JH, Falkow S: Construction and characterization of new cloning vehicles. II. A multipurpose cloning system. Gene 2:95, 1977.

26. Itakura K, Hirose T, Crea R, Riggs AD, Heyneker HL, Bolivar F, Boyer HW: Expression in *Escherichia coli* of a chemically synthesized gene for the hormone somatostatin. Science 198:1056, 1977.

27. Sanger F, Coulson AR, Hong GF, Hill DF, Peterson GB: Nucleotide sequence of bacteriophage lambda DNA. J Mol Biol 162:729, 1982.

28. Panasenko SM, Cameron JR, Davis RW, Lehman IR: Five hundredfold overproduction of DNA ligase after induction of a hybrid lambda lysogen constructed *in vitro.* Science 196:188, 1977.

29. Murray NE, Bruce SA, Murray K: Molecular cloning of the DNA ligase gene from bacteriophage T4. II. Amplification and preparation of the gene product. J Mol Biol 132:493, 1979.

30. Collins J, Hohn B: Cosmids: A type of plasmid gene-cloning vector that is packageable *in vitro* in bacteriophage lambda heads. Proc Natl Acad Sci USA 75:4242, 1979.

31. Palmiter RD, Brinster RL: Germ-line transformation of mice. Annu Rev Genet 20:465, 1986.

32. Chupeau M-C, Bellini C, Guerche P, Maisonneuve B, Vastra G, Chupeau Y: Transgenic plants of lettuce (*Lactuca sativa*) obtained through electroporation of protoplasts. BioTechnology 7:503, 1989.

33. Hammer RE, Pursel VG, Rexroad CE, Wall JR, Bolt DJ, Ebert KM, Palmiter RD, Brinster RL: Production of transgenic rabbits, sheep and pigs by microinjection. Nature 315:680, 1985.

34. Simons JP, Wilmut I, Clark AJ, Archibald AL, Bishop JO, Lathe R: Gene transfer into sheep. BioTechnology 6:179, 1988.

35. Clark AJ, Bessos H, Bishop JO, Brown P, Harris S, Lathe R, McClenaghan M, Prowse C, Simons JP, Whitelaw CBA, Wilmut I: Expression of human anti-hemophilic factor IX in the milk of transgenic sheep. BioTechnology 7:487, 1989.

36. Brinster RL, Chen HY, Trumbauer M, Senear AW, Warren R, Palmiter RD: Somatic expression of herpes thymidine kinase in mice following injection of a fusion gene into eggs. Cell 27:223, 1981.

37. Brinster RL, Chen HY, Warren R, Sarthy A, Palmiter RD: Regulation of metallothionein–thymidine kinase fusion plasmids injected into mouse eggs. Nature 296:39, 1982.

38. Palmiter RD, Brinster RL, Hammer RE, Trumbauer ME, Rosenfeld MG, Birnberg NC, Evans RM: Dramatic growth of mice that developed from eggs microinjected with metallothionein–growth hormone fusion genes. Nature 300:611, 1982.

39. DeGreef W, Delon R, DeBlock M, Leemans J, Botterman J: Evaluation of herbicide resistance in transgenic crops under field conditions. BioTechnology 7:61, 1989.

40. Lawson C, Kaniewski W, Haley L, Rozman R, Newell C, Sanders P, Tumer NE: Engineering resistance to mixed virus infection in a commercial potato cultivar: Resistance to potato virus X and potato virus Y in transgenic Russet Burbank. BioTechnology 8:127, 1990.

41. Delannay X, LaVallee BJ, Proksch RK, Fuchs RL, Sims SR, Greenplate JT, Marrone PG, Dodson RB, Augustine JJ, Layton JG, Fischhoff DA: Field performance of transgenic tomato plants expressing the *Baccilus thuringiensis* var. *Kurstaki* insect control protein. BioTechnology 7:1265, 1989.

42. Fitts R, Diamond M, Hamilton C, Neri M: DNA–DNA hybridization assay for detection of *Salmonella* spp. in foods. Appl Environ Microbiol 46:1146, 1983.

43. Echeverria P, Seriwatana J, Chityothin O, Chaicunpa W, Tirapat C: Detection of enterotoxigenic *Escherichia coli* in water by filter hybridization with three enterotoxin gene probes. J Clin Microbiol 16:1086, 1982.

44. Holben WE, Jansson JK, Chelm BK, Tiedje JM: DNA probe method for the detection of specific microorganisms in the soil bacterial community. Appl Environ Microbiol 54:703, 1988.

45. Barkay T, Fout DL, Olson BH: Preparation of a DNA gene probe for detection of mercury resistance gene in gram-negative bacterial communities. Appl Environ Microbiol 49:686, 1985.

46. Sayler GS, Shields MS, Tedford ET, Breen A, Hooper SW, Sirotkin KM, Davis JW: Application of DNA–DNA colony hybridization to the detection of catabolic genotypes in environmental samples. Appl Environ Microbiol 49:1295, 1985.

47. Cox KH, DeLeon DV, Angerer LM, Angerer RC: Detection of mRNAs in sea urchin embryos by *in situ* hybridization using asymmetric RNA probes. Dev Biol 101:485, 1984.

48. Rigby PWJ, Dieckmann M, Rhodes C, Berg P: Labelling deoxyribonucleic acid to high specific activity *in vitro* by nick translation with DNA polymerase I. J Mol Biol 113:237, 1977.

49. Feinberg AP, Vogelstein B: A technique for radiolabelling DNA restriction endonuclease fragments to high specific activity. Anal Biochem 132:6, 1983.

50. Downing RG, Dubbleby CJ, Villems R, Broda P: An endonuclease cleavage map of the plasmid pWWO-8, a derivative of the TOL plasmid of *Pseudomonas putida* mt-2. Mol Gen Genet 168:97, 1979.

51. Butler ET, Chamberlain MJ: Bacteriophage SP6–specific RNA polymerase I isolation and characterization of the enzyme. J Biol Chem 257:5772, 1982.

52. Sambrook J, Fritsch EF, Maniatis T: In Molecular Cloning: A Laboratory Manual, 2nd ed. Cold Spring Harbor, NY: Cold Spring Harbor Laboratory, 1989.

53. Langer PR, Waldrop AA, Ward DC: Enzymatic synthesis of biotin-labeled polynucleotides: Novel nucleic acid affinity probes. Proc Natl Acad Sci USA 78:6633, 1981.

54. Leary JJ, Brigati DJ, Ward DC: Rapid and sensitive colorimetric method for visualizing biotin-labeled DNA probes hybridized to DNA or RNA immobilized on nitrocellulose: Bioblots. Proc Natl Acad Sci USA 80:4045, 1983.

55. Durrant I, Benge LCA, Sturrock C, Devenish AT, Howe R, Roe S, Moore M, Scozzafava G, Proudfoot LMF, Richardson TC, McFarthing KG: The application of enhanced chemiluminescence to membrane-based nucleic acid detection. BioTechniques 8:564, 1990.

56. DeFlaun MF, Paul JH, Davis D: Simplified method for dissolved DNA determination in aquatic environments. Appl Environ Microbiol 52:654, 1986.

57. Fuhrman JA, Comeau DE, Hagstrom A, Chan AM: Extraction from natural planktonic microorganisms of DNA suitable for molecular biological studies. Appl Environ Microbiol 54:1426, 1988.

58. Paul JH, Carlson DJ: Genetic material in the marine environment: Implications for bacterial DNA. Limnol Oceanogr 29:1091, 1984.

59. Paul JH, Jeffrey WH, DeFlaun M: Particulate DNA in subtropical oceanic and estuarine planktonic environments. Mar Biol 90:95, 1985.

60. Paul JH, Myers B: Fluorometric determination of DNA in aquatic microorganisms by use of Hoechst 33258. Appl Environ Microbiol 43:1393, 1982.

61. Somerville CC, Knight IT, Straube WL, Colwell RR: Simple, rapid method for direct isolation of nucleic acids from aquatic environments. Appl Environ Microbiol 55:548, 1989.

62. Ogram A, Sayler GS, Barkay T: The extraction and purification of microbial DNA from sediments. J Microbiol Methods 7:57, 1987.

63. Torsvik VL: Isolation of bacterial DNA from soil. Soil Biol Biochem 12:15, 1980.

64. Tsai Y-L, Olson BH: Rapid method for direct extraction of DNA from soil and sediments. Appl Environ Microbiol 57:1070, 1991.

65. Steffan RJ, Goksoyr J, Bej AK, Atlas RM: Recovery of DNA from soils and sediments. Appl Environ Microbiol 54:2908, 1988.

66. Stahl DA, Flesher B, Mansfield HR, Montgomery L: Use of phylogenetically based hybridization probes for studies of ruminal microbial ecology. Appl Environ Microbiol 54:1079, 1988.

67. Weller R, Ward DM: Selective recovery of 16S rRNA sequences from natural microbial communities in the form of cDNA. Appl Environ Microbiol 55:1818, 1989.

68. Ward DM, Weller R, Bateson MM: 16S rRNA sequences reveal numerous uncultured microorganisms in a natural community. Nature 345:63, 1990.

69. Tsai Y-L, Park MJ, Olson BH: Rapid method for direct extraction of mRNA from seeded soils. Appl Environ Microbiol 57:765, 1991.

70. Barkey T, Liebert C, Gillman M: Hybridization of DNA probes with whole-community genome for detection of genes that encode microbial responses to pollutants: *mer* genes and Hg^{2+} resistance. Appl Environ Microbiol 55:1574, 1989.

71. Steffan RJ, Atlas RM: Solution hybridization assay for detecting genetically engineered microorganisms in environmental samples. BioTechniques 8:316, 1990.

72. Torsvik V, Goksoyr J, Daae FL: High diversity in DNA of soil bacteria. Appl Environ Microbiol 56:782, 1990.

73. Torsvik V, Salte K, Sorheim R, Goksoyr J: Comparison of phenotypic diversity and DNA heterogeneity in a population of soil bacteria. Appl Environ Microbiol 56:776, 1990.

74. Ferguson RL, Buckley EN, Palumbo AV: Response of marine bacterioplankton to differential filtration and confinement. Appl Environ Microbiol 47:49, 1984.

75. Saiki RK, Scharf S, Faloona F, Mullis KB, Horn GT, Erlich HA, Arnheim N: Enzymatic amplification of β-globin genomic sequences and restriction site analysis for diagnosis of sickle cell anemia. Science 230:1350, 1985.

76. Rotbart HA: Enzymatic RNA amplification of the enterovirus. J Clin Microbiol 28:438, 1990.

77. Bej AK, Steffan RJ, DiCesare J, Haff L, Atlas RM: Detection of coliform bacteria in water by polymerase chain reaction and gene probes. Appl Environ Microbiol 56:307, 1990.

78. Starnbach MN, Falkow S, Tompkins LS: Species-specific detection of *Legionella pneumophila* in water by DNA amplification and hybridization. J Clin Microbiol 27:1257, 1989.

79. Pollard PR, Johnson WM, Lior H, Tyler SD, Rozee KR: Detection of the aerolysin gene in *Aeromonas hydrophila* by the polymerase chain reaction. J Clin Microbiol 28:2477, 1990.

80. Ridgeway HF, Phipps DM, Leddy M: In situ and On Site Bioreclamation: An International Symposium. San Diego, March 19–21, 1991.

81. Rochelle PA, Olson BH: A simple technique for the electroelution of DNA from environmental samples. Biotechniques 11:xxx, 1991.

82. Barkay T, Olson BH: Phenotypic and genotypic adaptation of aerobic heterotrophic sediment bacterial communities to mercury stress. Appl Environ Microbiol 52:403, 1986.

83. Jansson JK, Holben WE, Tiedje JM: Detection in soil of a deletion in an engineered DNA sequence by using DNA probes. Appl Environ Microbiol 55:3022, 1989.

84. Knight IT, Shults S, Kaspar CW, Colwell RR: Direct detection of *Salmonella* spp. in estuaries by using a DNA probe. Appl Environ Microbiol 56:1059, 1990.

85. Walia S, Khan A, Rosenthal N: Construction and applications of DNA probes for detection of polychlorinated biphenyl-degrading genotypes in toxic organic contaminated soil environments. Appl Environ Microbiol 56:254, 1990.

86. Diels L, Mergeay M: DNA probe-mediated detection of resistant bacteria from soils highly polluted by heavy metals. Appl Environ Microbiol 56:1485, 1990.

87. Collins DM, Gabric DM, de Lisle GW: Identification of two groups of *Mycobacterium paratuberculosis* strains by restriction endonuclease analysis and DNA hybridization. J Clin Microbiol 28:1591, 1990.

88. Timberlake WE: Developmental gene regulation in *Aspergillus nidulans*. Dev Biol 78:497, 1980.

89. Mehdy MC, Ratner D, Firtel RA: Induction and modulation of cell type-specific gene expression in *Dictyostelium*. Cell 32:763, 1983.

90. Jiang X, Estes MK, Metcalf TG, Melnick JL: Detection of hepatitis A virus in seeded estuarine samples by hybridization with cDNA probes. Appl Environ Microbiol 52:711, 1986.

91. Varveri C, Ravelonandro M, Dunez J: Construction and use of a cloned cDNA probe for the detection of plum pox virus in plants. Phytopathology 77:1221, 1987.

92. Roy BP, Abou Haidear MG, Sit TL, Alexander A: Construction and use of cloned cDNA biotin and ^{32}P-labeled probes for the detection of papaya mosaic potexvirus RNA in plants. Phytopathology 78:1425, 1988.

93. Lakshman DK, Hiruki C, Wu XN, Leung WC: Use of [^{32}P]RNA probes for the dot-hybridization detection of potato spindle tuber viroid. J Virol Methods 14:309, 1986.

94. Chityothin O, Sethabutr O, Echeverria P, Taylor DN, Vongsthongsri V, Tharavanij S: Detection of heat-stable enterotoxigenic *Escherichia coli* by hybridization with an RNA transcript probe. J Clin Microbiol 25:1572, 1987.

95. Haun G, Gobel U: Oligonucleotide probes for genus-, species- and subspecies-specific identification of representatives of the genus *Proteus*. FEMS Microbiol Lett 43:187, 1987.

96. Steffan RJ, Atlas RM: DNA amplification to enhance detection of genetically engineered bacteria in environmental samples. Appl Environ Microbiol 54:2185, 1988.

97. Wilson KH, Blitchington R, Hindenach B, Greene RC: Species-specific oligonucleotide probes for rRNA of *Clostridium difficile* and related species. J Clin Microbiol 26:2484, 1988.

98. Hahn D, Starrenburg MJC, Akkermans ADL: Oligonucleotide probes that hybridize with rRNA as a tool to study *Frankia* strains in root nodules. Appl Environ Microbiol 56:1342, 1990.

99. Chapman NM, Tracy S, Gauntt CJ, Fortmueller U: Molecular detection and identification of enteroviruses using enzymatic amplification and nucleic acid hybridization. J Clin Microbiol 28:843, 1990.

100. Neill RJ, Fanning GR, Delahoz F, Wolff R, Gemski P: Oligonucleotide probes for detection and differentiation of *Staphylococcus aureus* strains containing genes for enterotoxins A, B, and C and toxic shock syndrome toxin 1. J Clin Microbiol 28:1514, 1990.

101. DeLeon R, Shieh C, Sobsey MD: Detection of enteroviruses and hepatitis A virus in environmental samples by gene probe and polymerase chain reaction. Proc Water Qual Technol Conf 18:833, 1991.

102. Wilde J, Eiden J, Yolken R: Removal of inhibitory substances from human fecal specimens for detection of group A rotaviruses by reverse transcriptase and polymerase chain reaction. J Clin Microbiol 28:1300, 1990.

103. Allard A, Girones R, Juto P, Wadell G: Polymerase chain reaction for detection of adenoviruses in stool samples. J Clin Microbiol 28:2659, 1990.

10

INNOVATIONS IN BIOLOGICAL PROCESSES FOR POLLUTION CONTROL

B.E. RITTMANN

Department of Civil Engineering, University of Illinois, Urbana, Illinois 61801

1. INTRODUCTION TO BIOLOGICAL PROCESSES

Environmental biotechnology for pollution control uses a biological process to remove pollutants from contaminated water, waste water, or other media. A biological process is an engineered system that accumulates desired microorganisms to a sufficiently large mass that the pollutant concentration is lowered to meet some treatment standard during the time that the microorganisms are in contact with the material being treated. Therefore, innovation in environmental biotechnology means finding better ways to identify and accumulate desired microorganisms to accomplish a treatment goal.

Environmental biotechnology today is fundamentally an engineering application of microbial ecology (1). Process controls are used to select the kinds and amounts of different microorganisms in the process, thereby creating an ecological system that accomplishes the treatment goal. Those controls are on substrate supply, cell retention, and process "loading."

The first process control, substrate supply, is very powerful, because broad classes of bacteria can be selected or eliminated by controlling which electron donors and acceptors are available as substrates. Most traditional waste water pollutants are either electron donors or acceptors; thus selection of the desired

Environmental Microbiology, pages 265–286, © 1992 Wiley-Liss, Inc.

microorganisms can be achieved by controlling the supply of the other essential substrate. For example, NH_4^+ is an electron donor that is a common waste water pollutant. When O_2 is supplied as an electron acceptor, nitrifying bacteria are able to gain energy and grow through the aerobic oxidation of NH_4^+ to NO_3^-. Another common electron-donor pollutant is organic matter, normally measured as biochemical oxygen demand (BOD). When O_2 is supplied as an electron acceptor, aerobic heterotrophic bacteria grow well through the aerobic oxidation of the carbon in BOD to CO_2. On the other hand, if no electron acceptors are supplied, a consortium of fermenting microorganisms, including the methane-producing archaebacteria, can be selected. In this case, the BOD is converted to CH_4, which evolves from the liquid as a gas. Finally, NO_3^- is a common pollutant that has the role of an electron acceptor. It can be reduced to N_2 gas by denitrification when an appropriate electron donor is supplied; thus selection for denitrifying bacteria requires a supply of electron donors, such as organic matter, reduced sulfur, or H_2.

Once the proper microorganisms are selected by controlling substrate availability, their mass in the process must be built up. Accumulation of a sufficiently large biomass is accomplished through the second process control, cell retention. The simplest form of retention utilizes a large reactor volume. However, a large liquid volume with a low microorganism concentration often is expensive and frequently is prohibited by size restraints. Thus cell-retention strategies that allow small liquid volumes with high cell concentrations are essential. The two most common cell-retention strategies are 1) sedimentation combined with recycle and 2) attachment.

When bacteria flocculate into aggregates of 50–200 μm in size, called *flocs,* they can be concentrated by sedimentation in a quiescent clarifier (2). This concentrated stream of cells is recycled back to the main reactor, where the appropriate substrates are supplied. Because the clarified effluent exits the system containing very little biomass, the biomass is concentrated by sedimentation and recycle. Cell-recycle systems are called *activated sludge* systems, because the sludge (or solid matter) in the reactor is "activated" by being enriched with active bacteria that are captured by sedimentation and recycled. These systems select for cells that are flocculent.

The second retention strategy is attachment (2). Here, the process utilizes a high amount of surface area on media such as rocks, sand, or plastic. Bacteria attach to the surfaces to form biofilms, which are layer-like aggregates of bacteria and extracellular polymers. The water flows past the biofilms, which extract the pollutants from the liquid while remaining fixed within the process. Microorganisms performing the same reactions can be selected and accumulated by cell recycle and by biofilm retention.

In addition to accumulating a large biomass, cell retention has an important selection function that augments substrate supply. Very high efficiency cell re-

**TABLE 10.1. Common Examples of Biological Processes
in Environmental Biotechnology[a]**

Process designation	Electron donor	Electron acceptor	Cell retention
Aerated lagoons	BOD	O_2	Volume
Conventional activated sludge	BOD and/or NH_4^+	O_2	Cell recycle
Trickling filters	BOD and/or NH_4^+	O_2	Biofilm
Rotating biological contactors	BOD and/or NH_4^+	O_2	Biofilm
Denitrification	BOD, H_2, or S^{2-}	NO_3^-	Cell recycle or biofilm
Anaerobic filter	BOD	BOD or CO_2	Biofilm
Anaerobic sludge digestion	BOD (in form of sludge)	BOD or CO_2	Volume
Anaerobic upflow sludge blanket reactor	BOD	BOD or CO_2	Biofilm on self-forming granules
Composting	BOD (in form of sludge or other solid matter)	O_2	Volume and cell recycle

[a]BOD, biochemical oxygen demand.

tention allows the accumulation of microorganisms that have a very long cell retention time, or sludge age (θ_x). Because some microorganisms, such as nitrifiers and methanogens, grow much more slowly than others, they can form stable populations only when the cell retention time is quite long, such as 10–15 days or longer. These slow-growing microorganisms are not selected if the cell retention is poor, even when their substrates are available. Thus cell retention can be used to select for or against certain slow-growing species (2).

The final control on a biological process is process *loading*. Simply stated, the concept of process loading is to provide enough contact time between the microorganisms and the pollutants that the pollutant concentration is lowered to meet the treatment standard. Loading involves a kinetic analysis of the process. Such an analysis is beyond the scope of this chapter; however, a concise summary has been published by Rittmann *et al.* (1). Briefly, the process loading for the cell-recycle system centers on achieving the proper cell retention time (θ_x), while the process loading for biofilm processes involves specification of a rate of substrate loading per unit surface area, called the *substrate flux* (J).

Table 10.1 provides some common examples of biological processes used in environmental biotechnology. The electron-donor and electron-acceptor substrates, as well as the cell-retention mechanisms, are listed. Note the wide variety of processes that utilize the same substrates.

TABLE 10.2. Trends in Government Regulations Stimulating Innovation

1. Greater reliability in the removal of traditional wastewater pollutants, such as BOD, suspended solids, N, and P
2. Removal from wastewater of specific and often hazardous organic components found in a mixture of many components; examples include particular chlorinated solvents, volatile organic compounds, and other industrial chemicals
3. Removal of specific and group contaminants to very low concentrations, often $<1\,\mu g$/liter
4. Removal of effluent "toxicity," as measured by various bioassays
5. Treatment of new media, including contaminated drinking water, groundwater, aquifers, and soil

2. DRIVING FORCES FOR INNOVATION

Although biological processes have been used for many decades (2,3), innovations are now occurring at an increasing rate. Three forces are driving the rapid pace of innovation: regulations, economic pressures, and scientific and technological advances.

Governments in the industrialized countries are imposing much more sweeping and stringent regulations on what must be treated and how well. Of course, the details vary greatly from country to country. Nevertheless, five key trends can be identified and are summarized in Table 10.2. While trend 1 forces greater reliability for traditional applications, trends 2 through 5 are, for the most part, requirements that biological processes accomplish treatments that have not been traditional objectives. In other words, the scope of what biological treatment must be able to achieve is being extended to include much greater specificity toward contaminants (trends 2 and 4), greater stringency in terms of concentration (trend 3), and greater breadth of applications (trend 5). In some cases, existing processes are unable to meet the new regulatory needs.

A second driving force is economic pressure. Unlike most of biotechnology, which is oriented toward producing a valuable product (*e.g.*, insulin, growth hormone, chemicals), environmental biotechnology mainly provides a service that is required by regulation. Therefore, much of the economic pressure ultimately is generated by the regulatory changes. When new regulations are imposed, existing technologies often can provide the service, but only in a cost-ineffective manner. Thus new technologies must be developed to provide cost-effective service, although existing technologies already are workable.

The final driving force comprises scientific and technological advancements. Regulatory and economic forces yield new and improved technologies only when the scientific and technological foundations are present. Fortunately, scientific research and technological development can provide the base on which to build new environmental biotechnology. The next sections discuss key trends in research and development leading to innovation.

3. RESEARCH INNOVATIONS

Although the leading research in environmental biotechnology is highly diverse, one overarching theme serves to unify much of it today: Research is uncovering, explicitly describing, and exploiting much more *structure* in all aspects of biological processes. To uncover structure means to differentiate in greater detail the specialized roles of different substrates, different parts of cells, different cells within a mixed community, and different locations within a process. Hence structure is described at the levels of molecules, cells, communities, and reactor space.

3.1. Structure in the Concept of Substrate

Traditionally, a substrate is a dissolved electron donor that is growth-rate limiting (4,5). Sometimes the concept is extended to include dissolved electron acceptors. The key features of the traditional substrate are that it is a dissolved molecule and that it is growth-rate limiting. Table 10.3 contrasts the traditional substrate with four other kinds of substrates that are important for understanding several of the innovative biological processes.

Colloidal organic material is particulate substrate. Although this material ultimately is an electron donor, it differs from the traditional electron donor in that it must first be captured and hydrolyzed before it is available for uptake and oxidation by the bacteria (6–8). Therefore, physico/chemical phenomena of particle transport, adhesion, and extracellular enzymatic hydrolysis are added steps that can affect the rate of particle removal and electron-donor utilization. Because sewage (9) and many other waste waters have more colloidal than soluble BOD, the sequential steps of capture, hydrolysis, and oxidation of particulate substrates must be understood.

Internally stored substrates are becoming increasingly important for the selection of bacteria that form good flocs and remove phosphorus (10–13). Under conditions of high concentrations of soluble organic electron donors (*e.g.*, acetate) and a deficiency of electron acceptors, certain strains of bacteria are able to absorb the electron donor and store it internally as nondissolved polymers, such as polyhydroxybutyrate (PHB). This storage material can be hydrolyzed and oxidized when external electron-donor substrates are depleted but electron acceptors are present. Hence strains that store internal substrates can be selected when the microorganisms are exposed alternately to feast and famine conditions

TABLE 10.3. Characteristics of Five Different Kinds of Substrates Important in Biological Process Innovations

1. Traditional
   a. Dissolved electron donor (and sometimes acceptor)
   b. Growth-rate limiting
2. Particulate
   a. Not dissolved, but colloidal
   b. Must be hydrolyzed before becoming available for uptake as an electron donor
   c. Frequently must be captured by physicochemical mechanisms before hydrolysis can occur
3. Internal storage
   a. Present internally, usually as solid inclusion body
   b. Serves as a store of electrons and energy
   c. Cyclically built-up and depleted as environmental conditions change
4. Secondary
   a. Contributes negligibly toward cell growth and maintenance; therefore
   b. Not growth-rate limiting
   c. Requires consumption of a primary substrate to create and sustain cell mass
   d. Can be driven to very low concentrations
5. Produced
   a. Produced as a natural result of cell metabolism
   b. Consumed by the same or other bacteria
   c. Hard-to-biodegrade fractions tend to accumulate

with respect to electron donors. Such selection has practical importance, because bacterial strains that store polymers also form good flocs and can accumulate larger than normal amounts of phosphorus. In the latter case, the extra phosphorus storage is in the form of polyphosphate, another internal storage polymer that seems to store energy for use under strictly fermenting conditions (10).

Attention to the biodegradation of organic compounds present at very low concentrations has led to the differentiation between primary and secondary substrates. A primary substrate normally conforms to a traditional nutrient in that it is the growth-rate-limiting electron donor. On the other hand, while a secondary substrate is an electron donor, its concentration is too low to sustain a steady-state biomass through its utilization alone (14–17). Therefore it is consumed by biomass grown and sustained through utilization of a primary substrate, which means that secondary substrates are not growth-rate limiting. Their removal is determined by their intrinsic kinetics and by the mass of active biomass. However, the active biomass is controlled by utilization of primary substrate, rather than secondary substrate (15,17).

The final class of substrates is called *produced*, because they are formed and released as part of cell metabolism. Soluble materials released by microorgan-

**TABLE 10.4. Important Examples of Cell
Structural Differentiation**

1. Internal substrates
   a. Internal storage, such as PHB and poly-P (see Table 10.3.)
   b. NADH/NAD
   c. ATP/ADP
2. Genetic content
   a. Immobile and essential
   b. Mobile and accessory

isms are called *soluble microbial products* (SMPs) (18,19). A portion of SMP is formed directly in proportion to substrate utilization and is termed substrate *utilization-associated products* (UAPs). A second portion of SMP is formed through basal metabolism and decay of active biomass and is called *biomass-associated product* (BAP). UAP and BAP are produced substrates, because much of the originally produced material is biodegradable (19,20). Thus significant carbon and electron flows are channeled through SMP. It is not yet known to what degree SMP biodegradation is by the same cells that produce the SMP or by specialized bacteria that have high affinity for low concentrations of the natural polymers that comprise cells and SMP.

3.2. Structure Within the Bacterial Cell

Table 10.4 lists environmentally important examples of structural differentiation within bacterial cells. The two broad classes of structure are internal substrates and genetic content. Within the former class, internal storage materials were discussed in the previous section.

Microbial reactions occur through multistep pathways that gradually oxidize or reduce substrates (21), producing carriers of electrons (NADH) and energy (ATP) that act as internal food sources. The kinetics of utilization for external substrates can be affected by the status of internal substrates. For example, a decrease in the internal concentration ratio NADH/NAD can increase the rate of electron-donor utilization (22), while a decrease in the internal concentration ratio ATP/ADP can increase electron-acceptor utilization (23). Thus the kinetic expressions normally used to describe the rates of substrate utilization in terms of external concentration (*e.g.*, the Monod function) have parameters that are not constant, but depend on the internal concentration. The effects of internal substrates on the utilization rates of external substrates appear to be most important when electron donor and acceptor are simultaneously at subsaturating concentrations.

The second major class of cell structure involves the genetic capability of the microorganisms. The genes encoding all the cell functions are stored and replicated as DNA. Most of the DNA within a bacterial cell is contained on the chromosome, which is characterized as being (mainly) immobile and essential. Immobile means that the chromosome is not transferred across its cell's boundary. When a bacterium divides into two daughter cells, the chromosome is replicated once so that each daughter cell has one chromosome. *Essential* means that the chromosomal DNA encodes functions that are required for the normal growth and maintenance of the cell.

Many bacteria contain DNA other than the chromosome. These other elements—including plasmids, transposons, and viral DNA—generally are much smaller than the chromosome, are mobile, and are accessory. The most important ones seem to be the plasmids, which are discussed here. The plasmids are mobile, which means that they can be transferred from one cell to another independently of cell replication. The most common transfer mechanism seems to be conjugation, in which the plasmid is replicated, independently of cell division, as it is transferred from a donor cell to a recipient cell. Plasmids also are characterized as accessory, which means that the functions encoded are not essential for the routine operation of the cell. Although plasmid DNA is accessory, it codes for some very interesting and useful reactions (1,24): included are biodegradation of halogenated and other xenobiotic organic chemicals, as well as resistances to antibiotics and heavy metals.

That some of the genes allowing biodegradation of or resistance to hazardous chemicals are found on plasmids has profound implications for innovative treatment. It suggests that the traditional process-control strategies, which control substrate availability and cell retention, may not be adequate to ensure selection of bacteria that degrade hazardous chemicals to very low concentrations. Because traditional control strategies select microorganisms according to traits normally encoded on the chromosome, they do not necessarily select for the presence of plasmids.

Exactly what selects for the presence of plasmids is not well known (1,25). Strong advantages conferred by a plasmid sometimes increase retention (26). For example, having resistance to an antibiotic is highly advantageous when the antibiotic is present at an inhibitory level. In addition, preliminary evaluation of conjugation kinetics suggests that high cell concentrations and the input of plasmid-containing cells (*i.e.*, bioaugmentation) can increase the content of plasmid-containing cells in a biological process (25). The importance of genetic structure in environmental biotechnology is covered in more detail by B.H. Olsen, this volume.

TABLE 10.5. **Key Areas of Research on the**
Structure of the Microbial Community

1. Floc-forming versus filamentous heterotrophs
2. Heterotroph and autotroph coexistence
3. Gene-containing versus gene-free

3.3. Structure Within the Microbial Community

The microbial community is the next larger level at which structure is being better elucidated. Unlike many other areas of biotechnology, environmental processes almost always involve mixed cultures. Control of the ecological balance in the microbial communities is an important research issue, summarized in Table 10.5.

The competition between floc-forming and filamentous heterotrophs in activated sludge processes is central to community control. In order that cell retention be successful in activated sludge, the bacteria must aggregate into large and dense flocs. When the flocs contain too many heterotrophic bacteria that grow in long filaments extending outside the primary floc, the sludge settles and compacts very poorly (2,24,27). This situation of too much filament is called sludge *bulking* and can result in serious deterioration of effluent quality and failure of cell retention.

The floc-forming and filamentous bacteria utilize the same electron-donor and -acceptor substrates. However, research has identified the more subtle conditions that favor floc formers (2,28). Filaments are favored by three main conditions: 1) high BOD loading with low dissolved oxygen concentration, 2) long cell retention times, or 3) the input of reduced sulfur. To a certain degree, sludge bulking can be mitigated by avoiding each of the three conditions. Unfortunately, completely eliminating all conditions favoring filaments is not always possible. In such a (common) case, a strategy that gives a competitive edge to the floc formers is necessary.

Certain floc-forming heterotrophs are able to produce internal storage substrates (see Section 3.1.) when the electron-donor concentration is very high (2,12). If the floc formers rapidly sequester most of the electron donor as storage material, they can gain a competitive advantage over the filaments, which do not store electron donor and are therefore starved. Ecological control over community structure is achieved by regularly exposing the activated sludge to periods of very high BOD loading; the location of this high load is called a *selector tank* (2,12,13).

The second community issue described in Table 10.5 is coexistence of heterotrophic and autotrophic bacteria. Heterotrophs use organic carbon as the carbon source, while autotrophs must reduce the carbon at the CO_2 oxidation state during synthesis. The impact of this difference in carbon source is that the energy and electron needs for synthesis of autotrophic cells are much greater than for heterotrophs (2,29). Thus the yield and specific growth rate of autotrophs are substantially smaller than for heterotrophs. Even when all substrates are amply available, autotrophs are at a competitive disadvantage because of their relatively slow growth rates. The most sharply pointed competition is between aerobic heterotrophs and nitrifying bacteria (2).

Although differing in their carbon sources, heterotrophs and autotrophs frequently compete in other ways. For instance, often they utilize the same electron acceptor. However, the most common form of competition is for desirable space in flocs or biofilms. Slow-growing autotrophs can be pushed toward less advantageous locations by fast-growing heterotrophs that can quickly grow into prime locations and keep out or push out autotrophs (30–32). Prime locations usually are those that have the highest concentrations of electron-donor and -acceptor substrates.

Since simultaneous existence of heterotrophs and autotrophs often is desired and sometimes is required for adequate treatment, the issue of competition is critical. Preliminary research shows that slow-growing species tend to be relegated to the deep portions of biofilms (30–32). Microorganisms in these layers usually are at a disadvantage, because substrate concentrations are reduced because of diffusion gradients. In addition, substrates that heterotrophs also use may be completely depleted in the outer layers of the biofilm. On the other hand, being deep in the biofilm protects the autotrophs from detachment and can enhance maintenance of slow-growing species in a mixed biofilm community (32). The ecological competition for substrates and space is a new research topic that offers significant scientific and practical insights.

The final issue in Table 10.5 concerns genetic advantages of specific populations. While ecological selection according to chromosome-encoded genes determines the strains that exist in the community, control of plasmid distribution can alter the genetic capabilities of different bacteria within the same strain (25). Thus gene distribution within strains provides a level of community structure that can be superimposed on the normal ecological structure of mixed populations.

3.4. Structure Within the Process Space

The process space provides an even larger scale of structure. The process level deals with interactions among the biological structures and nonbiological phenomena. In many cases, process-level structure makes it possible to implement the necessary controls on the substrate, cell, and community structures. Table 10.6 lists three research areas that are yielding improved understanding and

TABLE 10.6. Examples of Process-Level Structure

1. Establishment of macro- and microzones within the process, *e.g.*,
    a. Different electron acceptors: aerobic, anoxic, and anaerobic areas
    b. High and low concentrations of electron donor
2. Separation devices
    a. Improved clarifiers
    b. Membranes
3. Exploitation of nonbiodegraditive removal mechanisms
    a. Adsorption
    b. Volatilization

control of process structure. In all cases, nonbiological phenomena are crucial, but the connections to the previous levels of structure are important.

The establishment of zones within processes is an outstanding example of how a process-level structure yields better performance. The zones are physically distinct places or times within a process. As shown in point 1 of Table 10.6, these zones are characterized by having different electron acceptors or widely differing concentrations of electron donors.

One of the most common cases of electron-acceptor zoning is one-sludge denitrification of waste waters that contain BOD and total Kjeldahl nitrogen (TKN). The BOD and the NH_4^+–N in the TKN are electron-donor substrates that must be oxidized and removed. Oxidation of NH_4^+ yields NO_3^-, which is a possible electron acceptor that often must be removed. The concept of one-sludge denitrification depends on use of O_2 as the electron acceptor to produce NO_3^- from NH_4^+. Subsequently, the NO_3^- is used as electron acceptor to oxidize BOD. The BOD remaining must be oxidized aerobically, or with O_2 as electron acceptor.

Zoning is required, because the denitrification reactions stop if the denitrifying bacteria are exposed to nontrivial concentrations of dissolved oxygen. On the other hand, nitrification ceases if the dissolved oxygen concentration in contact with the nitrifiers is too low. Figure 10.1 illustrates several means to carry out one-sludge denitrification. Figure 10.1a illustrates classic predenitrification (33,34), in which denitrification of influent BOD and NO_3^- occurs in an unaerated anoxic reactor. The effluent from the anoxic reactor, which contains the BOD not removed by denitrification and NH_4^+, flows to the aerobic reactor, in which the remaining BOD is aerobically oxidized and NH_4^+–N is nitrified to NO_3^-. Most of the NO_3^- produced in the aerobic reactor is recycled back to the anoxic reactor to allow denitrification there. In the case of predenitrification, macrozones are created by aerating specific tanks. The one sludge (*i.e.*, one mixed community) cycles through the two zones continuously, being exposed alternately to aerobic or anoxic conditions.

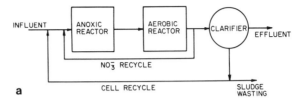

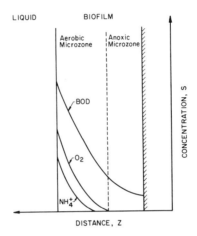

Fig. 10.1. Several means to create zones for one-sludge denitrification. **a**: predenitrification macrozones. **b**: Floc or biofilm microzones (a biofilm is illustrated).

Figure 10.1b illustrates that the same result can be created in microzones within biofilms (30,31) or flocs (33). In the case of microzones, the community does not cycle between aerobic and anoxic conditions. Instead, the microzones are established more or less permanently at different locations within the biofilm or floc. Careful control of the dissolved oxygen concentration in the bulk liquid (33) allows it to be completely consumed in the outer part of the film or floc. Then, an inner microzone is anoxic, consuming the NO_3^- produced in the aerobic microzone, as well as BOD not consumed in the aerobic layer.

A second example of zoning involves the use of a strictly anaerobic tank at the beginning of a process. Figure 10.2a illustrates how an anaerobic tank can be placed ahead of a predenitrification system for the purpose of selecting for floc-forming bacteria that store PHB and polyphosphate (11). In this example, known as the UCT process (11), anoxic and aerobic zones are used to achieve

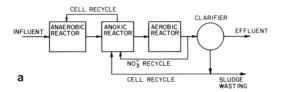

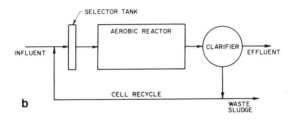

Fig. 10.2. Examples of addition of specialized zones ahead of the normal process. **a**: Addition of an anaerobic tank before predenitrification in order to effect enhanced P removal. **b**: Addition of a selector tank in order to control sludge bulking.

predenitrification, while the anaerobic zone is used to select ecologically for bacteria that store excess phosphates as polyphosphate.

Figure 10.2b shows another type of zoning scheme: one in which a selector tank is used to create a zone of high electron-donor concentration (12). The selector tank creates ecological pressure to select for floc-forming heterotrophs and against filamentous heterotrophs. Again, a specialized zone is created by process-level structuring.

The second major example in Table 10.6 describes separation devices. In all activated-sludge-type systems, cells must be captured and recycled in order to build up the activated sludge. Separation devices involve physical mechanisms for solids–liquid separation. Today, much more attention is being given to the physical phenomena of solids separation, because the lack of reliable and high-efficiency solids removal is a major obstacle to meeting some of the new regulations (see Table 10.2).

The traditional separation device is a gravity clarifier having a dual purpose. The first is to remove virtually all suspended solids (*i.e.*, the microorganisms) from the effluent. The second is to thicken the underflow sludge for recycling and wasting. Relatively recent research advances in the area of solids-flux theory (2,35,36) link the two functions and provide a rational basis for design and

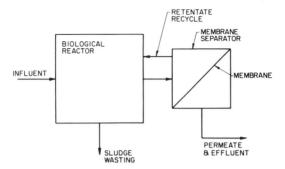

Fig. 10.3. Schematic representation of membrane filtration combined with a biological process.

operation of the clarifier portion of activated sludge units. In addition, new emphasis on activated-sludge flocculation in the aeration tank (33) and in the input stilling well (37) is leading to better designs that enhance floc formation and both aspects of clarifier performance.

An innovative alternative for solids separation with biological processes is membrane separation, which uses thin polymer skins. The skin has small pores of controlled size. For microfiltration, the pore size is $0.02–0.1$ μm, and for ultrafiltration the size is $0.001–0.02$ μm (38). Utilization of microfiltration membranes could replace gravity clarification, providing a perfect retention device for bacteria, most of which are larger than 0.2 μm in diameter. Thus microfiltration could provide 100% retention and 100% reliability for removal of almost all of the suspended solids in biological-process effluents. This innovation could go a long way toward meeting the requirements listed in Table 10.2. Figure 10.3 shows a typical scheme for employing a membrane separator with a biological reactor.

Further improvement in the quality and consistency of effluents could be achieved by use of ultrafiltration membranes, which are designed to retain moderate- to large-sized molecules ($1,000–100,000$ MW). Since a significant portion of effluent chemical oxygen demand (COD) is comprised of SMPs having a molecular weight of $1,000–10,000$ (18,19), ultrafiltration could significantly reduce the total effluent load of organic matter, including soluble and suspended. The innovation of ultrafiltration should prepare effluents for many types of reuse, an important requirement in arid regions.

Application of membrane separation as part of a biological process introduces at least two research questions. The first concerns fouling of the membrane. Fouling is observed as an increasing pressure drop to sustain permeation through the membrane. If the decrease in permeation rate is too high, the process becomes economically unattractive because of high energy costs. Membranes coupled to bioreactors can foul by four mechanisms: 1) plugging of the pores by strained particles, 2) accumulation of a "polarization gel" of retained solutes, 3) accumulation of a "cake" of retained particles, and 4) reduction of pore size by adsorption of permeating solutes. Solution of the fouling problem is essential before membrane processes can have a widespread role in environmental bio-technology.

The second research question is directed primarily toward ultrafiltration. The retention of moderate- to large-sized molecules causes accumulation of elevated concentrations in the biological process, potentially affecting gas transfer, aggregation, and microbial kinetics. Little is known about the specific effects, but solute build-up is a major problem when ultrafiltration is employed.

The third example listed in Table 10.6 is exploitation of nonbiodegradative removal mechanisms during biological treatment. The two most likely processes are volatilization to a gas phase and adsorption to a solid phase.

When a compound has a Henry's constant (H) greater than about 10^{-3} atm-m^3/mol (17,39,40), it can be significantly volatilized from typical biological processes as long as other mechanisms are not consuming the compound more rapidly. In aerobic processes, the volatilization rate is controlled by the aeration rate, which depends on the BOD and TKN inputs and on the aeration efficiency. The volatilization rate in anaerobic processes is controlled by the gas-evolution rate (CH_4 in methanogenesis and N_2 in denitrification), both of which are determined by the BOD input. Whether one wants to maximize removal by volatilization or minimize volatilization to prevent air pollution, the physical process is controlled most strongly by the input substrate loading and the type of process-level structuring.

Organic compounds with a large octanol-water partition coefficient (*e.g.*, larger than $1,000$ m^3H_2O/m^3 oct.) adsorb to biomass (17,40). Adsorption is a physico/chemical mechanism that competes with biodegradation and can reduce the degree of biodegradation. It also concentrates the compounds, many of which are considered hazardous, into the sludge, which eventually must be disposed of. Because of these factors, adsorption of biodegradable compounds is not usually beneficial.

On the other hand, adsorption of poorly biodegradable compounds can be beneficial when these compounds adversely affect effluent quality or process performance by inhibition of microbial processes. Adsorption to biomass can be accentuated by operating the system with a short cell retention time, in which case the adsorbed compounds are removed most rapidly with the wasted cell solids (40). Adsorption can be enhanced still further by adding a strong adsorbant, such as activated carbon. Activated carbon adsorbs and accumulates adsorbable and poorly degraded components (41). Easily degraded organics tend to be decomposed immediately or after a period of adsorption and bioregeneration (42). Thus accentuated sorption is most valuable for mitigation of adverse effects of adsorbable and poorly degradable organic components. In other words, adsorption physicially sequesters troublesome components so that the biological process is highly effective for biodegradation of the more degradable chemicals.

4. TECHNOLOGICAL INNOVATIONS

A technological innovation can yield a new reactor type, a totally new process, or even a new way of approaching a traditional process. Table 10.7 lists several innovations in biological treatment. A detailed discussion is provided for one of the technological advances, high-rate biofilm processes.

A high-rate biofilm process achieves its status by combining dense biofilms with attachment media having a high specific surface area (2). The biofilm accumulation per unit reactor volume is great, which allows the reactor to have a high volumetric loading, a short liquid detention time, and a compact size. The economic advantages of compact processes are obvious. In addition, compactness offers significant opportunities for use in small spaces, for retrofits, and for portable processes.

The greatest potential for high-rate processing is with fluidized-bed biofilms. Small ($<$2 mm) particles of sand, coal, activated carbon, or resins are fluidized by the upward flow of water or by gas turbulence. Biofilm is attached to the

---

TABLE 10.7. Examples of Technological Innovations

1. Anaerobic waste water treatment: The direct treatment of moderate- to high-strength waste waters by methanogenic consortia occurs in a range of biofilm and sludge-blanket reactors (43, 44)
2. Biological drinking-water treatment: Removal of low concentrations of biodegradable material by a range of biofilm processes (45) reduces disinfectant demand and problems of microbially produced tastes, odor, and corrosion
3. High-rate biofilm processes: Excellent cell retention is combined with a high specific surface area to make volumetric loadings very high
4. Biofilm/activated sludge systems: Adding biofilm attachment media to activated sludge increases biomass accumulation and process stability
5. Membrane bioreactors: Microorganisms are held inside the membrane, while substrates are fed with the wastewater and across the membrane (46)
6. On-line control with expert systems: Automation of on-line control with expert systems increases reliability when influent loadings vary greatly and when processes have short liquid detention times (47)
7. *In Situ* bioreclamation of contaminated aquifers: Biologically active zones, in which an increased mass of microorganisms biodegrades the contaminants in place, are stimulated by injecting required nutrients and substrates into the aquifer (48)

---

particles. Bed fluidization allows the use of small particles having a very high specific surface area, but without pore clogging by the biofilm. Fluidization also results in relatively rapid external mass transport of substrates to the biofilm.

Figure 10.4 illustrates the three main methods for creating bed fluidization. In the traditional liquid-fluidized bed, the upward velocity is controlled, usually via an effluent recycle flow, in such a manner that the bed is fluidized (or expanded) by 10% to several hundred percent of its unexpanded height. This expansion opens up the pores and increases porosity. Use of effluent recycle makes control of bed fluidization independent of influent loading.

Air-lift liquid fluidization is brought about through injection of air into the center section of a reactor having an annular baffle. The liquid flows up through the center and down the outside of the baffle. The upward flow of the liquid causes bed fluidization. Any medium that escapes the inner section is recycled to the bottom of the fluidized section. Control of bed fluidization is achieved by gas-flow control, which also is independent of influent loading.

Gas–turbulence fluidization does not take place in a columnar reactor. Instead, gas is introduced in a shallower tank. The turbulence caused by the gas causes random liquid velocities that fluidize the medium.

At least in concept, any fluidization approach can be combined with any biological reaction. In practice, the air-lift and gas-turbulence methods have been used mostly with aerobic processes, which require aeration. However, anaerobic processes could also be operated in these modes, as long as the gas phase was recirculated for fluidization. The traditional liquid-fluidized bed has been used with anaerobic (44), denitrifying (49), and aerobic (50) reactions. In most cases, gases are produced (CH_4, N_2) or required (O_2); hence the traditional liquid fluidized bed often is a three-phase process.

Because of the very high specific surface area of particles smaller than 1 mm, high volumetric loads can be achieved when the surface loads (substrate fluxes) are modest. Under these circumstances, the biofilm loss rate can be very low (51) and the density very high (51,52). These characteristics can enhance effluent quality by making the effluent suspended solids low and by allowing good biofilm mixing, which allows the effluent substrate concentration to decrease below S_{min}, the minimum concentration able to sustain steady-state biomass (2,53). The effect is accentuated further when the surface texture of the support medium is highly irregular, as with granular activated carbon (54).

Fluidized-bed technology still requires improved knowledge and technology in four areas. The first is flow distribution. When the bed diameter goes beyond the pilot scale, equal distribution of liquid and gas velocities is essential for stable bed fluidization and to preclude short-circuiting. In addition, bed and/or liquid oxygenation is still a major problem in high-rate aerobic systems (2).

Biomass distribution and control are essential (2). Too much biomass leads to bed stratification and loss of medium to the effluent. On the other hand, too little

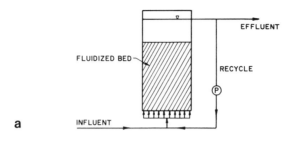

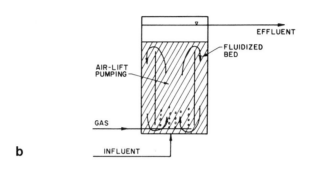

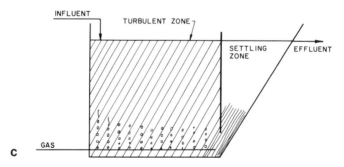

Fig. 10.4. Schematic diagrams of fluidized-bed biofilm processes. **a**: Traditional liquid fluidized. **b**: Air-lift liquid fluidized. **c**: Gas-turbulence fluidized.

accumulation, often from excessive detachment of slow-growing microorganisms, results in process failure. Thus work is needed to limit growth when surface loadings are high for fast-growing cells, and to minimize detachment of slow-growing microorganisms, particularly when surface loads are kept low in order to provide for excellent effluent quality.

Effective and reliable biofilm processes will be routine only after the mechanistic models of biofilm kinetics (*e.g.*, refs. 5,55) are translated into engineering practice. The insight and rigor of mechanistic models must be expressed with a few intuitively understandable parameters that are critical for characterizing system performance. Because current biofilm models appear to be complicated and difficult to use in practice, engineers will not benefit from them until they are presented in a context appropriate for engineering design and analysis. Although much remains to be achieved in translation for design, one recent attempt (56) using normalized loading curves is an important step in the right direction.

5. CONCLUSIONS

Environmental biotechnology is unique, because it is a service-oriented field that is driven by regulatory pressures. Today, the pressures for greater reliability, removal of specific contaminants, low effluent concentrations, control of toxicity, and new media to be treated are creating excellent incentives for innovation in biological treatment for pollution control.

Research is providing the knowledge needed to redefine what biological treatment can do. Key to research success is the explicit study and exploitation of greater structure in all levels of biological systems: substrates, cells, communities, and the process space itself.

Many technological advances are being implemented now or will be implemented soon. The interplay of needs, scientific understanding, and technology developments will determine what biological processes will be successful. Only those processes that meet a need and are effectively developed from sound scientific principles will survive and become the traditions of the next generation.

REFERENCES

1. Rittmann BE, Smets BF, Stahl DA: Genetic capabilities of biological processes, part I. Environ Sci Technol 24:23, 1990.

2. Rittmann BE: Aerobic biological treatment. Environ Sci Technol 21:128, 1987.

3. McCarty PL: One hundred years of anaerobic treatment. In Hughes DE, *et al.* (eds): Anaerobic Digestion 1981. Amsterdam: Elsevier, 1991, p 3.

4. Lawrence AL, McCarty PL: Unified basis for biological process design. J Sanitary Eng Div ASCE 96:757, 1970.

5. Rittmann BE, McCarty PL: Model of steady-state-biofilm kinetics. Biotechnol Bioeng 22:2343, 1980.

6. Henze M, Grady CPL Jr, Gujer W, Marais GVR, Matsuo T: A general model for single-sludge wastewater treatment systems. Water Res 21:505, 1987.

7. Rittmann BE, Sprouse GB: Fate of particulate organic matter in fixed-film anaerobic processes. Proc GRUTTEE conf Processus et Procedes Conduisant a la Methanisation, Meze, France, 1987 p 1.

8. Sprouse GB, Rittmann BE: Colloid removal in a fluidized-bed biofilm reactor. J Environ Eng 116:314, 1990.

9. Levine AD, Tchobanoglous G, Asano T: Characterization of the size distribution of contaminants in wastewater: Treatment and reuse implications. J Water Pollut Control Fed 57:805, 1985.

10. Comeau Y, Hall KJ, Hancock REW, Oldham WK: Biochemical model for enhanced biological phosphorous removal. Water Res 20:1511, 1986.

11. Meganck MTJ, Faup GM: Enhanced biological phosphorous removal. In Wise DL (ed): Biotreatment Systems, vol III. Boca Raton, FL: CRC Press, Inc., 1988, p 111.

12. Chudoba J, Cech JS, Farkac J, Grau P: Control of activated sludge filamentous bulking. Experience verification of a kinetic selection theory. Water Res 19:191, 1985.

13. Daigger GT, Robbins MH, Marshall BR: The design of a selector to control low F/M filamentous bulking. J Water Pollut Control Fed 57:220, 1985.

14. Kobayashi H, Rittmann BE: Microbial removal of hazardous organic compounds. Environ Sci Technol 16:170A, 1982.

15. Namkung E, Stratton RG, Rittmann BE: Predicting removal of trace organic compounds by biofilms. J Water Pollut Control Fed 59:670, 1987.

16. Namkung E, Rittmann BE: Modeling bisubstrate removal by biofilms. Biotechnol Bioeng 29: 269, 1987.

17. Namkung E, Rittmann BE: Estimating volatile organic compound (VOC) emissions from publically owned treatment works (POTWs). J Water Pollut Control Fed 59:670, 1987.

18. Namkung E, Rittmann BE: Soluble microbial products (SMP) formation kinetics by biofilms. Water Res 20:795, 1986.

19. Rittmann BE, Bae W, Namkung E, Lu CJ: A critical evaluation of microbial product formation in biological processes. Water Sci Technol 19:517, 1987.

20. Gaudy AF Jr, Blachly JR: A study of the biodegradability of residual COD. Proc 39th Purdue Industrial Waste Conf, 1984.

21. Lehninger AL: Biochemistry, 2nd ed. New York: Worth Publishers, Inc., 1975.

22. Matin A, Gottschal JC: Influence of dilution rate on NAD(P) and NAD(P)H concentrations and ratios in a *Pseudomonas* sp. grown in continuous culture. J Gen Microbiol 94:333, 1976.

23. Wilson DF, Erecinska M, Drown G, Silver IA: Effects of oxygen tension on cellular energetics. Am J Physiol 233:233, 1977.

24. Sherratt D: Control of plasmid maintenance. Symp Soc Gen Microbiol 39:239, 1986.

25. Smets BF, Rittmann BE, Stahl DA: Genetic capabilities of biological processes, part II. Environ Sci Technol 24:162, 1990.

26. Hardman DJ, Gowland PC, Slater JH: Large plasmids from soil bacteria enriched on halogenated alkanoic acids. Appl Environ Microbiol 51:44, 1986.

27. Sezgin M: The role of filamentous microorganisms in activated sludge settling. Prog Water Technol 12:97, 1980.

28. Strom PF, Jenkins D: Identification and significance of filamentous microorganisms in activated sludge. J Water Pollut Control Fed 56:449, 1984.

29. McCarty PL: Energetics and bacterial growth. In Faust SP, Hunter JV (eds): Organic Compounds in Aquatic Environments. New York: Marcel Dekker, Inc., 1971, p 495.

30. Wanner O, Gujer W: A multispecies biofilm model. Biotechnol Bioeng 29:314, 1986.

31. Kissel JC, McCarty PL, Street RL: Numerical simulation of mixed-culture biofilm. J Environ Eng 110:393, 1984.

32. Manem J: Interactions Between Heterotrophic and Autotrophic Bacteria in Fixed Film Biological Processes Used in Drinking Water Treatment. PhD dissertation, Department of Civil Engineering, University of Illinois, Urbana, 1988.

33. Rittmann BE, Langeland WE: Simultaneous denitrification with nitrification in single-channel oxidation ditches. J Water Pollut Control Fed 57:300, 1985.

34. Barnard JL: Biological denitrification. Water Pollut Control (G.B.) 72:705, 1973.

35. Laquidara WD, Keinath TM: Influence of three factors on clarification in the activated sludge process. J Water Pollut Control Fed 55:1331, 1983.

36. Dick RI: Analysis of the performance of final setting tanks. Trib Cebedeau 33:359, 1980.

37. Parker DS: Assessment of secondary clarification design concepts. J Water Pollut Control Fed 55:349, 1983.

38. Cheryan M: Ultrafiltration Handbook. Lancaster, PA: Technomic Publishing Company, Inc., 1986.

39. McCarty PL: Organics in water: An engineering challenge. J Environ Eng 106:1, 1980.

40. Rittmann BE, Jackson D, Storck SL: Potential for treatment of hazardous organic chemicals with biological processes. In Wise DL (ed): Biotreatment Systems, vol III. Boca Raton, FL: CRC Press, Inc., 1988, p 15.

41. Schultz JR, Keinath TM: Powdered activated carbon treatment process mechanisms. J Water Pollut Control Fed 56:143, 1984.

42. Chang HT, Rittmann BE: Verification of the model of biofilm on activated carbon. Environ Sci Technol 21:280, 1987.

43. McCarty PL, Smith DP: Anaerobic wastewater treatment. Environ Sci Technol 20:1200, 1986.

44. Jewell WJ: Anaerobic sewage treatment. Environ Sci Technol 21:14, 1987.

45. Rittmann BE, Huck PM: Biological treatment of public water supplies. CRC Crit Rev Environ Control 19:119, 1989.

46. Van Brunt J: Immobilized mammalian cells: The genetic way to productivity. Biotechnology 4:505, 1986.

47. Patry G (ed): Dynamic Modeling and Expert Systems in Wastewater Engineering. Ann Arbor, MI: Lewis Publishers, 1981.

48. Thomas JM, Ward CH: In situ biorestoration of organic contaminants in the subsurface. Environ Sci Technol 23:760, 1989.

49. Jeris JS, Beer C, Mueller JA: High rate biological denitrification using a granular fluidized bed. J Water Pollut Control Fed 46:2118, 1974.

50. Tanaka H, Uzman S, Dunn IJ: Kinetics of nitrification using fluidized sand bed reactor with attached growth. Biotechnol Bioeng 23:1683, 1981.

51. Wang YT, Suidan MT, Rittmann BE: Kinetics of methanogens in an expanded-bed reactor. J Environ Eng 112:155, 1986.

52. Trinet F, Heim R, Amar D, Chang HT, Rittmann BE: Study of biofilm and fluidization of bioparticles in a three-phase liquid-fluidized-bed reactor. Water Sci Technol 23:1347, 1991.

53. Rittmann BE: Comparative performance of biofilm reactor types. Biotechnol Bioeng 24:1341, 1982.

54. Chang HT, Rittmann BE: Comparative study of biofilm kinetics on different absorptive media. J Water Pollut Control Fed 60:362, 1988.

55. Saez PB, Rittmann BE: An improved pseudo-analytical solution steady-state-biofilm kinetics. Biotechnol Bioeng 32:379, 1988.

56. Heath MS, Wirtel S, Rittmann BE: Simplification design of biofilm processes using normalized loading curves. J Water Pollut Control Fed 62:185, 1990; see also Res J Water Pollut Control Fed 63:90–92, 1991.

BIOREMEDIATION OF ORGANIC CONTAMINANTS IN THE SUBSURFACE

E.J. BOUWER

Department of Geography and Environmental Engineering, The Johns Hopkins University, Baltimore, Maryland 21218

1. INTRODUCTION

Over the past few decades, the extensive production and use of synthetic organic compounds as organic solvents, fuels and fuel additives, pesticides, plasticizers, pigments, dyes, plastics, and chemical feedstocks has led to a wide distribution of these compounds throughout the environment. These synthetic chemicals enter the environment via losses during production, transport, and storage; dispersive losses in use of product; improper waste disposal; and accidental spills. Well water contaminated with chloroform, carbon tetrachloride, trichloroethylene, tetrachloroethylene, 1,1,1-trichloroethane, dichloroethylenes, methylene chloride, and dibromochloropropane has been widely encountered (1). Other compounds commonly observed in groundwater at trace levels include benzene, toluene, xylenes, phthalates, chlorinated aromatics, polynuclear aromatic hydrocarbons, and pesticides. A total of 39 pesticides have been detected in groundwater from 34 states or provinces with concentrations in the 0.1 to 5 μg/liter range (2). The most commonly found pesticides in groundwater fall into two classes: 1) soil fumigants and nematocides, such as EDB, 1,2-dichloropropane, and DBCP; and 2) herbicides, such as alachlor, atrazine, and 2,4-D. The presence of such organic contaminants in the subsurface has helped focus na-

Environmental Microbiology, pages 287–318, © *1992 Wiley-Liss, Inc.*

tional attention to our vast but limited groundwater resource. Groundwater supplies 25% of the fresh water for drinking, industrial, and agricultural purposes in the United States. About 50% of all Americans rely on groundwater, much of it minimally treated or untreated, as a primary source of drinking water (3). Approximately 75% of U.S. cities derive their water supplies, in total or in part, from groundwater.

An understanding of the processes that affect the fate of organic contaminants is needed in order to make sound decisions about their hazards to the public and environment and to develop suitable mitigation techniques. Important processes that influence the fate of organic contaminants in aquatic environments include sorption, volatilization, abiotic transformation (chemical or photochemical), and biotransformation. Sorption and volatilization do not destroy contaminants; rather, they concentrate them or transfer them to another medium. Abiotic chemical transformations involving organic contaminants are often slow, and photochemical reactions are insignificant in the subsurface. There is increasing recognition that bacteria are present in the subsurface and that they are capable of biotransforming many organic contaminants.

There is an urgent need to clean up polluted groundwater supplies and soils. Clean-up methods often employ flushing the subsurface with water so that contaminants can dissolve and be pumped to the surface for aboveground treatment. Because organic contaminants generally sorb to soils, they are not readily leached from the soil, and such aboveground treatment systems are generally inefficient and slow. Furthermore, most aboveground technologies involve physicochemical processes that simply sequester the contaminants or transfer them to another environmental medium. Bioremediation offers a potentially more effective and economical clean-up technique through partial or complete destruction of the contaminants. The process stimulates the growth of indigenous or introduced microorganisms in regions of subsurface contamination and thus provides direct contact between microorganisms and the dissolved and sorbed contaminants for biotransformation. Use of bioremediation for treatment of subsurface regions is challenging because these regions are difficult to characterize and the introduction of chemicals and microorganisms is not easy. Coupling this with our limited understanding of factors controlling biotransformation pathways and reaction rates of many organic contaminants of concern makes establishing the utility of *in situ* bioremediation an important scientific and engineering problem. This chapter addresses some important issues concerning the transformation of organic contaminants that can be applied to the problem of environmental contamination as well as to the development of engineered treatment processes for subsurface clean up. These include a discussion of relevant biotransformation processes, the role of electron acceptor, treatment scenarios, biotransformation rate, and reaction stoichiometry.

2. BIOTRANSFORMATION OF ORGANIC COMPOUNDS

Many organic contaminants are subject to attack by microorganisms (4–6). Biochemical pathways for some chemical classes are fairly well known, such as alkanes, alkylbenzenes, phenols, and alcohols. Biotransformations of organic contaminants are facilitated by enzymes during normal metabolic functions of microorganisms. The advantages of biotransformation include complete mineralization of contaminants to innocuous products and the observation that microbially mediated enzymatic reactions are generally faster than those of the same reactions in the absence of microorganisms.

2.1. Mechanisms

Biotransformations are driven by the ultimate goal of increasing the size and mass of microbial populations. Microorganisms must transform environmentally available nutrients to forms that are useful for incorporation into cells and synthesis of cell polymers. In general, cells utilize reduced forms of nutrients for these synthesis reactions. Reducing nutrients requires energy and a source of electrons. An electron donor is essential for growing cells; energy is made available for cell growth when the electron donor transfers its electrons to a terminal electron acceptor. Organic contaminants are often metabolized as the electron donors. Following is an example of a biotransformation in which an organic contaminant typified as benzene (C_6H_6) is oxidized to innocuous compounds and supports microbial growth:

$$C_6H_6 + 7.5\ O_2 \rightarrow 6\ CO_2 + 3\ H_2O \tag{1a}$$

$$C_6H_6 + 1.5\ HCO_3^- + 1.5\ NH_4^+ \rightarrow 1.5\ C_5H_7O_2N + 1.5\ H_2O \tag{1b}$$

In reaction 1a, the transfer of electrons between the organic contaminant benzene (electron donor) and O_2 (electron acceptor) provides energy for synthesis of cellular material ($C_5H_7O_2N$) from the benzene carbon (reaction 1b). By this process, a portion of an organic contaminant serves as a primary energy source that is converted to end products, and a portion of the contaminant carbon is synthesized into biomass. When used as primary substrates, organic contaminants are often completely degraded or detoxified. Examples of some xenobiotics that can serve as primary substrates include phenol, chlorophenols, chlorobenzoates, alkanes, benzene, toluene, xylene, and chlorobenzenes.

Because trace concentrations of organic contaminants are typically present in the environment (*e.g.*, at the μg/liter level), they frequently cannot support microbial growth as sole electron donors. In this situation, the organic contaminant can still be transformed by a microbial population that obtains the majority of its energy and carbon from a different compound that serves as the primary substrate. This metabolism is sometimes referred to as *secondary utilization* since the organic contaminants available at trace levels become secondary substrates and contribute negligibly to the energy and carbon needed for synthesis of new cell mass (7). Note that a secondary substrate does not have to share the same enzymatic pathways as the primary substrate, but the microorganism must be capable of biotransforming both compounds.

There are many organic contaminants that are biotransformed in the environment for which no microorganisms able to use them as sources of energy or carbon can be isolated. In this special case of secondary utilization, the organic contaminant cannot support any growth or maintenance since the growth yield is zero. This process, cometabolism, has been recently defined as "the transformation of a non-growth substrate in the obligate presence of a growth substrate or another transformable compound" (8). In cometabolism, enzymes produced by the microorganisms to metabolize the primary substrate can interact with an organic contaminant and bring about its transformation in a fortuitous manner. In many documented cases of cometabolism, the organic contaminant is an analog of the growth-supporting substrate. The enzymes that function in the initial sequences of metabolic pathways and transform organic contaminants are often nonspecific. However, subsequent enzymes are specific for intermediates of the primary substrate with the result that partially transformed intermediates of the organic contaminant accumulate, and the organic contaminant supplies neither energy nor cell carbon.

The distinctions among cometabolism, secondary utilization, and fortuitous biotransformation are difficult to discern and are perhaps unimportant in assessing the fate of organic contaminants in the presence of microorganisms. More importantly, if microorganisms cannot obtain sufficient energy from the transformation of the organic contaminant, then an appropriate primary energy substrate must be present for the microbial community to grow and function. Therefore, bioremediation of contaminants at trace levels may involve the addition of exogenous primary energy substrates.

2.2. Subsurface Microbial Ecology

Bacteria are the predominant microorganisms found in subsurface ecosystems, although fungi and protozoa have also been observed (9–12). Until recently, however, the presence and activity of microorganisms below the top few meters of soil were believed to be either extremely limited or nonexistent because of oligotrophic nutrient conditions and because of results from an early study suggesting that microbial numbers declined rapidly with depth (13). This inaccurate belief resulted from a general lack of scientific interest in this area, combined with inadequate or inappropriate methods for detecting subsurface microbial activity (14). Newly developed techniques in microbial ecology have indicated that microorganisms are present and active at considerable depths in the

TABLE 11.1. Classes of Organic Contaminants Known To Be Biotransformed by Subsurface Microorganisms

Phenols	Alkylbenzenes
Benzoates	Nitro-substituted aromatics
Halogenated aliphatics and aromatics	Polycyclic aromatics
Phthalate esters	PCBs
Phenoxyacetic acid herbicides	Some pesticides

subsurface (12, 15, 16). Deeper regions of the unsaturated zone and uncontaminated, shallow water-table aquifers contain relatively uniform populations of around 10^6–10^7 organisms per gram of dry subsurface material. Acridine orange direct counting with respiratory activity measured by 2–(p–iodophenyl)–3–(p–nitrophenyl)–5–phenyltetrazolium chloride (INT) reduction or ATP extraction indicated that between 1% and 10% of the total cell count was metabolically active (16). Research conducted by the U.S. Department of Energy's Subsurface Science Program (1986–1988) showed that similar numbers of diverse, metabolically active microorganisms exist in subsurface sediments of the Savannah River Plant at depths of more than 500 m (17).

Morphologically, subsurface bacteria differ very little from terrestrial bacteria, except with respect to size (<1 μm). Gram-positive forms are more prevalent in many uncontaminated soils (18). Many bacteria indigenous to the subsurface lack cytoplasmic constituents and contain polyhydroxybutyrate (PHB)-like storage inclusions, suggesting that they have acclimated to their oligotrophic surroundings (19). Because of their smaller size, subsurface bacteria have a higher surface-to-volume ratio, which allows them to take up and utilize nutrients effectively from dilute solutions. These bacteria are predominantly attached to soil surfaces because of the high specific surface area of soils and oligotrophic or copiotrophic nutrition.

The metabolic capabilities of subsurface microorganisms are quite diverse and similar to terrestrial counterparts. Biotransformation of many types of organic contaminants appears possible. Classes of organic compounds known to be biotransformed by subsurface microorganisms are listed in Table 11.1 In some instances, the compounds are detoxified via mineralization or formation of innocuous intermediates and products. Occasionally the organic contaminants are transformed into new compounds that are of similar or greater toxicity, such as the production of vinylchloride from trichloroethylene in reducing environments and the formation of nitrosamines via activation of secondary amines.

3. ROLE OF ELECTRON ACCEPTORS

The metabolic activity of subsurface microorganisms alters chemical speciation in the vicinity of the cell. The energy for microbial growth is obtained via

TABLE 11.2. Electron Acceptors in Biotransformation Processes

Microbial Process	Electron Acceptor	Reaction[a]	Free energy change (ΔG°) at pH 7 (kcal/equivalent)
Aerobic respiration	O_2	$CH_2O + O_2(g) = CO_2(g) + H_2O$	-29.9
Denitrification	NO_3^-	$CH_2O + 0.8\ NO_3^- + 0.8\ H^+$ $= CO_2(g) + 0.4\ N_2(g) + 1.4\ H_2O$	-28.4
Mn(IV) reduction	Mn(IV)	$CH_2O + 2\ MnO_2(s) + 2\ HCO_3^- + 2\ H^+$ $= CO_2(g) + 2\ MnCO_3(s) + 3\ H_2O$	-23.3
Fe(III) reduction	Fe(III)	$CH_2O + 4\ FeOOH(s) + 4\ HCO_3^- + 4\ H^+$ $= CO_2(g) + FeCO_3(s) + 7\ H_2O$	-10.1
Sulfate reduction	$SO_4^=$	$CH_2O + 0.5\ SO_4^= + 0.5\ H^+$ $= CO_2(g) + 0.5\ HS^- + H_2O$	-5.9
Methanogenesis	CO_2	$CH_2O + 0.5\ CO_2(g)$ $= CO_2(g) + 0.5\ CH_4(g)$	-5.6

[a]Oxidation and reduction half-reactions for the combinations listed were obtained from Stumm and Morgan (77).

oxidation of organic compounds, hydrogen, or reduced inorganic compunds of iron, nitrogen, or sulfur. Terminal electron acceptors, which include oxygen under aerobic conditions, and nitrate, Mn(IV), Fe(III), sulfate, and carbon dioxide under anaerobic conditions, are required for these oxidations. The growth of microorganisms and corresponding redox reactions affect the surrounding chemical environment, altering compound concentrations, pH, and the oxidation-reduction potential of the system. The metabolic activity results in transformations of inorganic and organic materials, some of which are organic contaminants, as already discussed.

Microorganisms preferentially utilize electron acceptors that provide the maximum free energy during respiration. Of the common electron acceptors, oxygen provides the most free energy during electron transfer (Table 11.2). Use of nitrate, Mn(IV), Fe(III), sulfate, and carbon dioxide yields less energy during electron transfer according to the order listed in Table 11.2. In the subsurface there is limited capacity for mixing, and rates of oxygen replenishment from the atmosphere are extremely slow following consumption by aerobic reactions. The coupling of mass transport and reaction in the subsurface results in spatial gradients of electron acceptor concentrations. Competition between heterotrophs deriving energy from oxidation of reduced organic carbon and nitrifiers deriving

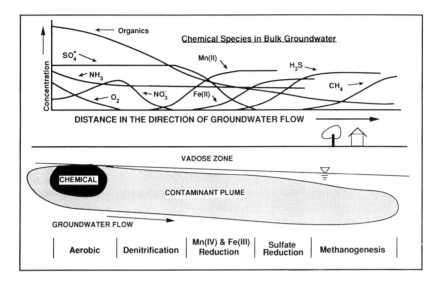

Fig. 11.1. Possible microbially mediated changes in chemical species and redox conditions in the direction of groundwater flow in the presence of organic contaminants.

energy from the oxidation of ammonia consumes oxygen. As oxygen is depleted, anoxic conditions follow with the sequential uitlization of nitrate (denitrification), Mn(IV) and Fe(III), sulfate (sulfate respiration or reduction), and carbon dioxide (methanogenesis) as electron acceptors by heterotrophs. The competition for available electron donors and acceptors can produce an ecological succession of microorganisms that accumulate along the groundwater flow path in the subsurface. The likely changes in bulk groundwater flow chemical composition are illustrated in Figure 11.1. The extent of overlap between the various redox regions in the subsurface is poorly understood. The importance of microbial reactions involving iron and manganese to organic contaminant biotransformations is unknown.

In many subsurface systems, the redox state is governed by microbial activity. Therefore, characterization of the geochemistry, speciation of redox components, and available electron donors and acceptors for respiration can provide a reliable indication of the microbiological conditions. Measurement of pH and concentrations of natural organic matter, oxygen, nitrogen, sulfur, iron and manganese species, methane, hydrogen, and trace metals are useful for this characterization. The relative distribution of these chemicals provides insight into the kinds of microorganisms present consistent with the chemical profiles illustrated in Figure 11.1. The chemical measurements are much less costly and time consuming to perform than is a detailed microbiological investigation that involves aseptic soil coring and sampling, enumeration of microorganisms, and culturing to assess active populations. The measurement of dissolved hydrogen concentration appears to correlate well with the predominant terminal electron-accepting reaction in anaerobic sediments (20). These authors suggest that steady-state H_2 concentrations are primarily dependent on the physiological characteristics of the microorganisms consuming the H_2 and will follow the order methanogens > sulfate-reducing > Fe(III)-reducing > Mn(IV)-reducing > nitrate-reducing. The uniqueness of the H_2 concentration associated with each of the electron acceptor conditions listed suggests that H_2 concentration can be used to identify the pertinent microbiological condition.

The redox environment is an important factor affecting microbial respiration and biotransformation of organic contaminants. Some compounds are only transformed under aerobic conditions; others require strongly reducing conditions, and still others are transformed in both aerobic and anaerobic environments. Therefore, knowledge of the spatial distribution of electron acceptors and other redox species within a region of subsurface contamination is important for identifying zones conducive to biotransformation of a particular organic contaminant. Examples of how the electron acceptor influences the biotransformation of some organic contaminants of concern are described in subsequent sections.

3.1. Aerobic Oxidation of Halogenated Solvents by Methanotrophs

Halogenated aliphatic compounds can be oxidized by methanotrophic bacteria under aerobic conditions. These bacteria cometabolize the halogenated compounds while using methane as a primary substrate. The initial enzyme responsible for methane utilization, methane-monooxygenase (MMO), is nonspecific and participates in the cometabolic transformations of a variety of other compounds, such as halogenated alkanes, alkenes, alcohols, cyclics, and aromatics (21). These compounds do not appear to be used by methanotrophs as a source of carbon or energy, but are oxidized in a fortuitous reaction secondary to methane oxidation. Trichloroethylene is mineralized in soil columns exposed to natural gas (22). Additional studies by Fogel et al. (23), Little et al. (24), Roberts et al. (25), and Leeson (26) using different methane-utilizing consortia grown on aquifer material have demonstrated that chlorinated methanes, ethanes, and ethylenes could be oxidized, and, in general, the rates of oxidation increased with a decrease in the number of chlorines on the molecule. For example, the chlorinated ethylenes used in order of decreasing rate were vinyl chloride, cis- and trans-dichloroethylene, 1,1-dichloroethylene, and trichloroethylene. Compounds saturated with chlorines, such as carbon tetrachloride and tetrachloroethylene, were not utilized by the methanotrophs. Epoxidation appears to be the first step in the oxidation of the chlorinated alkenes. The epoxides eventually yield aldehydes and acids prior to complete mineralization. A demonstration research project has confirmed that addition of methane and oxygen to an aquifer stimulated a native population of methanotrophs that was capable of substantially enhancing the transformation of halogenated aliphatic contaminants (25, 27). The prospect of stimulating methanotrophic activity for control of certain halogenated solvents in the subsurface appears good.

3.2. Biotransformation With Nitrate

The biodegradation of compounds with carboxyl functional groups attached to the aromatic ring (benzoate and analogs) under denitrification conditions has been well established (28, 29). However, it has long been thought that nitrate cannot serve as the terminal electron acceptor in hydrocarbon transformation. Recently it has been shown that nitrate can support the biotransformation of certain alkylbenzenes in subsurface material (30). m-Xylene was quantitatively (80%) oxidized to carbon dioxide at concentrations up to 0.4 mM (42 mg/liter) under conditions of denitrification. The m-xylene–adapted microorganisms biotransformed toluene, benzaldehyde, benzoate, m-toluylaldehyde, m-toluate, m-cresol, p-cresol, and p-hydroxybenzoate, but were unable to metabolize benzene, naphthalene, methylcyclohexane, and 1,3-dimethylcyclohexane (31). Bossert and Young (32) provide evidence for oxidation of p-cresol by a denitrifying bacterium. An acetate-supported biofilm with nitrate as electron acceptor was able to transform trace concentrations of hexachloroethane, bromoform, bromodichloromethane, and carbon tetrachloride (33).

The pathways for these transformations with nitrate are poorly understood. In the presence of molecular oxygen, aromatic and aliphatic hydrocarbons are initially oxygenated by mono- or dioxygenases and converted to catechols and alcohols, respectively, which are then amenable to subsequent mineralization (34). In the absence of molecular oxygen, oxygenases are inactive, and only those compounds with oxygen containing functional groups were thought to be mineralized (28). Aromatic and aliphatic hydrocarbons have no such activating substituent groups that facilitate hydration of the compounds. Studies under strongly reducing conditions of methanogenesis have provided insight into possible reactions in the absence of molecular oxygen. Degradation of (^{14}C)benzene and (^{14}C)toluene to $^{14}CO_2$ was reported in a methanogenic microbial culture originally enriched on ferulic acid (35). Trace amounts of p-cresol accumulated in the methanogenic culture growing in the presence of toluene. The oxygen incorporated into toluene came from water, as shown by using ^{18}O-labeled water and analyzing the produced p-cresol by mass spectrometry (36). Decomposition of alkylbenzenes in methanogenic aquifer material was reported by Wilson *et al.* (37). The methanogenic aquifer material incubated in the laboratory mineralized (^{14}C)toluene to $^{14}CO_2$.

Analogous to the methanogenic reactions above, there is some evidence that toluene biotransformation under denitrification conditions may proceed through benzyl alcohol and benzaldehyde or through p-cresol and p-hydroxybenzoate to benzoate before mineralization to CO_2 (31). Bouwer and Wright (33) found that 41% $\pm$ 3% of (^{14}C)carbon tetrachloride was mineralized to $^{14}CO_2$ in a denitrifying biofilm column reactor. However, bromoform was nearly stoichiometrically transformed to dibromomethane, and 14% of the carbon tetrachloride was transformed to chloroform. These latter end-products suggest that reduction reactions compete with mineralization to CO_2 in the absence of molecular oxygen.

3.3. Halogenated Compounds and Reductive Dehalogenation

Microbial reduction reactions involving organic contaminants increase in significance as environmental conditions become more reducing. For example, tetrachloroethylene was converted to trichloroethylene, dichloroethylene, and vinylchloride in methanogenic biofilm reactors (38, 39). A *Clostridium* sp. isolated from the methanogenic mixed-culture biofilm was able to dechlorinate 1,1,1-trichloroethane, chloroform, and carbon tetrachloride (40). Trichloroethylene transformation occurred in anaerobic soil with production of 1,2-dichloroethylene (41). Chlorophenols were metabolized in an active methanogenic aquifer by replacement of the halogen substituents with hydrogen atoms (42). Three of four chlorobenzoates, five of seven chlorophenols, and one of two chlorophenoxyacetate herbicides were reductively dehalogenated in methane-producing aquifer material (43). All three isomers of trichlorobenzene were reductively dechlorinated to chlorobenzene via dichlorobenzenes in columns wet-packed with anaerobic sediment from the river Rhine near Wageningen, The Netherlands (44). A study of PCB congener distribution in upper Hudson River sediments suggested that PCBs were being reductively dechlorinated by several populations of anaerobic bacteria (45). Sediments containing >50 ppm PCBs all showed losses of up to one-third of the chlorine originally present. Freon F113 appeared to have undergone transformation to one defluorinated and two dechlorinated products in anoxic groundwater at the Gloucester Landfill, Canada (46).

The above observed reductions of both halogenated aliphatic and aromatic compounds have been attributed to reductive dehalogenation. Here, the halogenated compound acts as an electron acceptor, and in this process a halogen is removed and is substituted with a hydrogen atom. Because of the highly electronegative character of halogen substituents, particularly on aliphatic compounds, many polyhalogenated compounds are relatively oxidized and serve as good electron acceptors, becoming reduced in the process (39). The more highly halogenated the compound, the faster the relative rate of reduction; while the less halogenated the compound, the faster the relative rate of oxidation. The stability of the carbon–halogen bond influences the reactivity of halogenated compounds.

Based on bond dissociation energies for methyl carbon–halogen and aromatic carbon–halogen bonds (tabulated by Morrison and Boyd [47]), the ease of reductive dehalogenation in a particular microbial redox environment generally follows the order I > Br > Cl >> F. The following three structural factors have also been found to influence the relative reactivity of halogenated aliphatics by reductive dehalogenation: 1) geminal and vicinal polyhalides are more reactive than simple monohalides; 2) an increase in the length of the carbon chain attached to the halogen atom decreases the reactivity; and 3) geminal halogens increase the reactivity over vicinal ones. The field observations cited above on organic contaminant biotransformations in reducing environments are consistent with these principles of relative reactivity.

Thermodynamic considerations can help indicate which organic contaminants are most susceptible to reductive dehalogenation in the presence of different electron acceptors being used by the microorganisms (39). A comparison between reduction potentials for electron acceptors in microbial respiration with acetate as electron donor is illustrated in Figure 11.2. Included in Figure 11.2 are the reduction potentials for reductive dehalogenation of hexachloroethane, ethylene dibromide, tetrachloroethylene, chlorobenzene, 1,2-dichlorobenzene, and pentachlorophenol. All of these reduction reactions appear to be energetically favorable under sulfate-reducing conditions. Reduction of ethylene dibromide and hexachloroethane even appears thermodynamically possible under aerobic respiration and denitrification.

The use of halogenated compounds as electron acceptors to support respiration might constitute an ecological advantage for subsurface microorganisms that ordinarily live where suitable electron acceptors are in short supply. The frequent occurrence of reductive dehalogenation reactions in anaerobic groundwaters suggests that the microorganisms involved are often present in the subsurface. Therefore, stimulation of their growth to increase rates of reductive dehalogenation in a bioremediation scheme is possible. Detoxification is likely if there is complete loss of halogens from the organic contaminant via a series of reductive dehalogenation reactions. However, reductive dehalogenation can also produce intermediate compounds that can have similar or greater associated health risks than the parent compound and remain a source of contamination. For example, reduction of tetrachloroethylene can lead to accumulation of vinylchloride (39), which is considered to have a significantly higher health risk (48).

3.4. Sequential Anaerobic/Aerobic Transformations of Halogenated Organic Compounds

Many examples of reductive dehalogenation of halogenated organic compounds under anaerobic microbial conditions have been described in a previous section. Halogenated compounds are relatively oxidized by the presence of halo-

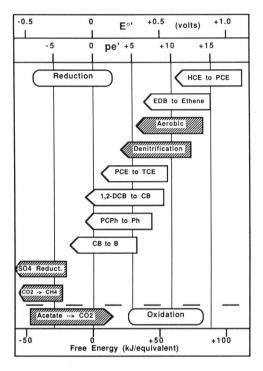

Fig. 11.2. Half-life reduction potentials for microbially mediated reactions. Base of arrows align with potential of half-reactions shown in volts. HCE, hexachloroethane; PCE, tetrachloroethylene; EDB, ethylene dibromide; TCE, trichloroethylene; CB, chlorobenzene; B, benzene; 1,2-DCB, 1,2-dichlorobenzene; PCPh, pentachlorophenol; Ph, phenol. Data were obtained from Stumm and Morgan (77) or computed as described by Vogel *et al.* (39).

gen substituents. The susceptibility to reduction reactions increases as the number of halogen substituents increases. Conversely, as the number of halogen substituents decreases on a given organic compound, reductive dehalogenation reactions become less rapid and are less likely to occur. Therefore, it is difficult to achieve complete loss of the halogen substituents by reductive dehalogenation under anaerobic conditions. Mono- and dihalogenated compounds tend to accumulate from the transformation of polyhalogenated organic compounds under reducing microbial conditions.

However, aerobic microorganisms are capable of transforming halogenated compounds, especially those with fewer halogen substituents. This oxidative process often results in complete mineralization to carbon dioxide and is mediated by three general mechanisms: incorporation of oxygen in the carbon–hydrogen bond, oxidation of a halogen substituent, and oxidation of a carbon–carbon double bond via epoxidation. With fewer halogen substituents, the more reduced the compound, and the more susceptible it is to oxidation. With removal of halogens, oxidation becomes more favorable than does reductive dehalogenation. Therefore, the combination of an anaerobic process followed by an aerobic reaction has promise for bioremediation of highly halogenated organic contaminants to innocuous products.

Examples of this combination of anaerobic and aerobic treatment have been tested in the laboratory. Tetrachloroethylene and trichloroethylene were transformed to dichloroethylene under methanogenic conditions in a 23 liter laboratory aquifer simulator containing contaminated soil and groundwater (49). A recirculation flow of glucose and nutrients was used to maintain methanogenic conditions. Oxygen was then introduced and the oxidation of dichloroethylene by methanotrophic bacteria was initiated. The sequential anaerobic/aerobic manipulations resulted in complete biotransformation of the tetrachloroethylene and trichloroethylene. Hexachlorobenzene, tetrachloroethylene, and carbon tetrachloride were dechlorinated to at least the dichlorinated products in a methanogenic biofilm column reactor fed acetate as the primary substrate (50). All of the reductive dechlorination products in the effluent of the methanogenic column were fed to an aerobic reactor seeded with settled sewage. The mono- and dichlorinated compounds were effectively utilized by the aerobic biofilm. The same sequential anaerobic/aerobic approach should be applicable for other halogenated organic contaminants. The dichlorobiphenyls, chlorobiphenyls, and biphenyl formed during reductive dechlorination of PCBs under reducing conditions (45) could be subsequently mineralized under aerobic conditions (51). Anaerobic reductive dechlorination of di-, tri-, and tetrachlorobenzenes yields chlorobenzene that can be completely mineralized under aerobic conditions (44).

3.5. Microbially Mediated Chemical Reactions

In the examples given above, microorganisms directly participated in the transformation of organic contaminants that are used as electron donors or acceptors. In addition, the environmental alterations that result from microbially mediated redox reactions at the microbial interface can lead to abiotic reactions, such as precipitation or dissolution of minerals and heavy metals, protonation or deprotonation of compounds and surfaces, and oxidation or reduction of compounds. For example, sulfate-reducing bacteria produce hydrogen sulfide, which is a strong nucleophile that can participate in substitution and addition reactions with several classes of organic functional groups. Hydrogen sulfide is also a strong reductant capable of reducing a variety of organic compounds. Cadmium, copper, lead, and zinc oxides are solubilized through microbial production of metabolites and acids that lower the pH (52). Much is yet to be learned about the interaction between chemical and biological processes involving organic contaminants.

4. TREATMENT SCENARIOS

The basic steps involved in subsurface bioremediation are 1) characterization of site hydrogeology and contamination, 2) removal of any separate immiscible phase, 3) assessment of biotransformation, 4) system design and operation, and 5) monitoring of system performance (53). Information regarding site geology and hydrology must be defined in order to determine properly the eventual location of the treatment system. Geological considerations should include stratigraphic effects such as horizontal extent of the aquifer and heterogeneity of the soil. Hydrogeological data include porosity, permeability, and groundwater velocity, direction, and recharge/discharge (54). In addition, hydraulic connection between aquifers, potential recharge/discharge areas, and water-table fluctuations must be considered. It is important initially to identify the contaminants present since the rates of biotransformation are compound specific.

The stimulation of microbial growth for biotransformation generally involves perfusion of nutrients and one or more electron acceptors through the formation. The process is most attractive when indigenous bacteria are used, as this avoids the significant problem of injecting and distributing a population of bacteria acclimated to the contaminants. Formations with hydraulic conductivities of 10^{-4} cm/sec or greater are most amenable to bioremediation. Other factors that are considered favorable for applying biotransformation in a cleanup operation are listed in Table 11.3. For comparison, unfavorable conditions for *in situ* bioremediation are also included in Table 11.3.

Feasibility studies for the biotransformation of the contaminants are usually conducted in the laboratory using subsurface material collected from the site. These experiments are conducted to establish the presence of microorganisms capable of biotransforming the organic contaminant(s), their nutrient and electron acceptor requirements, and the range of contaminant concentrations that are not completely inhibitory to the microorganisms. Sorption studies should be conducted to determine the extent and rates of partitioning of contaminants onto the aquifer solids. The greater the degree to which the contaminants are sorbed, the more difficult it is for the contaminant to come into contact with microorganisms, and the longer the time for clean up.

The appropriate treatment scenario will depend on the degree and type of contamination as well as its location. The most common means to contain and remediate contaminated groundwater is to extract the water and treat it at the surface, which is referred to as "pump-and-treat" technology (Fig. 11.3) (55). A conventional biological process can be employed in the aboveground treatment

TABLE 11.3. **Favorable and Unfavorable Chemical and Hydrogeological Site Conditions for Implementation of** *In Situ* **Bioremediation**[a]

Favorable factors
  Chemical characteristics
    Small number of organic contaminants
    Nontoxic concentrations
    Diverse microbial populations
    Suitable electron acceptor condition
    pH 6–8
  Hydrogeological characteristics
    Granular porous media
    High permeability ($K > 10^{-4}$ cm/sec)
    Uniform mineralogy
    Homogeneous media
    Saturated media
Unfavorable factors
  Chemical characteristics
    Numerous contaminants
    Complex mixture of inorganic and organic compounds
    Toxic concentrations
    Sparse microbial activity
    Absence of appropriate electron acceptors
    pH extremes
  Hydrogeological characteristics
    Fractured rock
    Low permeability ($K < 10^{-4}$ cm/sec)
    Complex mineralogy
    Heterogeneous media
    Unsaturated–saturated conditions

[a]Data obtained from Wagner *et al.* (78).

train to achieve bioremediation. Clean up of groundwater contamination typically proceeds slowly using pump-and-treat because of the effects of contaminant desorption and the effects of geologic complexity (56). The goal of reaching stringent health-based clean-up standards is difficult to achieve with pump-and-treat, and the ultimate cost of clean-up is high because pumping must be continued for a long time.

Stimulation of microbial processes *in situ* to achieve organic contaminant biotransformation may offer a better solution. If the organic contaminant is present as a separate immiscible phase, various pumping schemes should be used

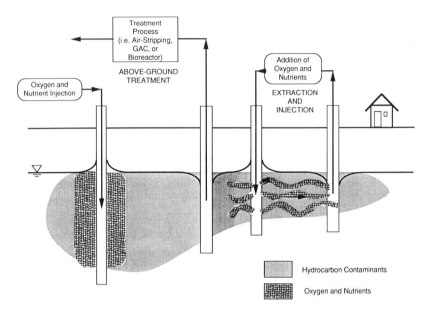

Fig. 11.3. Subsurface treatment using aboveground reactors, injection of oxygen and nutrients, or dynamic systems that involves injection and extraction wells.

to remove as much of the immiscible phase as possible prior to stimulation of biotransformation. A land farming approach can be implemented if the contamination is restricted to the soil near the land surface. The soil can be churned up, given appropriate nutrients and moisture, and incubated similar to a composting operation. Soil contaminated at shallow depths can be treated using infiltration galleys to allow nutrient-laden water to percolate through the soil. When contamination is located at greater depths, *in situ* bioremediation can be accomplished by simply infiltrating or injecting nutrients and oxygen into the contaminated subsurface to stimulate aerobic biotransformation in the contaminant plume (Fig. 11.3). Alternatively, the necessary growth factors could be injected downgradient from the contaminant plume to establish a biological treatment zone. Biotransformation of organic contaminants occurs as the plume percolates through the zone of microbial activity. If chemotactic bacteria are present, bacteria may slowly move upgradient in the direction of increasing organic contaminant concentrations. This will expand the zone of biotransformation and allow contact with sorbed and immiscible compounds. A dynamic system that includes injection and extraction wells and equipment for the addition and mixing of nutrients is often used to control flow and movement of nutrients and contaminants (Fig. 11.3). With bioremediaton, each pass of water laden with chemicals might stimulate subsurface microorganisms to transform a portion of the desorbed contaminant. Since the microorganisms colonize the soil surfaces, contaminants could be biotransformed as they desorb from the aquifer solids. Biotransformation reduces the solution contaminant concentration, thus enhancing the rate of desorption. Periodic sampling of the soil and groundwater is essential to determine the progress of the bioremediation.

5. ORGANIC CONTAMINANT BIOTRANSFORMATION RATES

Knowledge of the biotransformation rates in a bioremediation process is useful for determining the length of time required for meeting a treatment objective. Factors affecting the biotransformation rates of organic compounds used as primary substrates by bacteria are well known. Considerably less information is available about factors influencing rates of secondary utilization. Both primary and secondary utilization kinetics are described in the following sections.

5.1. Primary Substrate Utilization

The reaction rate for a xenobiotic that can serve as a primary energy source for bacterial growth is generally given by the Monod relationship:

$$\frac{dS}{dt} = -\frac{kXS}{K_s + S},$$
(2)

where S is the concentration of primary substrate, X is the concentration of active bacteria, k is the maximum specific substrate utilization rate, K_s is the Monod half-velocity coefficient, and t is time. The Monod relationship assumes a specific rate-limiting enzymatic process, a constant enzyme concentration, and no change in the species distribution of the microbial community. The applicability of these assumptions to microbial consortia and organic contaminant biotransformations in the subsurface needs to be critically examined.

At high substrate concentration, the biotransformation rate is at its maximum and zero order with respect to substrate concentration, but first order with respect to organism concentration:

$$\frac{dS}{dt} = -kX, \quad (S \gg K_s).$$
(3)

At substrate concentrations well below the K_s value, the biotransformation rate is first order with respect to both organism and substrate concentrations:

$$\frac{dS}{dt} = -\left(\frac{k}{K_s}\right)XS, \ (S<<K_s). \tag{4}$$

Kinetic models in the literature frequently refer to the ratio of k/K_s as the biodegradation second-order rate constant. Such a second-order model that assumes negligible change in active bacteria over the course of biotransformation is frequently useful for describing the rate of disappearance of organic contaminants at the low concentrations typical of natural environments (57). However, a model coupling equation 4 with change in microbial growth in response to primary substrate utilization was superior for describing p-cresol biodegradation (58).

In order for there to be net growth of the organisms consuming a primary substrate, the rate of growth must exceed the rate of decay (maintenance), as given by the following equation:

$$\frac{dX}{dt} = \frac{YkXS}{K_s + S} - bX, \tag{5}$$

where Y is the true yield coefficient and b is the first-order decay coefficient. The substrate concentration where growth just balances decay is termed S_{min}, the minimum substrate concentration, and is given by

$$S_{min} = \frac{K_s b}{Yk - b}. \tag{6}$$

The S_{min} concept implies that there is a minimum value under steady-state conditions to which a substrate concentration can be reduced when it is the sole source of energy. Consequently, if the organic contaminant enters the subsurface at a concentration below S_{min}, then there can be no net growth of organisms and the biotransformation rate will not increase. With the low microbial numbers in the subsurface, such compounds may not even be utilized or disappear slowly with times measured in months to years. However, when the organic contaminant concentration is greater than S_{min}, then there can be net growth, and both the microbial population and biotransformation rate increase with time.

Since organic contaminants are often present at trace levels in the subsurface, perhaps well below the S_{min} value, there is concern that they are likely to persist. However, it has been experimentally demonstrated that the organisms can grow if they are provided with a primary substrate at a concentration greater than their S_{min} requirement (59, 60). With a primary substrate sustaining growth, subsurface microorganisms may also be capable of simultaneously transforming an organic contaminant while using it as a secondary substrate.

5.2. Secondary Substrate Utilization

There is relatively little knowledge about factors affecting biotransformation rates by cometabolism or by secondary utilization. In some cases utilization of the organic contaminant occurs only while the primary substrate is being consumed. Primary substrate utilization is required to maintain the transforming enzymes in an active state. In the absence of a primary substrate, energy generated from storage compounds or through endogenous metabolism may also be adequate to produce active enzymes capable of transforming some organic contaminants.

Since the secondary substrate contributes negligibly to the energy needs of the cell, utilizaton of this compound is not directly coupled to microbial growth. Schmidt et al. (61) proposed several models for the kinetics of biodegradation of organic compounds not supporting growth. Some of the models involved exponential or logistic functions that were developed by curve-fitting rather than a mechanistic formulation.

An alternative approach to modeling the kinetics is to assume that the trace substrate is utilized by the biomass made from utilization of the primary substrate. The primary substrate can be a single organic compound or an aggregate of many compounds. The rate of secondary utilization is determined by coupling biofilm mass, which is supported by the primary substrate, with specific rate parameters for the secondary substrate (equation 2, 3, or 4). Insight into reaction kinetics using such an approach can be obtained by comparing the rate of organic contaminant biotransformation with the rate of primary substrate utilization. Relative utilizaton rates for several organic contaminants are given in Table 11.4 along with comparable rate data for the primary substrate, acetate, under both aerobic and methanogenic conditions as measured in continuous-flow acetate-supported biofilm columns adapted to the test compounds for at least 6 months. Here, reaction rates are indicated by the ratio of k/K_s. The k/K_s ratios for the secondary substrates were generally different from that of the acetate primary substrate. Some were higher and some were lower, but several were similar to that of acetate (within a factor of 3). These data indicate that secondary utilization may be faster than, the same as, or slower than primary substrate utilization. Generalization of these rate data to other organic contaminants and identification of the factors affecting their relative rates of utilization in multiple substrate systems are needed.

These k/K_s ratios may be important in estimating the half-lives ($t_{1/2}$) for organic contaminants in the subsurface using the expression

$$t_{1/2} = \frac{ln\ 2}{(k/K_s)X} .$$

(7)

TABLE 11.4. Measured Values of k/K_s and Corresponding Half-Lives in Days for Simultaneous Utilization of Acetate as Primary Substrate and Several Xenobiotics as Secondary Substrates Under Aerobic and Methanogenic Conditions[a]

	Aerobic culture				Methanogenic culture			
	Conc. (μg/liter)	k/K_s (L/mg/day)	$t_{1/2}$ (days)[b]	$t_{1/2}$ (days)[c]	Conc. (μg/liter)	k/K_s (L/mg/day)	$t_{1/2}$ (days)[b]	$t_{1/2}$ (days)[c]
Primary substrate								
Acetate	1 mg/liter	3.8	182	1.8	100 mg/liter	0.63	1,100	11
Secondary substrates								
Chlorobenzene	11	2.5	277	2.8				
1,2-Dichloro-benzene	10	10	69	0.7				
1,4-Dichloro-benzene	10	11	63	0.6				
1,2,4-Trichloro-benzene	11	5	139	1.4				
Ethylbenzene	9	35	20	0.2				
Styrene	8	50	14	0.14				
Naphthalene	14	40	17	0.17				
Chloroform					28	0.85	815	8.2
Carbon tetrachloride					17	2.1	330	3.3
1,1,1-Trichloro-ethane					18	0.46	1,500	15
Tetrachloro-ethylene					15	0.08	8,700	87

[a]Data from Bouwer and McCarty (79).
[b]Active microorganism concentration (X) = 0.001 mg/liter.
[c]Active microorganism concentration (X) = 0.1 mg/liter.

Equation 7 is likely to be adequate for large-scale subsurface contamination with low active microorganism concentration, slow groundwater movement, and biotransformation rates slow enough so that mass transfer limitation is insignificant. A summary of half-lives for trace organic contaminants as a fraction of the active biomass and the electron acceptor conditions is given in Table 11.4. The results indicate that, even with the low microbial populations indigenous to the subsurface in the range of 0.001 to 0.01 mg/liter, significant biotransformation of organic contaminants can occur. Elevation of the biomass concentration by one or two orders of magnitude in a bioremediation effort (0.1 mg/liter) by the addition of primary substrate, electron acceptor, or nutrients appears to result in half-lives between days and months. The half-life values also show how the electron acceptor condition markedly affects the biotransformation rates.

Simultaneous biotransformation of several trace substrates may offer a kinetic advantage that results in increased biomass accumulation and faster removal rates in comparison with the same concentration of a single substrate (62). This work suggests that biotransformation of a mixture of trace organic compounds enhances the stability of a mixed-culture system. The interaction between the primary substrate used for energy and organic contaminants that serve as secondary substrates needs more clarification. For example, LaPat-Polasko et al. (63) found that *Pseudomonas* sp. strain LP exhibited preferential utilization of methylene chloride over acetate regardless of which compound was the primary substrate. In addition, the presence of acetate stimulated the rate of methylene chloride utilization. Such interactions make formulation of reaction rates difficult, and reaction kinetics are likely to be more complex for secondary utilization of organic contaminants in comparison to the primary substrate alone.

There may be competition between the primary substrate and the organic contaminant as secondary substrate for an enzyme required for both reactions (64). The methane monooxygenase (MMO) enzyme that is used as the first step of methane oxidation by methanotrophs is an example of this kind of competition (65). This enzyme also transforms other compounds, such as trichloroethylene, in a cometabolic reaction. Methane as the primary substrate supplies the growth and maintenance energy requirements of the cell, and oxidation of methane generates reducing power that is essential for the MMO enzyme to function. The affinity of the MMO for methane is much greater than that for trichloroethylene. Consequently, increases in methane concentration reduce the number of enzyme sites available for trichloroethylene transformation, inhibiting trichloroethylene oxidation. However, some methane must be present to provide energy for growth and for the MMO to function. Therefore, an intermediate concentration of the methane will result in optimum biotransformation rate of the trichloroethylene.

This effect of methane concentration on trichloroethylene transformation by methanotrophs in batch cultures is shown in Figure 11.4. Headspace concentrations of methane ranged from zero to approximately 80% in oxygen (aqueous $[CH_4] = 17.4$ mg/liter). The initial trichloroethylene concentration was about 10 μg/liter. The maximum loss rate for trichloroethylene was observed with 16% methane (aqueous $[CH_4] = 3.5$ mg/liter). Both lower and higher values of methane in the headspace resulted in decreased rates of trichloroethylene transformation.

Enzyme kinetics for competitive inhibition can be used to model the interaction between methane and trichloroethylene for the MMO, which results in the following expression for substrate utilization rate (64):

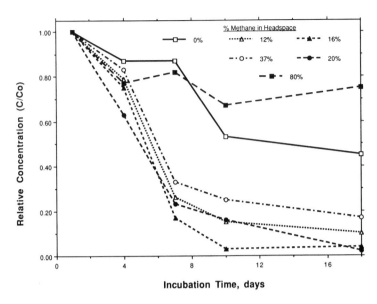

Fig. 11.4. Effect of methane concentration on trichloroethylene biotransformation by methanotrophs. Initial trichloroethylene concentration = 10 μg/liter. (Data from Leeson [26].)

$$\frac{dS}{dt} = -kX \frac{S}{S + K_s\left(1 + \dfrac{I}{K_i}\right)} .$$ (8)

Here I is the concentration of the inhibitor and K_i is the inhibition constant. In models of methane utilization rate, trichloroethylene becomes the inhibitor, and trichloroethylene utilization rate is modeled with methane as the inhibitor. For the batch data in Figure 11.4, the trichloroethylene concentrations were low compared with methane, and the dominant competitive effect was inhibition of trichloroethylene utilization by methane. The methane utilization was not influenced by the low trichloroethylene concentrations; therefore, Monod kinetics were appropriate for describing methane utilization. However, the competitive influence of methane on trichloroethylene biotransformation shown in Figure 11.4 was evaluated using equation 8. This yielded a K_i = 11 ± 14 mg/liter CH_4/liter (corresponds to 51% methane in the headspace) and k/K_s for trichloroethylene = 4.6 liter/g/day.

Many organic contaminants tend to sorb onto soil such that only a small fraction of the compound may actually be in the bulk water phase. Few studies exist to facilitate our understanding of how sorption affects biotransformation rates. There is evidence that sorbed compounds may either enhance or hinder these rates. Sorption can enhance the performance of microorganisms by removing from solution inhibitory compounds, thereby reducing their toxic effects on microbial growth. Also, retention of compounds through sorption can allow time for culture adaptation and ultimate biodegradation. The simplest approach is to assume that sorbed molecules are sequestered, and the biotransformation rate is determined by the solution concentration alone. This would reduce the overall reaction rate, since sorption reduces the solution concentration. The desorption rate of α-hexachlorocyclohexane from soil slurries appeared to limit the overall aerobic biotransformation rate (66). Modeling and experimental work with biofilms colonizing granular activated carbon (GAC) have demonstrated that, as biotransformation reduces the solution concentration, sorbed molecules can move into the solution phase and be transformed within the biofilm (59, 67). A better understanding of biotransformation rates in multisubstrate systems with sorption, and more fundamental studies on competitive interactions, are needed.

5.3. Solute Transport Modeling

A comprehensive model of bioremediation would involve the integration of the rate expressions (equations 2, 3, 4, 5, and 8) into solute transport models. A detailed coverage of this topic is beyond the scope of this chapter and has been covered elsewhere (68–70). A few noteworthy points are described here. One feature of the subsurface is that most of the microorganisms are attached to the soil particles and are not in the water phase. The rate of removal of the organic contaminant from the water phase is then described by a flux of the contaminant from the water to the microorganisms attached to the soil surfaces:

$$organic\ contaminant\ loss\ rate\ =\ -JA, \tag{9}$$

where J is the substrate flux to the attached microorganisms and A is the surface area of the attached microorganisms. The fixed-film models of Rittmann and McCarty (71, 72) provide relatively simple techniques to compute the flux J under both steady-state and nonsteady-state conditions.

The incorporation of equation 9 into a groundwater model with reaction terms presents no conceptual challenge to modeling subsurface transport; however, several difficulties can arise in this effort. The first occurs when large or spatially complicated subsurface systems are being modeled. The expressions for the flux J are nonlinear and require considerable computer time to obtain an accurate solution. A second difficulty is that several different components, such as electron donors and acceptors and biomass fractions, need to be considered to describe microbial growth accurately. These components must be properly combined to consider sequential and competitive utilization of various electron donors and acceptors, growth of the individual biomass fractions, and the interaction between primary and secondary substrates. Furthermore, the influence of electron acceptor condition on organic contaminant biotransformation must be included. The characteristic time and length scales for the microbial processes differ markedly from the values for physical transport processes, which creates numerical difficulties in the model solution. A fourth complication is that accumulation of biomass in the soil can lead to a reduction in permeability. This will alter flow paths and delivery of substrate to the attached growth. The mechanisms of clogging by microbial growth are not sufficiently developed to be able to establish accurate quantitative expressions that can be used in a model. The above difficulties are often accentuated when bioremediation strategies are to be modeled. The use of injection and extraction wells (Fig. 11.3) creates gradients in flow velocity and solute concentration. These gradients in turn induce localized and heterogeneous microbial activity. These limitations are severe enough that incorporation of biological processes into solute transport models is not presently common in field practice; successful application has largely been restricted to research studies.

6. BIOTRANSFORMATION STOICHIOMETRY

In subsurface bioremediation, nutrients and an electron acceptor often must be supplied in order to stimulate microbial growth. These chemicals are not among the chemical constituents available in the contaminated region, and growth is limited without them. In the engineering of a microbial treatment system for organic contaminants, the chemical needs of the microorganisms must be defined and are described by the reaction stoichiometry.

When the organic contaminant can be used as the sole source of energy for microbial growth, a stoichiometric equation for the overall biotransformation can be readily established using the thermodynamic model reported by McCarty (73). In this model, electrons from the electron donor (typically the organic contaminant) can be coupled with the electron acceptor to generate energy or can be used to synthesize biomass. The relative amount of the electron donor being oxidized for energy and being converted to biomass is established with an energy balance. The amount of energy released during oxidation of the electron donor must balance the amount required to synthesize the cell material. Several balanced equations for possible bioremediation reactions are shown in Table 11.5.

Such stoichiometric relationships established with the model can be used to determine the appropriate solution of nutrients and electron acceptors to flush throughout the zone of contamination. An example of stoichiometric calculations for toluene contamination follows. When hydrocarbons enter the subsurface, the

TABLE 11.5. Stoichiometric Relationships for Possible Bioremediation Reactions[a]

Reaction 1: aerobic biotransformation of toluene

C_7H_8 + 2.66 O_2 + 1.27 HCO_3^- + 1.27 NH_4^+ → 1.27 $C_5H_7O_2N$ + 1.93 CO_2 + 2.73 H_2O

Reaction 2: biotransformation of toluene with nitrate

C_7H_8 + 5.31 NO_3^- + 5.31 H^+ → 0.41 $C_5H_7O_2N$ + 2.45 N_2 + 4.94 CO_2 + 5.21 H_2O

Reaction 3: methane utilization by methanotrophs

CH_4 + 1.04 O_2 + 0.137 NO_3^- + 0.137 H^+ → 0.137 $C_5H_7O_2N$ + 0.315 CO_2 + 1.59 H_2O

Reaction 4: reductive dechlorination of trichloroethylene to dichloroethylene

C_2HCl_3 + 0.47 CH_3COO^- + 0.089 NH_4^+ + 0.619 H_2O → 0.089 $C_5H_7O_2N$ + $C_2H_2Cl_2$ + HCl + 0.117 CO_2 + 0.385 HCO_3^-

[a]$C_5H_7O_2N$ = empirical formula for biomass. All compounds were considered in aqueous phase, except CO_2, N_2, and CH_4 were taken as gaseous. Free energy of formation values for the organic compounds were obtained from Handbook of Organic Chemistry (80).

residual amount retained on the soil after drainage has been found to range between 0.5% and 4% by volume depending on the soil type (14). Medium to fine sand typically retains 1% to 2% by volume of the organic contaminant. For this sand with bulk density of 2,000 kg/m^3, about 13 kg of residual toluene (15 liters) will remain per m^3 of soil after removal of free product by pumping and draining. Toluene is readily biotransformed under aerobic conditions (5). According to reaction 1 in Table 11.5, 12 kg of oxygen and 2.5 kg of ammonia nitrogen would be required per m^3 of soil in order for the biotransformation of toluene to proceed properly and to be in balance. Other nutrients that are required in lesser quantities for bacterial growth are not included in the balanced equation. However, the phosphorus requirement is about one-sixth of that for nitrogen. Thus, this toluene biotransformation would require 0.42 kg of phosphorus per m^3 of soil. Bacterial biomass is represented by the empirical formula $C_5H_7O_2N$, and 20.3 kg of cells would be formed per m^3 of soil during the aerobic oxidation of the toluene. One approach for conveying these chemicals to the zone of toluene contamination is to inject water with dissolved oxygen, ammonia, and phosphorus so that it comes in contact with the toluene. To supply the 12 kg of oxygen required per m^3 of soil, 1,500 m^3 of water at air saturation (approx. 8 mg/liter) would be required or about 6,000 volume changes of water (soil porosity = 25%). Alternatively, pure oxygen could be used to saturate the water to an oxygen level of 40 mg/liter to reduce the volume of water required to 300 m^3 or 1,200 volume changes. The limited solubility of oxygen in water greatly limits the biotransformation of toluene in the contaminated soil. The rate of supply of oxygen effectively controls the rate of toluene biotransformation. For this aerobic process, large quantities of water must be pumped for a long time period at an enormous cost.

Hydrogen peroxide and nitrate are two possible chemical additives that could be used to increase the oxidant capacity of the contaminated soil to promote organic contaminant biotransformation. Peroxide is more than seven orders of magnitude more soluble in water than molecular oxygen. Henry's Law constants are 7.1×10^4 and 1.3×10^{-3} M/atm for hydrogen peroxide and molecular oxygen, respectively (74). Thus, much more oxidant can be delivered during well injection when hydrogen peroxide replaces oxygen. It is thermodynamically unstable with respect to molecular oxygen (O_2) and water and tends to disproportionate once in the ground:

$$H_2O_2 + H_2O_2 = O_2 + 2H_2O. \tag{10}$$

One mole of hydrogen peroxide is equivalent to 0.50 mol of molecular oxygen. The molecular oxygen produced by disproportionation is then available as an electron acceptor by bacteria for decomposition of organic contaminants. For the above case with toluene, 25.5 kg of hydrogen peroxide per m^3 soil would be required to supply the needed oxygen.

The oxidant capacity of hydrogen peroxide can be consumed by abiotic processes that do not involve production of molecular oxygen. Hydrogen peroxide is a powerful oxidant, capable of oxidizing Fe(II), Mn(II), H_2S, and other inorganic and organic reductants. Water is the ultimate product of oxidation by hydrogen peroxide. Consequently, its usefulness as a source of molecular oxygen for aerobic biotransformations may be limited in aquifers by other hydrogen peroxide–consuming reactions that compete with disproportionation. Under optimum conditions, disproportionation occurs slowly enough to provide a continual electron acceptor supply for bacterial utilization. If it occurs too quickly, evolution of oxygen gas can form bubbles that reduce its availability to bacteria and can lower aquifer permeability. Since the abundance of iron, manganese, organic matter, and other hydrogen peroxide–reacting species varies from one soil to another, the performance of bioremediation with hydrogen peroxide varies substantially from site to site. Modifications in treatment strategy may also be necessary to minimize hydrogen peroxide toxicity. One study of groundwater in sand columns showed significant toxicity toward inoculated bacteria at hydrogen peroxide concentrations above 1.5×10^{-2} M (75).

A second approach to increase oxidant capacity is to add nitrate. This will stimulate denitrifying bacteria that can possibly oxidize alkylbenzenes when using nitrate as the electron acceptor as described in a previous section. Nitrate is advantageous because it is inexpensive, is very soluble, is chemically stable, and is nontoxic to microorganisms. Many genera of denitrifying bacteria are commonly found in the subsurface, such as *Arthrobacter*, *Bacillus*, *Pseudomonas*, *Agrobacterium*, *Alcaligenes*, and *Flavobacterium* (76). Thus there is a good chance that denitrifying bacteria will already be present in the contaminated soil. Biotransformation rates with nitrate are generally similar to those with oxygen. Nitrate could be added in sufficient quantity so that the organic contaminant becomes the rate-limiting compound for microbial growth. This will greatly reduce the number of pore volumes of water that will need to be recirculated to decompose the organic contaminant and will shorten the time for clean up. Nitrate will also be easier to distribute throughout the contaminated soil in contrast with dissolved oxygen that will be depleted by the microorganisms within a short travel distance from an injection well. For the above example with toluene, 46.5 kg of nitrate per m^3 of soil would be required to supply the needed electron acceptor (reaction 2, Table 11.5). Since nitrate is a contaminant of concern in drinking water supplies (current maximum contaminant level is 10 mg/liter NO_3–N), controlled stoichiometric addition of nitrate will be needed to prevent nitrate pollution during a bioremediation operation.

The production of biomass in a bioremediation process is another engineering factor that must be considered. As a result of toluene biotransformation in the above example, 20.3 kg of biomass would be formed with oxygen and 6.5 kg of biomass would be formed with nitrate per m^3 of soil (Table 11.5). The microbial growth under aerobic conditions forms a larger mass than that of the original toluene. The biomass growth with either electron acceptor has the potential for reducing the soil permeability and interfering with water flow during injection of chemicals. About 80% of the biomass will be biodegradable, and if additional oxygen or nitrate is supplied, the biomass will decay to about 20% of the amounts given in Table 11.5.

If the organic contaminant is biotransformed as a secondary substrate, then a primary substrate must be supplied. Our knowledge of the chemical requirements for optimum biotransformation of secondary substrates is poor. Studies with methanotrophic bacteria that biotransform trichloroethylene using methane and oxygen as primary substrates indicated that the ratio of methane mass used to the mass of trichloroethylene oxidized varied between 100/1 and 1,000/1 (25, 26). Consequently, large quantities of methane would have to be injected into the contaminated soil system for treatment of even a relatively small amount of trichloroethylene contamination. The stoichiometry given by reaction 3 in Table 11.5 can be used to obtain the appropriate amounts of methane, oxygen, and nutrients that must be supplied for growth of the methanotrophs and the amounts of biomass and other products that will be formed.

Methanogenic conditions have been shown to be conducive for reductive dehalogenation of trichloroethylene. The stoichiometry given by reaction 4 in Table 11.5 indicates the amount of electron donor represented by acetate needed for the single reductive dehalogenation reaction of trichloroethylene to dichloroethylene. However, trichloroethylene does not appear solely to support growth under methanogenic conditions, and a primary substrate must be present to maintain a viable microbial population. The mass of acetate to trichloroethylene transformed was 100/1 in studies using acetate as the primary substrate for methanogenic bacteria (38). Both of these examples of secondary utilization reactions illustrate the need for a large quantity of primary substrate. More research is needed to learn how to minimize the chemical requirements for secondary utilization of organic contaminants.

7. CONCLUSIONS

Biotransformation of organic contaminants is an important fate process in the subsurface. Many different bacterial species are capable of transforming organic contaminants either by oxidative or reductive reaction pathways; examples include the oxidation of halogenated solvents by methanotrophs, the oxidation of alkylbenzenes by bacteria that use nitrate as the terminal electron acceptor, and the reductive dehalogenation of the halogenated aliphatic and aromatic compounds by anaerobic microorganisms. This has led to an interest in using *in situ* techniques for biotransformation of organic contaminants as an alternative to aboveground treatment systems that generally involve physicochemical processes that do not destroy the contaminants. The objective of bioremediation is to stimulate the growth of indigenous or introduced microorganisms in regions of subsurface contamination and thus provide direct contact between microorganisms and the dissolved and sorbed contaminants for biotransformation. The process typically entails perfusion of nutrients and one or more electron acceptors through the contaminated soil. For some organic contaminants that can readily be used as primary substrates, enhanced biotransformation is relatively simple to achieve. However, the supply of nutrients and appropriate electron acceptor can be difficult in tight and heterogeneous soils. Much less knowledge exists to stimulate biotransformation of organic contaminants as secondary substrates. Knowledge of the biotransformation rates is important for determining the length of time required for achieving clean-up goals. Clearly, more information about factors affecting interdependency between primary and secondary substrate utilization is needed. Research needs to be conducted on factors affecting the relative rates of utilization of organic contaminant mixtures. More information about the chemical requirements for *in situ* bioremediation of trace organic contaminants will help to optimize their biotransformation. Both basic laboratory studies and well-controlled field experiments are needed to solve the many problems associated with bioremediation efforts.

ACKNOWLEDGMENTS

This work was supported in part by the U.S. National Science Foundation, grant ECE-8451060, and by the U.S. Environmental Protection Agency, grant

R814383010. I thank Melinda Trizinsky for assistance in preparing the manuscript.

REFERENCES

1. Pye VI, Patrick R, Quarles J: Groundwater Contamination in the United States. Philadelphia: University of Pennsylvania Press, 1983.

2. Hallberg GR: Pesticide pollution of groundwater in the humid United States. Agric Ecosyst Environ 26:299–367, 1989.

3. Solley WB, Merk CR, Pierce RR: Estimated use of water in the United States in 1985. U.S. Geological Survey Circular 1004. Washington, DC: U.S. Government Printing Office, 1988.

4. Alexander M: Biodegradation of organic chemicals. Environ Sci Technol 18:106–111, 1985.

5. Atlas RM: Microbial degradation of petroleum hydrocarbons: An environmental perspective. Microbiol Rev 45:180–209, 1981.

6. Kobayashi H, Rittmann BE: Microbial removal of hazardous organic compounds. Environ Sci Technol 16:3, 170A–183A, 1982.

7. McCarty PL, Reinhard M, Rittmann BE: Trace organics in groundwater. Environ Sci Technol 15:40–51, 1981.

8. Dalton H, Stirling DE: Co-metabolism. Philos Trans R Soc Lond 297:481–496, 1982.

9. Atlas RM, Bartha R: Microbial Ecology: Fundamentals and Applications. Reading, MA: Addison-Wesley Publishing Company, 1981.

10. Ghiorse WC, Balkwill DL: Microbial characterization of subsurface environments. In Ward CH, Giger W, McCarty PL (eds): Ground Water Quality. New York: John Wiley & Sons, 1985, pp 387–401.

11. Ghiorse WC, Wilson JT: Microbial ecology of the terrestrial subsurface. Adv Appl Microbiol 33:107–172, 1988.

12. Wilson JT, McNabb JF, Balkwill DL, Ghiorse WC: Enumeration and characterization of bacteria indigenous to a shallow water-table aquifer. Ground Water 21:134–142, 1983.

13. Waksman SA: Bacterial numbers in soils, at different depths, and in different seasons of the year. Soil Sci 1:363, 1916.

14. Wilson JT, Leach LE, Henson M, Jones JN: *In situ* biorestoration as a groundwater remediation technique. Ground Water Monit Rev 6:56–64, 1986.

15. Balkwill DL, Ghiorse WC: Characterization of subsurface bacteria associated with two shallow aquifers in Oklahoma. Appl Environ Microbiol 50:580–588, 1985.

16. Webster JJ, Hampton GJ, Wilson JT, Ghiorse WC, Leach FR: Determination of microbial cell numbers in subsurface samples. Ground Water 23:17–25, 1985.

17. Fliermans CB: Microbial life in the terrestrial subsurface of southeastern coastal plain sediments. Haz Water Haz Materials 6:155–171, 1989.

18. Lee MD, Thomas JM, Borden RC, Bedient PB, Ward CH, Wilson JT: Biorestoration of aquifers contaminated with organic compounds. CRC Crit Rev Environ Control 18:29–89, 1988.

19. Poindexter JS: Oligotrophy: Feast and famine existence. Adv Microbial Ecol 5:63–89, 1981.

20. Lovley DR, Goodwin S: Hydrogen concentrations as an indicator of the predominant terminal electron-accepting reactions in aquatic sediments. Geochim Cosmochim Acta 52:2993–3003, 1988.

21. Haber C, Allen L, Zhao S, Hanson R: Methylotrophic bacteria: Biochemical diversity and genetics. Science 221:1147–1153, 1983.

22. Wilson JT, Wilson BH: Biotransformation of trichloroethylene in soil. Appl Environ Microbiol 49:242–243, 1985.

23. Fogel MM, Taddeo AR, Fogel S: Biodegradation of chlorinated ethenes by a methane-utilizing mixed culture. Appl Environ Microbiol 51:720–724, 1986.

24. Little CD, Palumbo AV, Herbes SE, Lidstrom ME, Tyndall RL, Gilmer PJ: Trichloroethylene biodegradation by a methane-oxidizing bacterium. Appl Environ Microbiol 54:951–956, 1988.

25. Roberts PV, Hopkins GD, Mackay DM, Semprini L: A field evaluation of *in situ* biodegradation of chlorinated ethenes: Part 1, methodology and field site characterization. Ground Water 28: 591–604, 1990.

26. Leeson A: Aerobic Biotransformations of Halogenated Aliphatic Compounds by a Methanotrophic Consortium. Ph.D. Dissertation, Department of Geography and Environmental Engineering, The Johns Hopkins University, 1990.

27. Semprini L, Roberts PV, Hopkins GD, McCarty PL: A field evaluation of *in situ* biodegradation of chlorinated ethenes: Part 2, results of biostimulation and biotransformation experiments. Ground Water 28:715–727, 1990.

28. Evans WC: Biochemistry of the bacterial catabolism of aromatic compounds in anaerobic environments. Nature 270:17–22, 1977.

29. Schennen U, Braun K, Knackmuss HJ: Anaerobic degradation of 2-fluorobenzoate by benzoate-degrading, denitrifying bacteria. J Bacteriol 161:321–325, 1985.

30. Zeyer J, Kuhn EP, Schwarzenbach RP: Rapid microbial mineralization of toluene and 1,3-dimethylbenzene in the absence of molecular oxygen. Appl Environ Microbiol 52:944–947, 1986.

31. Kuhn EP, Zeyer J, Eicher P, Schwarzenbach RP: Anaerobic degradation of alkylated benzenes in denitrifying laboratory aquifer columns. Appl Environ Microbiol 54:490–496, 1988.

32. Bossert ID, Young LY: Anaerobic oxidation of p-cresol by a denitrifying bacterium. Appl Environ Microbiol 52:1117–1122, 1986.

33. Bouwer EJ, Wright JP: Transformations of trace halogenated aliphatics in anoxic biofilm columns. J Contam Hydr 2:155–169, 1988.

34. Gibson DT, Subramanian V: Microbial degradation of aromatic hydrocarbons. In Microbial Degradation of Organic Compounds. Gibson DT (ed): New York: Marcel Dekker, Inc., 1984, pp 181–252.

35. Grbic-Galic D, Vogel TM: Transformation of toluene and benzene by mixed methanogenic cultures. Appl Environ Microbiol 53:254–260, 1987.

36. Vogel TM, Grbic-Galic D: Incorporation of oxygen from water into toluene and benzene during anaerobic fermentative transformation. Appl Environ Microbiol 52:200–202, 1986.

37. Wilson BH, Smith GB, Rees JF: Biotransformations of selected alkylbenzenes and halogenated aliphatic hydrocarbons in methanogenic aquifer material: A microcosm study. Environ Sci Technol 20:997–1002, 1986.

38. Bouwer EJ, McCarty PL: Transformations of 1- and 2-carbon halogenated aliphatic organic compounds under methanogenic conditions. Appl Environ Microbiol 45:1286–1294, 1983.

39. Vogel TM, Criddle CS, McCarty PL: Transformations of halogenated aliphatic compounds. Environ Sci Technol 21:722–736, 1987.

40. Gälli R, McCarty PL: Biotransformation of 1,1,1-trichloroethane, trichloromethane, and tetrachloromethane by a *Clostridium* sp. Appl Environ Microbiol 55:837–844.

41. Kloepfer RD, Easley DM, Haas BB Jr, Deihl TG, Jackson DE, Wurrey CJ: Anaerobic degradation of trichloroethylene in soil. Environ Sci Technol 19:277–280, 1985.

42. Suflita JM, Miller GD: Microbial metabolism of chlorophenolic compounds in ground water aquifers. Environ Toxicol Chem 4:751–758, 1985.

43. Gibson SA, Suflita JM: Extrapolation of biodegradation results to groundwater aquifers: Reductive dehalogenation of aromatic compounds. Appl Environ Microbiol 52:681–688, 1986.

44. Bosma TNP, van der Meer JR, Schraa G, Tross ME, Zehnder AJB: Reductive dechlorination of all trichloro- and dichlorobenzene isomers. FEMS Microbiol Ecol 53:223–229, 1988.

45. Brown JF Jr, Bedard DL, Brennan MJ, Carnahan JC, Feng H, Wagner RE: Polychlorinated biphenyl dechlorination in aquatic sediments. Science 236:709–712, 1987.

46. Lesage S, Jackson RE, Priddle MW, Riemann PG: Occurrence and fate of organic solvent residues in anoxic groundwater at the Gloucester Landfill, Canada. Environ Sci Technol 24: 559–566, 1990.

47. Morrison RT, Boyd RN: Organic Chemistry, 3rd ed. Boston: Allyn and Bacon, Inc., 1974.

48. Sayre IM: International standards for drinking water. J Am Water Works Assoc 80:53–60, 1988.

49. Dooley-Danna M, Fogel S, Findlay M: The sequential anaerobic/aerobic biodegradation of chlorinated ethenes in an aquifer simulator. International Symposium on Processes Governing the Movement and Fate of Contaminants in the Subsurface Environment. Stanford University, Stanford, CA, July 23–26, 1989.

50. Vogel TM, Fathepure BZ, Selig H: Sequential Anaerobic/Aerobic Degradation of Chlorinated Organic Compounds. International Symposium on Processes Governing the Movement and Fate of Contaminants in the Subsurface Environment. Stanford University, Stanford, CA, July 23–26, 1989.

51. Kohler H-PE, Kohler-Staub D, Focht DD: Cometabolism of polychlorinated biphenyls: Enhanced transformation of Arochlor 1254 by growing bacterial cells. Appl Environ Microbiol 54:1940–1945, 1988.

52. Francis AJ, Dodge CJ: Anaerobic microbial dissolution of transition and heavy metal oxides. Appl Environ Microbiol 54:1009–1014, 1988.

53. Thomas JM, Ward CH: In situ biorestoration of organic contaminants in the subsurface. Environ Sci Technol 23:760–766, 1989.

54. Freeze RA, Cherry JA: Ground Water. Englewood Cliffs, NJ: Prentice-Hall, Inc., 1979.

55. Mercer JW, Skipp DC, Giffin D: Basics of Pump-and-Treat Ground-Water Remediation Technology. Washington, DC: Robert S. Kerr Environmental Research Laboratory, U.S. Environmental Protection Agency, EPA/600/8-90/003, 1990.

56. Mackay DM, Cherry JA: Groundwater contamination: Pump-and-Treat remediation. Environ Sci Technol 23:630–636, 1989.

57. Paris DF, Steen WC, Baughman GL, Barnett JT Jr: Second-order model to predict microbial degradation of organic compounds in natural waters. Appl Environ Microbiol 41:603–609, 1981.

58. Suflita JM, Smolenski WJ, Robinson JA: Alternative nonlinear model for estimating second-order rate coefficients for biodegradation. Appl Environ Microbiol 53:1064–1068, 1987.

59. Bouwer EJ, McCarty PL: Removal of trace chlorinated organic compounds by activated carbon and fixed-film bacteria. Environ Sci Technol 16:836–843, 1982.

60. Rittmann BE, Brunner CW: The nonsteady-state-biofilm process for advanced organics removal. J Water Pollut Control Fed 56:874–880, 1984.

61. Schmidt SK, Simkins S, Alexander M: Models for the kinetics of biodegradation of organic compounds not supporting growth. Appl Environ Microbiol 50:323–331, 1985.

62. Namkung E, Rittmann BE: Modeling bisubstrate removal by biofilms. Biotechnol Bioeng. 29:269–278, 1987.

63. LaPat-Polasko LT, McCarty PL, Zehnder AJB: Secondary substrate utilization of methylene chloride by an isolated strain of *Pseudomonas* sp. Appl Environ Microbiol 47:825–830.

64. Bailey JE, Ollis DF: Biochemical Engineering Fundamentals and Applications. Reading, MA: Addison-Wesley Publishing Company, Inc., 1986.

65. Strand SE, Bjelland MD, Stensel HD: Kinetics of chlorinated hydrocarbon degradation by suspended cultures of methane-oxidizing bacteria. Res J Water Pollut Control Fed 62:124–129, 1990.

66. Bachmann A, de Bruin W, Jumelet JC, Rijnaarts HN, Zehnder AJB: Aerobic biomineralization of alpha-hexachlorocyclohexane in contaminated soil. Appl Environ Microbiol 54:548–554, 1988.

67. Chang HT, Rittmann BE: Mathematical model of biofilm on activated carbon. Environ Sci Technol 21:273–280, 1987.

68. National Research Council: Ground Water Models Scientific and Regulatory Applications. Washington, DC: National Academy Press, 1990.

69. McCarty PL, Rittmann BE, Bouwer EJ: Microbiological processes affecting chemical transformations in groundwater. In Bitton G, Gerba CP (eds): Groundwater Pollution Microbiology. New York: Wiley Interscience, 1984, pp 89–115.

70. Bouwer EJ, Cobb GD: Modeling of biological processes in the subsurface. Water Sci Technol 19:769–779, 1987.

71. Rittmann BE, McCarty PL: Model of steady-state biofilm kinetics. Biotechnol Bioeng 22: 2343–2357, 1980.

72. Rittmann BE, McCarty PL: Substrate flux into biofilms of any thickness. J Environ Eng Div, ASCE, 107:EE4, 831–849, 1981.

73. McCarty PL: Energetics and bacterial growth. In Faust J, Hunter JV (eds): Organic Compounds in Aquatic Environments. New York: Marcel Dekker, Inc., 1971, pp 495–531.

74. Seinfeld J: Atmospheric Chemistry and Physics of Air Pollution. New York: John Wiley & Sons, Inc., 1986.

75. Britton LN: Feasibility studies on the use of hydrogen peroxide to enhance microbial degradation of gasoline. Texas Research Institute. Washington, DC: American Petroleum Institute, Report 36, 1985.

76. Payne WJ: Denitrification. New York: John Wiley & Sons, Inc., 1981.

77. Stumm W, Morgan JJ: Aquatic Chemistry. New York: John Wiley and Sons, 1981.

78. Wagner K, Boyer K, Claff R, Evans M, Henry S, Hodge V, Mahmud S, Sarno D, Scopino E, Spooner P: Remedial Action Technology for Waste Disposal Sites, 2nd ed. Park Ridge, NJ: Noyes Data Corporation, 1986.

79. Bouwer EJ, McCarty PL: Utilization rates of trace halogenated organic compounds in acetate-supported biofilms. Biotech Bioengr 17:1564–1571, 1985.

80. Dean JA: Handbook of Organic Chemistry. New York: McGraw-Hill Book Company, 1987.

12

THE IMPORTANCE OF GENETIC EXCHANGE IN DEGRADATION OF XENOBIOTIC CHEMICALS

MICHAEL BOYLE

Division of Applied Sciences, Harvard University, Cambridge, MA 02138

1. INTRODUCTION

Certain synthetic chemicals, because of their toxicity and bioaccumulation, are an unacceptable burden on the environment (1). Of particular concern are recalcitrant compounds such as pentachlorophenol (PCP), 1,1,1-trichloro-2,2-bis(p-chlorophenol) ethane (DDT), 2,4,5-trichlorophenoxyl acetic acid (2,4,5-T), and polychlorinated biphenyls (PCBs). Although bacteria have evolved mechanisms to degrade complex natural compounds (*i.e.*, lignin, humic material), certain chemical structures were absent or rare in the environment before their manufacture and use as pesticides or industrial chemicals (2). Xenobiotic chemicals, because they often contain these novel structures, provide a unique opportunity to study the microbial evolution of new degradative pathways.

The development of new catabolic pathways requires environmental conditions that can support a large genetic pool. The transfer of xenobiotic-catabolizing genes across taxonomic barriers has been demonstrated in the laboratory under artificial selective pressure (3). The natural construction of a xenobiotic-

Environmental Microbiology, pages 319—333, © *1992 Wiley-Liss, Inc.*

degrading pathway through genetic exchange is an attractive yet unproven hypothesis and is the main focus of this chapter.

2. GENETIC EXCHANGE

Microbial adaptation to new and unusual substrates is illustrated by the decomposition of xenobiotic chemicals. Biochemical and genetic strategies to enhance degradation include enzyme induction, expression of cryptic (previously repressed) genes, population changes, mutation, and genetic transfer (1,4). In enrichment studies, the time required for chemical degradation could be a result of such biotic factors as the exhaustion of preferred nonxenobiotic substrates; the growth of a species originally present at low numbers to achieve an ecologically significant density; or the time required for gene mutation and rearrangement. An acclimation period lasting several weeks could accommodate genetic exchange as a mechanism of microbial adaptation to hazardous organic chemicals (5). Genetic information can be transferred between closely associated populations through phage-mediated transduction, transformation of naked DNA, and plasmid-assisted conjugation.

The transfer of DNA has been demonstrated to occur through transduction in natural environments. The freshwater *in situ* rate of transduction was recorded to be 8.8×10^{-5} transconjugates recipient^{-1} h^{-1} (6). In another *in situ* study with a donor to recipient ratio of 20, 1.8 per ml transductions were detected after 48 hours (7). The survival of infectious phage particles usually decreases in the presence of a natural microflora. Cells harboring one or more prophages could provide a rich source of transducing phages to the surrounding population if the infected hosts are induced to lyse by environmental or community conditions (7).

The ability to transform DNA is widely but not universally distributed among bacteria. The process involves the recognition and binding of DNA to the surface of a recipient cell; the active transport of DNA through the membrane of the competent cell; and the integration of the foreign DNA into the host replicon. The presence of extracellular DNA within the environment needs to be demonstrated for natural transformations to have any ecological significance. The residence time of naked DNA may be increased through physical and chemical protection conferred by soil, sediments, or humic materials. Particle-bound DNA was found to transform less efficiently, but was degraded by DNase to a lesser extent than soluble DNA (8). If DNA was continually released into a population through cell lysis or active extrusion, then the persistence of extracellular DNA would be less important for effective transformation (9,10).

Plasmids are circular strands of DNA that can replicate as separate entities from the host chromosome. In general, the size of the plasmids range from 1 kb (kilobase pairs), capable of carrying only a couple of genes, to over 500 kb.

Smaller plasmids are often maintained in multiple copies, up to 40 per organism. Certain plasmids can exchange genetic material from donor to recipient in a process called *conjugation*. Conjugative plasmids can mobilize part of the host chromosome, with transfer rates occurring at 10^{-4} to 10^{-8} per plasmid transferred. Nonconjugative plasmids may also be mobilized to a recipient bacterium if a conjugative plasmid co-resides in, or is introduced into, the host cell (a triparental mating). The lower copy number conjugative plasmids are usually larger to contain genes (*tra*) that code for the production of sex pili and surface characteristics that allow for the transfer of genetic material through cell to cell contact. The expression of the *tra* operon is usually repressed in these plasmids. "Promiscuous" or "epidemic" conjugative plasmids have a broad host range and can transfer DNA between different species and genera. Lack of sequence homology usually prevents interspecies plasmid DNA from being integrated into a host chromosome, but a number of different yet compatible plasmids can be stably maintained within an individual cell (11–13).

Plasmids often carry transposons and insertion sequences (IS) that are capable of translocating genes. Both transposons and IS can form copies of themselves and relocate to new nonhomologous sections of DNA. Unlike transposons, IS do not code for an easily recognizable trait. Transposons and IS cause mutations by inactivating genes into which they are inserted. Under optimal conditions transposition frequencies vary among transposons between 10^{-7} (for Tn10) and 10^{-1} (for Tn501) per cell (14). This intracellular exchange of mobile genetic elements can increase the flexibility of gene expression through gene recombination (15). The increase in mutation rate caused by gene movement may give a host an evolutionary advantage (16,17).

Genes can have a variety of copy numbers. If an increase in gene copies contributes to the amplification of enzymatic activity, then the phenotypic expression is determined by gene dosage. This can occur when a transposon-associated gene relocates from a low copy number to a high copy number replicon (*i.e.*, from the chromosome to a high copy number plasmid) (14). This gene duplication allows for mutations, deletion, and recombination on extra copies of DNA without the loss of an essential metabolic function. Mutational events such as base substitution of misaligned DNA readings (frameshifts) may enhance catabolic expression if they are also associated with the transfer of blocks of genes (18,19). Genetic transfer is discussed in more detail by Olson and Tsai, Chapter 9 in this volume.

3. MOLECULAR BREEDING

Genes encoding for the metabolism of pesticides, PCBs, and other xenobiotic compounds are often, but not always, located on plasmids. The expansion of

TABLE 12.1. Catabolic Plasmids Associated
With the Degradation of Halogenated
Aromatics.

Plasmid	Degradative compound
pAC21	p-Chlorobiphenyl
pKF1	p-Chlorobiphenyl
pSS50	p-Chlorobiphenyl
pJP2	2,4-D
pJP4	2,4-D, 3,chlorobenzoate
pRC10	2,4-D
pBR60	3-Chlorobenzoate
pWR1	3-Chlorobenzoate
pAC25	3-Chlorobenzoate
pAC27	Chlorobenzoate
pAC31	3,5-Dichlorobenzoate
TOL (pWWO)	Xylene, toluene
SAL	Salicylate

Adapted from Karns *et al.* (64).

substrate range through plasmid acquisition has produced bacteria capable of utilizing chlorobenzoates, chlorophenols, chlorotoluates, chloroaniline, and chlorobiphyenyls (Table 12.1) (20–25). The formation of catachols (dihydroxylbenzene) seems to be an important bacterial strategy to destabilize halogenated and nonhalogenated aromatic rings. When a catabolic pathway is "constructed," often by relaxing substrate specificity, it needs to be protected from intermediates that are counterproductive to further degradation (26). *Meta*-ring cleavage is a major degradative pathway for nonhalogenated catechols, but for chlorinated catechols this route can lead to the formation of persistent "dead end" or highly reactive "suicide" metabolites that inactivate the ring-cleaving enzyme (pyrocatechase) (27,28). Under optimal conditions haloaromatics can serve as the sole source of carbon and energy for organisms utilizing the more effective *ortho*-cleavage catabolism. The process of microbial adaption to a specific xenobiotic is proposed in Table 12.2.

Through genetic manipulation, Reineke and Knackmuss (29) constructed a catabolic pathway for the mineralization of a chlorinated aromatic substrate. The metabolic sequences are shown in Figure 12.1. The authors isolated a transconjugate able to use 4-chlorobenzoate as the sole source of carbon and energy from a mating between *Pseudomonas putida* (containing the broad specificity TOL plasmid) and *Pseudomonas* B13 (a 3-chlorobenzoate decomposer). They suggested that gene recruitment was responsible for the transposition of a 41.6 kb

**TABLE 12.2. The Metabolic Sequence for Microbial Adaptation
to Chlorophenol Degradation**

1. Prevention of *meta*-cleavage pathway from forming "dead end" products
2. Establishment of a chlorocatechol-assimilating sequence through plasmid-assisted acquisition of appropriate genetic information
3. Induction of high levels of phenol hydroxylase to degrade and detoxify chlorophenol

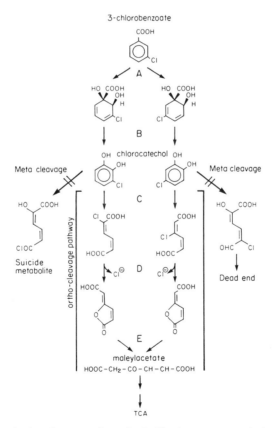

Fig. 12.1. The *ortho*-ring cleavage pathway for 3-chlorobenzoate proceeds through enzymes A, benzoate 1,2-dioxygenase; B, dihydrodihydroxybenzoate dehydrogenase; C, catechol 1,2-dioxygenase type II (pyrocatechase); D, chloromuconate cycloisomerase; and E, 4-carboxymethylenebut-2-en-4-olide hydrolase (20).

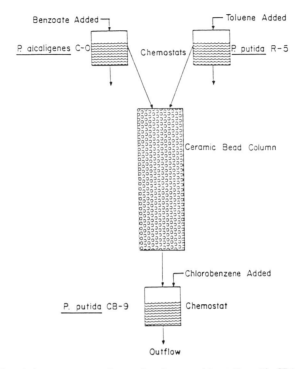

Fig. 12.2. The tri-chemostat system that produced a recombinant *P. putida* CB1-9 strain able to grow on chlorobenzenes unlike its parental *P. alcaligenes* C-0 and *P. putida* R-5 strains.

segment from the TOL plasmid to the recipient *Pseudomonas* B13 chromosome, resulting in an organism capable of using 4-chlorobenzoate by a modified *ortho*-ring cleaving pathway. The critical enzymatic steps for the catabolic conversion of similar haloaromatics are the ring-cleaving pyrocatechase, the dechlorinating cycloisomerase, and the hydrolase.

Kellogg *et al.* (30) described a plasmid-assisted program of strong yet gradual selective pressure that produced a 2,4,5-trichlorophenoxyacetic acid (2,4,5-T)–degrading consortium. A *Pseudomonas cepacia* (AC1100) isolated by Kilbane *et al.* (31) from a similar incubation was effective in removing 2,4,5-T in contaminated soil. Genetic transfer through conjugation was also demonstrated by Latorre *et al.* (21) between an aniline-metabolizing *Pseudomonas* sp. and the 3-chlorobenzoate degrader *Pseudomonas* sp. B13, which produced a chloroaniline degrader.

The time required for isolating a recombinant strain capable of degrading dichlorobenzene was decreased from months to weeks using a three-stage continuous culture system designed by Krockel and Focht (32). Two species (*P. putida* R5-3 and *P. alcaligenes* C-0), unable to grow on chlorobenzene, were cultured separately and then introduced at high cell density into a "reaction bed column" (Fig. 12.2). This column contained ceramic beads that increased the surface area for cell adsorption. The isolated recombinant strain (*P. putida* CB1-9) was able to grow on 1,4-dichlorobenzene as the sole energy source by utilizing genetic information supplied by both parental stains.

Although chlorinated biphenyls are persistent in the environment, there are bacterial consortia and co-cultures that are able to transform, cometabolize, and mineralize certain members of this class of compounds (33–37). The mineralization of mono- and dichlorobiphenyls by a single laboratory strain has also been reported (38,39). Plasmids that are associated with PCB degradation include the widely distributed pKF1, which specifies the conversion of p-chlorobiphenyls to chlorobenzoic acid, while the plasmids pAC27 and pAC31 code for the further utilization of mono- and dichlorobenzoates. The development of more than one dioxygenase degradative pathway and the fact that in anaerobic environments PCB can be reductively dehalogenated as either a substrate or as an electron acceptor both demonstrate the versatility of microbial adaption to these exotic compounds (35,40).

4. THE ECOLOGY OF GENETIC TRANSFER

A high density of plasmid-carrying bacteria, along with a low density of a competing natural microflora, seem to encourage *in situ* genetic transfer when cell to cell contact is either necessary, as in conjunction, or beneficial, as in transformation and transduction (7,12,41). The yield of plasmid-containing transconjugates often decreases when the parental density drops below 10^8 cells ml^{-1}. The environment (*e.g.*, redox potential, presence of adsorptive surfaces, temperature, available water) not only determines the size and diversity of microbial populations but also influences the transfer rate of genetic material (1,12). Some environmental and biological factors that would favor microbial adaptation to xenobiotics through the exchange of genetic material are presented in Table 12.3.

The importance of surfaces in promoting gene transfer is well documented in laboratory filter matings. *In situ* surfaces such as sewage flocs, sediment–water interface, and the root–soil surface (rhizoplane) may provide the sites for aggregation while also buffering the physicochemical environment (42,43,12).

In a soil column study, plasmid transfer was suspected as the mechanism conferring 3-chlorobenzoate–degrading ability to the indigenous microflora from inoculated *P. putida* and *P. aeruginosa* (44). In another soil system, a gene

TABLE 12.3. A Habitat Conducive to the Catabolic Adaptation to Xenobiotic Stress Through Genetic Transfer

1. High density of DNA recipient bacteria
2. A microniche-rich habitat that has a range of environmental parameters and xenobiotic concentration in close proximity
3. A variety of natural analogs that could support cometabolism and induce enzymatic activity
4. A variety of microorganisms that are catabolic-versatile (*i.e., Pseudomonans*-like)
5. Microorganisms able to transmit plasmids, especially those that carry traits for broad substrate utilization

promoter cassette was inserted into a broad host range plasmid (pFL67-2) and transferred its plasmid to members of the native microflora. The indigenous population was able to maintain this plasmid for at least 100 days without selective pressure (45). Van Veen (46) noted a rapid decline in the number of transconjugates with distance from the root surface in his inoculated soil-plant chambers. *Erwinia chrysanthemi* was found to transfer plasmid RP1 at a higher frequency in plants than *in vitro* (47,48). Although this is the exception to most *in situ* studies, it does indicate that the laboratory cannot always mimic the selective pressures present in microhabitat-dense systems.

In aquatic systems, surfaces and interfaces are important sites for nutrient and xenobiotic accumulation, as well as microbial degradation. Attached bacteria, unlike planktonic microorganisms, can also remain in polluted aquatic environments for sufficient time to respond genetically to selective pressure (49). Freshwater isolates have the capacity to act as recipients to broad-host-range plasmids (*i.e.*, R68 and R1162). Grabow *et al.* (50) noted that 3 of 10 donor strains transferred antibiotic resistance to *E. coli* at low frequency in dialysis bags submerged in a river (51). Conjugation was also detected in freshwater after 192 hours of coincubation at a frequency of 10^{-6} transconjugates cell^{-1}, which is 4 orders of magnitude less than under laboratory conditions (41).

In the absence of selective pressure, *in situ* transfer between mating pairs occurred at a frequency of 10^{-4} to 10^{-5} in membrane diffusion chambers submerged in a sewage treatment facility with and without sterile sewage (12,52,53). The acquisition of genetic material from the indigenous population to an introduced pseudomonad, as well as the mobilization of a nonconjugative plasmid, was also demonstrated in a laboratory-scale waste treatment system (42,54).

Usually, plasmids that do not confer a selective advantage are lost through competitive exclusion (55,56). Structural instability (*i.e.*, loss of part or all of the plasmid) has been observed for catabolic plasmids when grown on nonselecting media (56). However, even when selective pressure is absent, a plasmid may be maintained in a population providing a "community memory" of a prior successful strategy (57,58).

Cryptic genes are another example of the cell's ability to carry excess genetic information that may be activated under appropriate environmental conditions, or merely by chance. The "metabolic burden" associated with a cell harboring such genetic baggage, as determined by a decrease in growth rate, is caused by plasmid replicaton, mRNA translation, and foreign protein expression (15,59). If nonessential traits are harbored on plasmids within a few species, then the metabolic burden would be shared among members of the community (60). These plasmids would also confer an adaptive flexibility to a carrier population.

Stable communities can degrade xenobiotics without incorporating the pathway into a single species (61,62). In fact, a strain harboring the pSS50 plasmid that can cometabolize 4-chlorobiphenyl, with biphenyl as the primary carbon source, was out–competed in a chemostat study by a three-member consortium (39). Another form of synergistic attack is demonstrated by a stable population able to use 3-chlorobenzoate as the sole carbon and energy source enlisting dechlorinating, benzoate-oxidizing, and methanogenic bacteria (63). The close physical association among consortium members that facilitites substrate exchange would also be conductive to the transfer to genes regulating the optimization of energy transfer.

The development and association of microbial species capable of degrading a xenobiotic chemical appears to be a function of severity of selective pressure and time (64). An explanation given by Kellogg *et al.* (30) for the lack of haloaromatic-mineralizing microorganisms in nature is that the concentration most commonly found in contaminated sites provides insufficient available substrate to act as an incentive to evolve a degradation pathway. A threshold quantity of xenobiotic chemical is usually required to produce such an adaptive response (65).

5. GENE RECRUITMENT

Catabolic genes are usually clustered in operons on chromosomes and plasmids. This clustering allows the phenotypic expression coded by the DNA segment, such as dehalogenation, to be more easily amplified through gene dosage (56). In the TOL plasmid, the aggregated genes are positively regulated with the substrates of the pathway inducing transcription of the catabolic operon (66,67). Similarities in the clustering of chlorobenzoate-degrading genes, inducibility of their gene products, and the homology of these genes with others in plasmids (TOL and SAL) indicted to Karns *et al.* (64) that xenobiotic degradation pathways evolved primarily by recruitment of replication, copy number, and incompatibility genes from other replicons (30,68,69). Sequence homology between plasmids with the same catabolic phenotype suggests that in many cases the structural genes for xenobiotic degradation are conserved (70).

Pseudomonad-like organisms, in addition to other plasmid-carrying bacteria, are found at higher numbers in contaminated sites (49,70–72). The success of introducing novel catabolic genes into the gram-negative *Pseudomonas* was attributed by Clark (73) to this genus's regulation system that may have more flexible responses to novel substrates. Mutational events have been known to activate previously cryptic catabolic pathways within this nutritionally versatile bacterial group (67). The catabolic pathways of *Pseudomonas* appeared to Harayama and Timmis (67) to be composed of discrete functional units controlled by genetic modules that evolved through genetic transfer and rearrangement.

Acknowledging the difficulty of proving a negative hypothesis in ecology, *i.e.*, that genetic transfer does *not* contribute to bacterial adaptation to stress from xenobiotic chemicals, the evidence remains that genetic exchange *has evolved*—albeit with severe environmental and regulatory restrictions (17). One can speculate that the limiting factor in the expression of degradation of xenobiotic chemicals is the transfer rate of genes governing the various combinations of catabolic attack. The benefits to a cell being a recipient of only a portion of a catabolic pathway, as is the development of 5% of a bird's wing, is questionable because the intermediate forms do not yield energy—or fly. But the final utilization of a trait gives little indication of how it evolved (74). A portion of a catabolic pathway could detoxify xenobiotics with varying degrees of success from ''dead end'' products to dehalogenation. Such a nonspecific function could eventually develop into a mineralization pathway. The role of natural gene transfer in connecting partial components of a degradative pathway remains untested.

6. CONCLUSIONS

Conjugative plasmids, transformation, and phage-mediated transduction allow for a horizontal genetic connection among microbial communities. Such genetic exchange does not necessarily require an adherence to a ''multiorganism'' or a ''super-organism'' concept that conflicts with individual Darwinian evolution (75). But this link could temporarily broaden the genetic base from which the environment selects. Intracellular rearrangement and intercellular exchange of genes increases short-term microbial variability. This genome plasticity could be beneficial for community adaptation to xenobiotic pollution. But Terzaghi and O'Hara (15) noted a distinction between this short-term strategy dependent on genetic recombination and that long-term variation that becomes incorporated into a taxonomic coherent group of bacteria. The long-term evolutionary history of a genus is not usually maintained through plasmid-associated traits.

Xenobiotic-contaminated soils, sediments, and sludges provide vast numbers of microhabitats, each containing a population adapting to the novel chemical. This variety of environmental conditions could support a wealth of genetic challenges and adaptive strategies. Xenobiotic degradation may be limited by the rate of genetic transfer between, and rearrangement within, those niche-dense microbial habitats that can support a large and diverse genetic reservoir.

ACKNOWLEDGMENTS

The author thanks Steve DePalma for his clarifying suggesting during the preparation of this chapter.

REFERENCES

1. Alexander M: Biodegradation of chemicals of environmental concern. Science 211:132–138, 1981.

2. Siuda JF, Debernardis JF: Naturally occurring halogenated organic compounds. Lloydia 36: 197–243, 1973.

3. Don RH, Pemberton JM: Properties of six pesticide degradation plasmids isolated from *Alicaligenes parodoxus* and *Alicaligenes eutrophus*. J Bacteriol 145:681–686, 1981.

4. Wiggins BA, Jones SH, Alexander M: Explanation for the acclimation period preceding the mineralization of organic chemicals in aquatic environments. Appl Environ Microb 53:791–796, 1987.

5. Lappin HM, Geaves MP, Slater JH: Degradation of the herbicide Mecoprop [2-(2-methyl-4-chlorophenoxy) propionic acid] by a synergistic microbial community. Appl Environ Microbiol 49:429–433, 1985.

6. Morrison WD, Miller RV, Sayler GS: Frequency of F116-mediated transduction of *Pseudomonas aerugenosa* in a freshwater environment. Appl Environ Microbiol 36:724–730, 1979.

7. Saye DJ, Ogunseitan O, Sayler GS, Miller RV: Potential for transduction of plasmids in a natural freshwater environment: effect of plasmid donor concentration and a natural microbial community on transduction in *Pseudomonas aeruginosa*. Appl Environ Microbiol 53:987–995, 1987.

8. Aardema BW, Lorenz MG, Krumbein WE: Protection of sediment-adsorbed transformating DNA against enzymatic inactivation. Appl Environ Microbiol 46:417–29, 1983.

9. Stewart GJ, Carlson CA: The biology of natural transformation. Ann Rev Microbiol 40:211–35, 1986.

10. Trevors JT, Barkay T, Bourquin AW: Gene transfer among bacteria in soil and aquatic environments: A review. Can J Microbiol 33:191–198, 1987.

11. Hardy K: Bacterial Plasmids. American Society of Microbiology. Washington, DC: 1983.

12. Stotzky G, Babich H: Fate of genetically-engineered microbes in natural environments. Recomb. DNA Tech Bull 7:163–188, 1984.

13. Beringer JE, Hirsch PE: The role of plasmids in microbial ecology. In Klug MF, Reddy CA (ed): Current Perspectives in Microbial Ecology. American Society of Microbiology: Washington, DC, 1984, pp 63–70.

14. Eaton RW, Timmis KN: Genetics of xenobiotic degradation. In Klug MJ, Reddy CA (ed): Current Perspectives in Microbial Ecology, American Society of Microbiology, Washington, DC, 1983, pp 694–703.

15. Terzaghi E, O'Hara M: Microbial plasticity the relevance to microbial ecology. In Marshall K (ed): Advances in Microbial Ecology vol. 11 New York, Plenum Press: 1990, pp 431–460.

16. Chao L, Vargas C, Spear BR, Cox EC: Transposable elements as mutator genes in evolution. Nature 303:633–635, 1983.

17. Young JPW: The population genetics of bacteria. In Hopwood DA, Chater KF (ed): Genetic Bacterial Diversity. London, Academic: 1989, pp 417–439.

18. Okada H, Negoro S, Kimura H, Nakamura S: Evolutionary adaptation of plasmic-encoded enzymes for degrading oligomers. Nature 306:203–206, 1983.

19. Franz B, Aldrich T, Chakrabarty AM: Microbial degradation of synthetic recalcitrant compounds. Biotech Adv 5:85–99, 1987.

20. Weightman AJ, Don RH, Lehrbach PR, Timmis KN: The identification and cloning of genes encoding haloaromatic catabolic enzymes and the construction of hybrid pathways for substrate mineralization. In Omenn GS, Hollaender A (ed): Genetic Control of Environmental Pollutants. Basic Life Sciences vol. 28, New York, Plenum Press: 1984, pp 47–80.

21. Latorre J, Reineke W, Knackmuss H: Microbial metabolism of chloranilines: enhanced evolution by natural genetic exchange. Arch Microbiol 140:159–165, 1984.

22. Spain JC, Nishino SF: Degradation of 1,4-dichlorobenzene by a *Pseudomonas* sp. Appl Environ Microbiol 53:1010–1019, 1987.

23. Schraa G, Boone ML, Jetten MSM, van Neerven ARW, Colberg PJ, Zehnder AJB: Degradation of 1,4-dichlorobenzene by *Alcaligenes* strain A175. Appl Environ Microbiol 52:1374–1381, 1986.

24. de Bont JAM, Vorage MJAW, Hartmans S, van den Tweel WJJ: Microbial degradation of 1,3-dichlorobenzene. Appl Environ Microbiol 52:677–780, 1986.

25. Boyle M: The environmental microbiology of chlorinated aromatic decomposition. J Environ Microbiol 18:395–402, 1989.

26. Reineke W, Knackmuss H-J: Microbial degradation of haloaromatics. Ann Rev Microbiol 42:263–287, 1988.

27. Zeyer J, Wasserfallen A, Timmis KN: Microbial mineralization of ring-substituted anilines through as ortho-cleavage pathway. Appl Environ Microbiol 50:447–453, 1985.

28. Hartmann J, Reinke W, Knackmuss HJ: Metabolism of 3-chloro, 4-chloro, and 3,5 dichlorobenzoate by a pseudomonad. Appl Environ Microbiol 37:421–28, 1979.

29. Reineke W, Knackmuss H-J: Construction of haloaromatic utilizing bacteria. Nature 277:385–386, 1979.

30. Kellogg ST, Chatterjee DK, Chakrabarty AM: Plasmid assisted molecular breeding—new technique for enhanced biodegradation of persistent toxic chemicals. Science 214:1133–1135, 1981.

31. Kilbane JJ, Chatterjee DK, Chakrabarty AM: Detoxification of 2,4,5-trichlorophenoxyacetic acid from contaminated soil by *Pseudomonas cepacia*. Appl Environ Microbiol 45:1697–1700, 1983.

32. Krockel L, Focht DD: Construction of chlorobenzéne-utilizing recombinants by progenative manifestation of a rare event. Appl Environ Microbiol 53:2470–2475, 1987.

33. Adrians P, Kohler HPE, Kohler-Staub D, Focht DD: Bacterial dehalogenation of chlorobenzoates and coculture biodegradation of 4,4'-dichlorobiphenyl. Appl Environ Microbiol 55:887–892, 1989.

34. Brunner W, Sutherland FH, Focht DD: Enhanced biodegradation of polychlorinated biphenyls in soil by analog enrichment and bacterial inoculation. J Environ Qual 14:324–328, 1985.

35. Bedard DL, Wagner RE, Brennan MJ, Haberl ML, Brown JF Jr: Extensive degradation of Aroclors and environmentally transformed polychlorinated biphenyls by *Alicaligenes eutrophus* H850. Appl Environ Microbiol 53:1094–1102, 1987.

36. Furukawa K, Chakrabarty AM: Involvement of plasmids in total degradation of chlorinated biphenyls. Appl Environ Microbiol 44:619–626, 1982.

37. Shields MS, Hooper SW, Sayler GS: Plasmid-mediated mineralization of 4-chlorobiphenyl. J Bacteriol 163:882–889, 1985.

38. Parson JR, Sijm DTHM, van Laar A, Hutzinger O: Biodegradation of chlorinated biphenyls and benzoic acids by *Pseudomonas* strain. Appl Microbiol Biotechnol 29:81–84, 1988.

39. Pettigrew CA, Breen A, Corcoran C, Sayler GS: Chlorinated biphenyl mineralization by individual populations and consortia of freshwater bacteria. Appl Environ Microbiol 56:2036–2045, 1990.

40. Linkfield TG, Suflita JM, Tiedge JM: Characterization of the acclimation period before anaerobic dehalogenation of halobenzoates. Appl Environ Microbiol 55:2773–2778, 1989.

41. Gowland PC, Slater JH: Transfer and stability of drug resistance plasmid in *Escherichia coli* K12. Microb Ecol 10:1–13, 1984.

42. Mancini P, Fertels S, Nave D, Gealt MA: Mobilization of plasmid pHSV106 from *Escherichia coli* HB101 in a laboratory-scale waste treatment facility. Appl Environ Microbiol 53:665–671, 1987.

43. McPherson P, Gealt MA: Isolation of indigenous wastewater bacterial strains capable of mobilizing plasmid pBR325. Appl Environ Microbiol 51:904–909, 1986.

44. Pertsova RN, Kunc F, Golovleva LA: Degradation of 3-chlorobenzoate in soil by pseudomonad carrying biodegradative plasmids. Folia Microbiol 29:242–247, 1984.

45. Henschke RB, Schmidt RJ: Plasmid mobilization from genetically engineered bacteria to members of the indigenous soil microflora *in situ*. Curr Microbiol 20:105–110, 1990.

46. Van Veen JA: Ecology of genetically engineered organisms in soil and risk assessment procedures/regulations in the Netherlands. Boston, MA: AAAs National Convention Abstracts, p 55, 1988.

47. Reanney DC, Roberts WP, Kelly WJ: Genetic interactions among microbial communities. In Bull AT, Slater JH (ed): Microbial Interactions and Communities. New York, Academic Press: 1982, p 287–322.

48. Lacey GH: Genetic studies with plasmic RP1 in *Erwinia Chrysanthemi* strains pathogenic on maize. Phytopathology 68:1323–1330, 1978.

49. Burton NF, Day MJ, Bull AT: Distribution of bacterial plasmids in clean and polluted sites in a south Wales river. Appl Environ Microbiol 44:1026–1029, 1982.

50. Grabow WOK, Prozesky OW, Burger JS: Behavior in a river and dam of coliform bacteria with transferable and non-transferable drug resistance. Water Res 10:717–782, 1975.

51. Genthner FJ, Chatterjee P, Barkay T, Bourquin AW: Capacity of aquatic bacteria to act as recipients of plasmid DNA. Appl Environ Microbiol 54:115–117, 1988.

52. Altherr MR, Kasweck KL: *In situ* studies with membrane diffusion chambers of antibiotic resistance in Escherichia coli. Appl Environ Microbiol 44:838–843, 1982.

53. Mach PA, Grimes DJ: R-plasmid transfer in a wastewater treatment plant. Appl Environ Microbiol 44:1395–1403, 1982.

54. McClure NC, Weightman AJ, Fry JC: Survival of *Pseudomonas putida* UWC1 containing cloned catabolic genes in a model activated-sludge unit. Appl Environ Microbiol 55:2627–2634, 1989.

55. Slater JH, Godwin D: Microbial adaptation and selection. In Ellwood DC, Hedger JN, Latham MJ, Lynch JM, Slater JH (eds): Contemporary Microbial Ecology. New York: Academic Press, 1980, pp 137–160.

56. Hardman DJ, Gowland PC, Slater JH: Large plasmids form soil bacteria enriched on halogenated alkanoic acids. Appl Environ Microbiol 51:44–51, 1986.

57. Boumam JE, Lenski RE: Evolution of a bacteria/plasmid association. Nature 335:351–352, 1988.

58. Jain RK, Sayler GS, Wilson JT, Houston L, Pacia DP: Maintenance and stability of introduced genotypes in groundwater aquifer material. Appl Environ Microbiol 53:996–1002, 1987.

59. Bentley WE, Mirjalili N, Andersen DC, Davis RH, Kompala DS: Plasmid-encoded protein: The principal factor in the "metabolic burden" associated with recombinant bacteria. Biotech Bioeng 35:668–681, 1990.

60. Rittman BE, Smets BF, Stahl DA: The role of genes in biological processes. Environ Sci Technol 24:23–29, 1990.

61. Bull AT: Biodegradation: Some attitudes and strategies of microorganisms and microbiologist. In Ellwood DC, Latham MJ, Hedger JN, Lynch JM, Slater JH (eds): Contemporary Microbial Ecology. London, Academic Press: 1980. pp 107–136.

62. Slater JH, Godwin D: Microbial adaptation and selection. In Ellwood DC, Hedger JN, Latham MJ, Lynch JM, Slater JH (eds): Contemporary Microbial Ecology. New York: Academic Press, 1980, pp 137–160.

63. Suflita JM, Horowtz A, Sheldon DR, Tiedje JM: Dehalogenation: A novel pathway for the anaerobic biodegradation of haloaromatic compounds. Science. 218:1115–1117, 1982.

64. Karns JS, Kilbane JJ, Chatterjee DK, Chakrabarty AM: Microbial biodegradation of 2,4,5-tirchlorophenoxyacetic acid and chlorophenols. In Omenn GS, Hollaender A (eds): Genetic Control of Environmental Pollutants. Basic Life Sciences Vol. 28. New York: Plenum Press, 1984, pp 3–21.

65. Spain JC, Van Veld PA: Adaptation of natural communities to degradation of xenobiotic compounds; effect of concentration, exposure time and inoculum and chemical structure. Appl Environ Microbiol 45:428–435, 1983.

66. Ghosal D, You I.-S, Chatterjee DK, Chakrabarty AM: microbial degradation of halogenated compounds. Science 228:135–142, 1985.

67. Harayama S, Timmis KN: Catabolism of aromatic hydrocarbons by *Pseudomonas*. In Genetics of Bacterial Diversity, Hopwood DA (ed): San Diego, CA: Academic Press, 1989, pp 151–175.

68. Chatterjee DK, Chakrabarty AM: Genetic rearrangement in plasmids specifying total degradation of chlorinated benzoic acids. Mol Gen Genet 188:279–285, 1982.

69. Chatterjee DK, Chakrabarty AM: Genetic homology between independently isolated chlorobenzoate-degradative plasmids. J Bact 153:532–534, 1983.

70. Sayler GS, Hooper SW, Layton AC, Henry King JM: Catabolic plasmids of environmental and ecological significance. Microbial Ecol 19:1–20, 1990.

71. Hada HS, Sizemore RK: Incidence of plasmids in marine *Vibrio* spp. isolated from an oil field in the northwestern Gulf of Mexico. Appl Environ Microbiol 44:199–202, 1981.

72. Ogunseitan OA, Tedford ET, Pacia D, Sirotkin KM, Sayler GS: Distribution of plasmids in groundwater bacteria. J Ind Microbiol 1:311–317, 1987.

73. Clarke PH: Evolution of new phenotypes. In Klug MJ, Reddy CA (eds): Current Perspectives in Microbial Ecology. Washington, DC: American Society of Microbiology, 1984, pp 71–78.

74. Gould SJ: An Urchin in the Storm. New York: W.W. Norton & Co. p 233, 1985.

75. Regal P: The ecology of evolution: implication of the individualistic paradigm. In Halvorson HO, Pramer D, Rogul M (eds): Engineered Organisms and the Environment: Scientific Issues. Philadelphia, PA: American Society of Microbiology, 1985, pp 11–19.

13

MICROBIAL CONTROL OF PLANT DISEASES

ALEX SIVAN

The Institute for Agriculture and Applied Biology, The Institutes for Applied Research, Ben–Gurion University of the Negev, Beer-Sheva, Israel

ILAN CHET

Faculty of Agriculture, The Hebrew University of Jerusalem, Rehovot, Israel

1. BIOCONTROL AS AN ALTERNATIVE TO CHEMICAL PESTICIDES

During the past three decades crop production has been greatly influenced by the dramatic increase in the use of pesticides. Indeed, these compounds have played an extremely important role in obtaining the maximal yield potential of almost every commercial crop by reducing yield losses caused by plant pathogens and other pests. However, because of their residual toxic effects the continuous use of these potentially hazardous chemicals is imposing an increasing environmental threat. Pesticide contamination of food, water reservoirs, and soil has become a fact of life, and Rachel Carson's message of 30 years ago, set out so clearly in her book *Silent Spring,* that the continuous use of pesticides would lead to an environmental disaster, has become a reality. Recently, Ritter (1) reported that over 70 pesticides including soil fumigants have been detected in groundwater in 38 states in the United States. Thirty two of these pesticide detections were attributed to point sources or misuse. A landmark study published by the U.S. Environment Protection Agency (EPA) indicates that, in the United States alone, 3,000–6,000 cancer cases were induced annually by pes-

Environmental Microbiology, pages 335–354, © 1992 Wiley-Liss, Inc.

ticide residues on foods and another 50–100 by exposure to pesticides during application. Pesticides on foods pose a similar cancer risk to indoor air pollution (mainly tobacco smoke), and both are considered to be the largest environmental manmade cancer risks that can be quantified (2). Moreover, pesticide contamination of the environment is harmful to wildlife and to nontarget and other beneficial organisms.

As a result of increased awareness of the environmental hazards caused by synthetic pesticides, on the part of both the public and the authorities, some of these chemicals have been removed from the market. For example, several chemical seed protectants have been banned from use in the United States, including mercury-containing fungicides, while others such as the commonly used broad-spectrum fungicide Captan are being reevaluated.

Some of the commonly used traditional fungicides act by affecting different metabolic processes in the target pathogens, and they are thus able to exhibit a broad range of activity against several pathogens. Since most of the hazardous fungicides belong to this group of chemicals, during the last 20 years pesticide manufacturers have focused on the development of new pesticides that have more specific modes of action and that are thus safer to nontarget organisms. However, such pesticides are more likely to induce resistance in the target population (3), and their ''life-spans'' may thus be much shorter than the traditional broad-range protectants. This may have a significant economic effect, since the development costs of a new specific pesticide are now estimated to be about U.S. $30,000,000. These and other reasons have resulted in increased efforts on the part of academic institutions and commercial companies during the past decade to develop alternative nontoxic means of plant disease control. Among these approaches, biological control of plant pathogens is becoming increasingly important.

Biocontrol of plant pathogens may possess advantages—other than their non-hazardous nature—over conventional pesticides. Most fungicides have only a temporary effect and usually require repeated applications during the growth season. Biological control agents have the ability to reproduce, to establish themselves in the soil ecosystem, and to colonize seeds, spermosphere, rhizosphere, rhizoplane, and foliage (3). Furthermore, biocontrol strategies are highly compatible with the sustainable agriculture practices that are required for conserving natural resources of agriculture. In their comprehensive book on biological control, Cook and Baker (3) have indicated that ''the need for a sustainable agriculture will be met in part by wider use of biological control of plant pathogens.''

Biological control may be achieved by a number of strategies, both direct and indirect. Indirect strategies include the use of organic soil amendments that enhance the activity of indigenous microbial antagonists against a specific pathogen. Another indirect approach, cross-protection, involves the stimulation of plant self-defense mechanisms against a particular pathogen by prior inoculation of the plant with a nonvirulent strain that is taxonomically related to the pathogen. Successful cross-protection that results in induced resistance has been documented for viruses (4,5), fungi (6), and bacterial pathogens (7).

The direct approach involves the introduction of specific microbial antagonists to soil or to planting material (3). These antagonists have to proliferate and establish themselves in the appropriate ecological niche in order to be active against the pathogen. This approach provides an excellent paradigm for biocontrol and is the focus of this chapter.

2. MICROBIAL BIOCONTROL AGENTS

Introduction of naturally occurring antagonists to the pathogen's habitat was the first biocontrol approach to be investigated, and it is probably the most thoroughly studied one. However, the use of introduced biocontrol agents of plant pathogens lags far behind that of plant insect pests. The first means of controlling an insect pest by an introduced natural insect enemy was tried as long ago as 1888 (8), whereas the first attempt to introduce microbial inoculants for the control of a plant pathogen was performed only 70 years ago (9). This experiment involved the application of a mixture of fungi to control the damping-off of pine seedlings caused by *Pythium debaryanum*. Research on the exploitation of this approach over the past seven decades and especially in the last 15 years has led to a better understanding of the physiology, ecology, and mode of action of biocontrol agents that will facilitate the commercialization of biopesticides in the near future.

2.1. Bacteria

About 20 genera of bacteria have been shown to be potential biocontrol agents for numerous plant diseases; most of these bacteria have been tested against soil-borne plant pathogens (10,11). However, only a few strains gave consistent results under field conditions (11). It is not within the scope of this chapter to review all the potential bacterial antagonists but rather to mention briefly those bacteria that have attracted the greatest attention because of their consistently satisfactory performance under field conditions.

Probably the best example of the utilization of an introduced microbial antagonist for the biocontrol of an economically important plant pathogen was the pioneering work with *Agrobacterium radiobacter* strain 84 for the control of *A. tumefaciens* (12,13). *A. tumefaciens* is the causal agent of crown gall in a large number of crops throughout the world and was even considered one of the three most destructive agricultural pathogens in California (14). This was the first case of a successful commercialization of a biocontrol agent.

Bacillus subtilis strain A13 was the next bacterial antagonist to be commercialized (11). In contrast to most bacterial biocontrol agents, which are gram negative, gram-positive bacilli produce endospores that provide them with a high tolerance to heat and desiccation. Therefore, they have a significant ecological advantage over other bacterial antagonists. *B. subtilis* A13 has been shown to increase yields of carrots and oats (15) apparently by suppression of major and minor pathogens and possibly also by direct stimulation of plant growth (16,17).

During the past decade fluorescent rhizobacteria from the genus *Pseudomonas* have raised much interest as microbial antagonists for a wide variety of plant pathogens (11). Studies with *P. fluorescens* and *P. putida* initiated by Burr *et al.* (18) showed the potential of these bacteria to improve potato growth. The antagonists were later shown to increase yields of sugarbeets (19) and radishes significantly, the latter by up to 144% (20). These effects of growth promotion and yield increase correlated with *in vitro* inhibition of a wide range of soil-borne plant pathogens by these bacteria (19). Similarly, Kloepper and Schroth (21) demonstrated the inability to increase yields of noninhibitory mutants that colonized the roots.

One of the most intensively studied biocontrol systems using rhizobacteria was generated after the isolation of *P. fluorescens* strains from a soil naturally suppressive to *Gaeumannomyces graminis* var. *tritici,* the causal agent of take-all disease of wheat (22). Again, these were fluorescent rhizosphere-colonizing bacteria that, when applied as seed treatments in a soil conducive to the disease, provided significant disease control. A combination of strains 2-79 and 13-79 of *P. fluorescens* provided biocontrol of take-all in wheat superior to that of each of the strains alone (11). Moreover, in a commercial-scale experiment a combination of these strains with a third strain of *P. fluorescens* resulted in a further improvement of the biocontrol (23).

2.2. Fungi

The number of genera producing promising biocontrol agents is far lower for fungi than for bacteria. The most thoroughly studied fungal antagonists belong to the genus *Trichoderma*. Until now, only two fungal antagonists have been introduced commercially. The first is *Peniophora (Phlebia) gigantea,* which is used for the control of *Heterobasidium (Fomes) annosum* root rot on pines, and the second is *Trichoderma viride,* which is marketed for the control of several tree diseases. *Trichoderma* spp. (especially *T. harzianum* and *T. viride*) exhibit considerable variability among strains with respect to their biocontrol activity, host range, physiological and biochemical properties, and their ecological and environmental adaptability (24,25).

The mycoparasitic nature and the biocontrol potential of *Trichoderma* spp. was first demonstrated by Weindling (26). However, during the following four decades relatively little research was done on *Trichoderma* spp. In fact, the first biocontrol experiment with this antagonist, performed under natural field conditions, was reported as recently as 1972 (27). Since then, more and more field trials have shown the practical potential of *Trichoderma* spp. as a biocontrol agent for various soil-borne plant pathogens in vegetable, field, and ornamental crops (24,28).

Potential biocontrol agents from the genus *Trichoderma* may be isolated from the natural habitats of the target pathogens (*i.e.*, infested soil or plants). Indeed, this approach has led to the isolation of several *T. harzianum* strains effective in the control of the pathogens *Rhizoctonia solani* (29), *Sclerotium rolfsii* (30), *Pythium aphanidermatum* (31), *Macrophomina phaseolina* (32), *Fusarium oxysporum,* and *F. culmorum* (33,34) under natural field conditions.

Biological control of wilt pathogens that penetrate into the vascular system of their hosts (*e.g., F. oxysporum* or *Verticillium dahliae*) and that may attack the plant at various growth stages is considered to be less feasible than control of plant pathogens that induce damping-off. In the latter case the host is susceptible to the pathogen only for a limited time, at the early growth stage of the plant, after which it is not susceptible to infection. Thus, for biocontrol of damping-off, protection is required for only a limited period of time. This might be the reason for the slower progress in developing introduced antagonists (including fungi) for the biocontrol of wilt pathogens. Marois *et al.* (35) showed that a mixture of fungal antagonists consisting of three strains of *T. harzianum,* one of *Aspergillus ochraceus,* and one of *Penicillium fumiculosum* was effective in controlling fusarium crown rot of tomato when applied to fumigated soil. A strain of *T. harzianum* (T-35) exhibited high biocontrol efficacy under natural soil conditions in the greenhouse against fusarium wilts caused by *F. oxysporum* f. sp. *vasinfectum* and f. sp. *melonis* on cotton and melon, respectively, and against fusarium seedling blight in wheat caused by *F. culmorum* (33). The same strain was further tested under field conditions in commercial plots where a severe disease caused by either fusarium wilt of melon or fusarium crown rot of tomato had been recorded for few consecutive years. Application of T-35 to the root zone of tomato or melon seedlings, transplanted into these plots, reduced the incidence of the disease significantly, and melon and tomato yields increased by 33% and 18%, respectively (34,36).

3. MECHANISMS INVOLVED IN BIOCONTROL

Antagonistic interactions among microorganisms in nature include parasitism or lysis, antibiosis, and competition. These microbial interactions serve as the basic mechanisms by which biocontrol agents might operate. Elucidation of mechanisms involved in biocontrol activity is considered to be one of the key factors in the development of useful biocontrol agents. Of the numerous biocontrol agents examined, only a few have been subjected to a thorough analysis of the mechanisms involved in the suppression of the pathogen. However, all three mechanisms of biocontrol have been described in a number of fungal and bacterial antagonists. We focus, in this section, on *Trichoderma* as a model system: We describe those mechanisms that are thought to contribute to its biocontrol activity and that may well be involved in other biocontrol systems.

3.1. Mycoparasitism

Mycoparasitism is defined as a direct attack of a parasite on a fungal thallus. Some examples of destructive mycoparasitism have recently been reviewed (24,37,38). In *Trichoderma,* several consecutive steps are involved in its destructive mycoparasitism on other plant pathogenic fungi.

The mycoparasitic process is apparently initiated by attraction or chemotropism. Directed growth of *Trichoderma* toward its host was observed by Chet *et al.* (39), although no specific chemoattractants other than amino acids and simple sugars have thus far been identified (Barak and Chet, unpublished data).

The second step in the mycoparasitic activity of *Trichoderma* involves recognition that appears to play a major role in mycoparasitism. The fact that there is distinct variability between the ability of various strains to parasitize a range of host fungi has led to the hypothesis that this specificity might be derived by molecular or biochemical determinants. Indeed, recognition between *T. harzianum* and two of its major hosts, *R. solani* and *S. rolfsii,* was found to be controlled by two different lectins present on the host hyphae. *R. solani* carries a lectin that binds to galactose and fucose residues on the *Trichoderma* cell walls as well as to type O erythrocytes (40). The activity of a second lectin isolated from *S. rolfsii* was shown to be inhibited by D-glucose or D-mannose residues, apparently present on the cell walls of *T. harzianum*. This lectin was further shown to agglutinate conidia of a mycoparasitic strain of *T. harzianum* but did not agglutinate two nonmycoparasitic strains (41).

Following recognition, *Trichoderma* hyphae grow along or coil around the host hyphae, producing appresoria or hook-like structures at certain attachment sites (42). Presumably, this is the last step before excretion of lytic enzymes (*i.e.,* β-1,3-glucanase and chitinase) and penetration (42). These enzymes could be induced either by laminarin and chitin or by purified fungal cell walls (43,44,45). Recently, Sivan and Chet (45) found that the failure of strain T-35 of *T. harzianum* to parasitize colonies of *F. oxysporum* is not because of its strain's inability to produce or excrete cell wall–degrading enzymes (*i.e.,* β-1,3-glucanase and chitinase). *In vitro* tests showed that this strain is an effective mycoparasite of *R. solani* and *P. aphanidermatum* and is capable of producing high levels of β-1,3-glucanase and chitinase. Higher levels of these enzymes were induced by the cell walls of *R. solani* and *S. rolfsii* than by the cell walls of *F. oxysporum*. However, treatment of the *Fusarium* cell walls with 2 N NaOH, trypsin, or protease prior to their incubation with lytic enzymes obtained from strain T-35 resulted in a significant increase in the release of monomers from these cell walls. A similar increase was not obtained with cell walls of *R. solani* or *S. rolfsii*. These results suggest that the cell wall proteins of *Fusarium* interfere with the activity of β-1,3-glucanase and chitinase, thus increasing their resistance to lysis (45). However, the hypothesis that lack of recognition between *Fusarium* and *Trichoderma* prevented mycoparasitism has still to be studied.

3.2. Antibiosis

Although various *Trichoderma* strains possess the ability to excrete toxic or inhibitory metabolites (31,46), there is not sufficient evidence for their contribution to pathogen suppression and disease reduction *in situ*. Wright (47) reported that a gliotoxin-producing strain of *T. viride* (later thought to be *Gliocladium virens*) was more effective in controlling *Pythium* damping-off than a gliovirin-producing strain. However, a third strain that produced neither of these antibiotics gave inferior results compared with the antibiotics-producing strains.

Howell and Stipanovic (48) combined a biochemical approach with an analysis of mutants to demonstrate that the antibiotic glovirin, produced by *G. virens*, is involved in the biocontrol of *P. ultimum*. This antibiotic was shown to kill the pathogen, since mycelia that had been exposed to the antibiotic, washed, and transfered to a fresh medium did not grow. The role of gliovirin in the biocontrol of *P. ultimum* was demonstrated by the work of Howell and Stipanovic (48), who showed that a mutant that did not produce gliovirin was unable to control *Pythuim* damping-off. Moreover, another mutant that overproduced gliovirin provided similar control to that of the wild type, although it exhibited a much lower growth rate.

Claydon *et al.* (49) have identified volatile alkyl pyrones produced by *T. harzianum* that were inhibitory to a number of fungi *in vitro*. When these metabolites were added to a peat–soil mixture, they reduced the incidence of *R. solani*–induced damping-off on lettuce. Recently, Ordentlich and Chet (unpublished data) isolated a novel inhibitory substance, 3-(2-hydroxypropyl),4-(2,4-hexadienyl),2(5H)-furanone, produced by *T. harzianum* that was found to suppress growth of *F. oxysporum* and may be involved in the biocontrol of fusarium wilts.

3.3. Competition

Competition for limiting amounts of nutrients (*i.e.,* carbon, nitrogen, and iron) and its role in the biocontrol of plant pathogens has been studied for many years, with emphasis on bacterial biocontrol agents (11,50), whereas very little attention has been paid to fungal antagonists. Recently, we found that a strain of *T. harzianum* (T-35) that controls *Fursarium* spp. on various crops may utilize competition for nutrients and rhizosphere colonization (51).

The potential of a microorganism applied as a seed treatment to proliferate and establish along the developing rhizosphere has been termed *rhizosphere competence* (52). When conidia of T-35 were applied to soil enriched with chlamydospores of *F. oxysporum* f. sp. *melonis* and *F.* sp. *vasinfectum* and amended with low levels of glucose and asparagine, the ability of the chlamydospores to germinate was reduced. This inhibitory effect could be reversed if an excess of glucose and asparagine or of seedling exudates was added to the soil. After its application as seed treatment, this strain effectively colonized the rhizosphere of melon and cotton and prevented colonization of these roots by *F. oxysporum*. Thus competition for carbon and nitrogen in the rhizosphere as well as rhizosphere competence might be involved in the biocontrol of *F. oxysporum* by *T. harzianum* strain T-35 (51).

The biocontrol activity of *Trichoderma* strains (as well as of other biocontrol agents) may be conferred by more than one exclusive mechanism. In fact, it seems advantageous for a biocontrol agent to suppress a plant pathogen using multiple mechanisms. More research efforts should be directed to the application of this approach.

4. BIOLOGICAL BARRIERS TO THE DEVELOPMENT OF MICROBIAL PESTICIDES

Today, despite the increasing interest in studying the basic and applied aspects of biological control of plant pathogens, the commercial use of biocontrol is very limited (53). Development of reliable biocontrol systems will depend on the accurate assessment of the problems and on the limiting factors that interfere with them. Currently, most of the reports in this field deal with successes and do not carefully describe cases in which previously screened antagonists have failed to provide control of a pathogen. A careful analysis of such cases might be critical for the determination of problems associated with effective development.

One of the most prominent problems associated with biocontrol is apparently the inconsistent performance of microbial antagonists under field conditions (11). Several reasons can be attributed to this phenomenon.

4.1. Screening Techniques

It is essential that control agents be screened under conditions approximating as closely as possible the natural habitat in which they are expected to be active

(53). However, since this requires complex and time-consuming *in vivo* bioassays, indirect screening techniques are often used. These usually rely on antagonistic phenotypes on plates (*i.e.,* mycoparasitism and antibiosis). Although these methods are fast, they are capable only of reflecting the potential of an antagonist to be used as a biocontrol agent and can sometimes be misinterpreted.

4.2. Ecological Factors

Since 1 g of soil can contain up to 10^9 bacteria and 10^5 fungi, an introduced microbial antagonist will have to overcome the competition from the indigenous soil microflora. For instance, in soils with a low iron level, some *Trichoderma* strains are sensitive to competition from fluorescent pseudomonads, resulting in growth inhibition and reduction in biocontrol activity (54). Soil pH may also be an important factor determining the success or failure of introduced antagonists. Indeed, Chet and Baker (55) showed that *Trichoderma* is more active under low pH conditions. Competition as well as other antagonistic interactions (*e.g.,* antibiosis and parasitism) between the introduced biocontrol agent and the pathogen may also take place in the rhizosphere. Rhizosphere colonization by introduced biocontrol agents appears to be a key factor in the biocontrol of soilborne plant pathogens (11). Thus, factors affecting root colonization should be determined and introduced as an additional screening tool for biocontrol agents.

4.3. Mechanisms of Control

As noted earlier in this chapter, it may be impossible to project the actual mode(s) of action under field conditions from data obtained under laboratory conditions. Elucidation of the mechanism should include genetic studies to identify and isolate genes involved in biocontrol. These could be further used to complement mutants lacking the respective mechanistic phenotype (56); this could lead to a better understanding of mechanisms and thus to more effective exploitation of antagonists.

4.4. Problems Associated With Formulation and Large-Scale Production

The development of an effective, stable, and cost-effective formulation that is compatible with agricultural equipment and agrotechnical methods is critical for the successful commercialization of introduced antagonists (57). Formulations should suit the macro- and microhabitats in which control agents are to be applied and should ensure a long shelf life of the biopesticide. Furthermore, because of the competition with soil microflora, the formulation should also provide a selective advantage to the introduced antagonist over the resident microorganisms. This advantage is often conferred by a food base that is suitable for the antagonist. However, food bases should be carefully selected, since readily available nutrients (*e.g.*, simple sugars and amino acids) may favor the pathogen. An excellent example of this problem is the attempt to control *Phytophthora cinnamomi* by application of *T. harzianum* formulated in clay granules impregnated with molasses; nutrients that leaked from this preparation were found to favor the pathogen and to cause disease (58).

The control activity of antagonists may be reduced after large-scale fermentation. The Biotechnology General Company (BTG, Rehovot, Israel) has developed a liquid fermentation process that produces 3×10^8 conidia/ml after 60 hours. However, it is not clear whether these conidia are as effective as those produced by aerial mycelium (28).

5. ENHANCEMENT OF BIOCONTROL PERFORMANCE

Although antagonistic microorganisms are not expected to have the biocidal effect of pesticides (3), they should provide consistent control at a level that will be commercially acceptable to the grower. Two main approaches for enhancing biocontrol efficacy are discussed. The first deals with modified formulations for tailoring the optimal microenvironment for the introduced antagonist, and the second focuses on genetic manipulation of the biocontrol agent for enhancing the expression of traits required for biocontrol activity.

5.1. Improved Formulation and Application Techniques

Since the soil ecosystem is biologically balanced, it is expected that most introduced microorganisms will fail to establish. Therefore, application of antagonists to soil requires a suitable food base to provide a carrier and a substrate for the biocontrol agent. However, this type of application requires large amounts of microbial biomass and is rarely, if ever, economical. Application of bioprotectants on the planting material, such as the seed-coating technique, has been proven to be as effective as broadcast application and, because it requires low amounts of inoculum, is expected to be much more economical than the broadcast application methods. Moreover, seed treatments with rhizosphere-competent strains may result in secondary deployment of these biocontrol agents along the root (11,51,52,59), which may prevent infection by root rot and wilt pathogens (34,51).

The density of the antagonist propagules applied to the seeds may affect both root colonization and disease control. Thus determination of this optimal inoculum density may be important for reducing inoculum concentration without losing biological efficacy. A microbial seed treatment to various crops may require different concentrations of the biocontrol agent. Figure 13.1 shows that

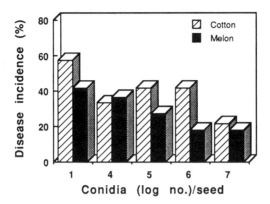

Fig. 13.1. The effect of increasing concentrations of conidia of *T. harzianum* strain T-35 on cotton and melon seeds on biological control of fusarium wilt.

different concentrations of the microbial antagonist are required for control of fusarium wilt in cotton and melons.

Another approach for enhancing seed treatment efficacy is to modify the seed coat and spermosphere microenvironments to favor proliferation of the biocontrol agent selectively over other competing microorganisms, including the pathogen. Nelson *et al.* (60) reported that 25 of 88 compounds tested for their ability to favor growth of *T. koningii* while not supporting the growth of *P. ultimum* significantly increased biocontrol activity when incorporated with the antagonist on the seed coat. These compounds included fatty acids, polysaccharides, alcohol sugars, and detergents.

Modification of the microenvironment for improving biocontrol activity can be achieved on the seed surface by combining the microbial antagonist with a seed osmoconditioner, a method known as seed priming. One of these methods is known as solid matrix priming (SMP) (61). Incorporation of *T. harzianum* with a solid matrix primer improved the root colonization by the antagonist, biocontrol activity against *P. ultimum* and *F. graminearum,* and plant growth response of peas and cucumbers (62).

5.2. Genetic Manipulations of Biocontrol Agents

Although a large variety of effective control agents can be isolated from natural habitats, genetic modification of these microorganisms may result in the improvement or acquisition of certain traits that are important in biocontrol activity. One of the simplest methods evaluated for improving biocontrol agents is mutation by chemical mutagens or irradiation. In the past this technique has been utilized for producing biocontrol strains resistant to fungicides such as benomyl (63). Some of the benomyl-resistant mutants of *Trichoderma* spp. differed from their respective wild types in biocontrol activity, growth characteristics, sporulation, survival in soil, and rhizosphere competence (52,64).

Recombinant DNA procedures may be very useful for improving biocontrol agents by cloning or deleting certain genes in desired microorganisms. However, this approach depends on the identification and isolation of the genes involved in biocontrol. Thus far, this approach has been used mainly for elucidating control mechanisms. In an excellent study, Thomashow and Weller (56) demonstrated that phenazine antibiotics were essential for the control of take-all disease in wheat by *Pseudomonas fluorescens*. These researchers not only deleted activity of the genes in question but also were able to clone back these genes into deficient mutants and to restore the antibiotic production and activity.

In another study aimed at proving the involvement of the chitinase of *Serratia marcescens* in suppressing *S. rolfsii,* Shapira *et al.* (65) developed a system for expression of the *S. marcescens* chitinase in *Escherichia coli*. A DNA fragment carrying the chitinase gene (*chi* A) and the oLpL bacteriophage lambda promoter were subcloned into the plasmid pBR322 to yield a new plasmid, pLCHIA. When *E. coli* was transformed with this plasmid it produced and secreted high amounts of chitinase. The transformed *E. coli* or the *S. marcescens* chitinase it produced controlled *S. rolfsii* on beans when applied to soil infested with this pathogen.

In the absence of sexual recombination, protoplast fusion may be utilized for inducing parasexuality and genetic recombination between different strains. This technique was first used for the production of asexual hybrids in animal and plant cells and was later employed in fungi. Stasz *et al.* (66) developed a procedure for fusing protoplasts from different auxotrophic mutant strains of *Trichoderma*. Thus far, the use of this technique for improving control of disease has been demonstrated only in *Trichoderma* (62,67,68). Although several *Trichoderma* strains that are effective agents have been isolated, there is quite a large variability between different strains in terms of their host range, biocontrol attributes, and their ability to grow under low temperatures (which may be essential for the eradication of some pathogens). Thus, combining desired traits from different strains into one is required in order to produce superior strains (53). Some of these attributes are likely to be polygenic; for example, mycoparasitism consists of several steps (*i.e.,* recognition, attachment, and excretion of lytic enzymes). Hence, asexual hybridization via protoplast fusion may have an advantage over mutation or transformation because of its potential to incorporate and combine large DNA sequences from different wild types into the resulting hybrid.

In *Trichoderma,* protoplast fusion resulted in progeny that were shown to be superior to either of the prototrophic (wild-type) parental strains in their biocontrol ability and their growth at 10°C and at 25°C (62). One of these improved strains of *T. harzianum* (strain 1295-22) was further analyzed and found to be a better rhizosphere competent than its wild-type parental strains T12 and T95 (68). Similarly, Pe'er and Chet (67) reported the improved control of *R. solani* damping-off in cotton by an isolate of *T. harzianum* obtained by fusing protoplasts of two auxotrophic mutants. This strain also showed enhanced mycoparasitic activity compared with the parental strains.

Although protoplast fusion in *Trichoderma* was expected to result in nuclear fusion and haploidization, yielding recombinant strains (*i.e.*, parasexualily), thus far this process does not seem to occur in *Trichoderma*. Protoplast fusion progeny of different strains or species of the fungus resulted in vegetative incompatibility, which was demonstrated by a low complementation frequency for fusion between complementary auxotrophic parental strains and by the recovery of progeny that were unstable and slow growing and that exhibited a wide range of colony morphologies. These progeny were shown to be either very imbalanced heterokaryons or homokaryons that differed markedly from the parental strains (69). A partial elimination of this vegetative incompatibility could be obtained by insertion of nuclei of *T. harzianum* isolated from one strain into protoplasts of another, suggesting that the cytoplasmic components might be involved, to some extent, in this incompatibility (70). Although not all the postfusion genetic events are completely understood in the *Trichoderma* protoplast, fusion has proven to be a useful tool in obtaining more effective strains.

6. INTEGRATED CONTROL

Although biocontrol has achieved some remarkable successes, it is well recognized that at least for the time being, if it is to be part of large-scale agriculture, it should be combined with other control practices, leading to integrated pest management (IPM). This section focuses on the experience gained in the integration of *Trichoderma* with chemical fungicides or with physical methods for plant disease control.

6.1. Chemical–Biological Control

Because of their relatively high tolerance to some fungicides, *Trichoderma* biocontrol agents are good candidates to be combined with fungicides. This phenomenon is illustrated in Figure 13.2. *Trichoderma,* as well as other biocontrol agents, may be applied simultaneously with sublethal doses of fungicides. This may improve disease control, avoiding much of the risk imposed by the application of the chemicals at standard doses. In keeping with this approach, Hadar *et al.* (29) showed that a combination of a sublethal dose of pentachlo-

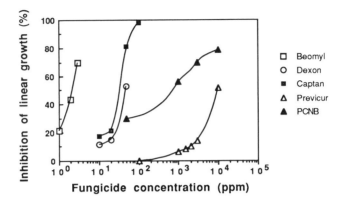

Fig. 13.2. Relative tolerance of *T. harzianum* strain T-315 to some fungicides.

ronitrobenzene (PCNB) together with *T. harzianum* reduced *R. solani* root rot of eggplants from 40% (unprotected) to 13%, while *T. harzianum* alone reduced it to 26%. Similarly, improved control of bean blight caused by *S. rolfsii* was obtained following a combination of *T. harzianum* with PCNB (28). The natural resistance of *T. harzianum* strain T-315 to several fungicides, including prothiocarb (Fig. 13.2), which is being used for the control of *Pythium* spp., was exploited by combining this fungicide with T-315 for controlling *P. aphanidermatum* on gypsophila cuttings. The combined treatment was superior to the application of either the chemical or T-315 alone. This example of integrated control is illustrated in Figure 13.3.

Another means of combining chemical and biological agents is to apply fungicide-resistant mutants together with the relevant chemical. Baker and Scher (71) applied a mutant strain of *T. harzianum* (T-95), resistant to benomyl, to a rooting mixture of carnation infested with *R. solani*. When the antagonist or benomyl was applied alone to the rooting mixture, a reduction of about 50% in rhizoctonia rot incidence was obtained, and a combined treatment of both gave 100% control. The same strain was further analyzed and found to be an effective rhizosphere competent (52).

Recently, Ordentlich *et al.* (72) reported that a combined treatment of *T. harzianum* and Captan resulted in significant control of verticillium wilt of potatoes caused by *V. dahaliae* under field conditions and also facilitated a 17% increase in the potato yield.

Alternatively, the biocontrol agent may be introduced after soil fumigation to prevent reinfestation with the pathogen. Application of *T. harzianum* after soil fumigation with a sublethal dose of 200 kg/ha of methylbromide resulted in significant control of rhizoctonia damping-off of carrots and had a similar effect

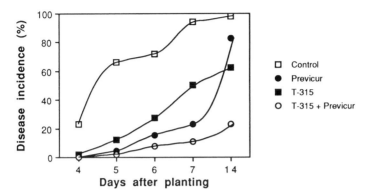

Fig. 13.3. Integrated control of *P. aphanidermatum* on gypsophila by *T. harzianum* strain T-315 and prothiocarb.

on growth, yield, and disease control to the full recommended dose of the fumigant (500 kg/ha) (73). Moreover, application of *Trichoderma* to a fumigated soil had a prolonged effect that resulted in prevention of its reinfestation by *R. solani* throughout three consecutive growth cycles of beans in the same soil. This combination has been shown to be effective for controlling soil-borne pathogens such as *F. oxysporum* f.sp. *radicis-lycopersici* (the causal agent of fusarium crown rot in tomato) that produce airborne microconidia (34).

6.2. Physical–Biological Control

Solar heating (soil solarization) as a means for controlling soil-borne plant diseases was developed in Israel (74). This method involves mulching of the soil with transparent polyethylene sheets that trap the incoming solar radiation and kill propagules of pathogens as well as weed seeds. The combination of biocontrol agents with this method may extend its effect and may prevent reinfestation during the growth season.

Soil solarization may also result in natural biocontrol of propagules exposed to sublethal heat and thereby weakened. Such propagules are more susceptible to indigenous soil microorganisms that tolerate the heat levels generated by soil solarization (75).

When a strain of *T. harzianum* was introduced into solar-heated soil, the control of *R. solani* and *S. rolfsii* on potatoes was improved compared with solar heating alone. Furthermore, this combination reduced the inoculum density of *R. solani* and delayed its subsequent build-up in field plots and under greenhouse conditions (76). Because of its safety, solar heating may well be used in combination with biocontrol agents as a substitute for fumigants used in soil disinfestation and in integrated control.

7. CONCLUSIONS

Because of their relatively high efficiency, biocontrol agents, including *Trichoderma* spp., can probably serve as suitable alternatives to some commercial fungicides (especially broad-spectrum chemicals). *T. harzianum* has already been shown to be as effective as the fungicides benomyl and Captan in the control of damping-off (62,71). However, more studies are required on improving screening techniques and elucidating the molecular basis of mechanisms. This, in turn, will enable genetic manipulation and production of superior biocontrol strains and a better understanding of the ecological factors that interfere with biocontrol activity. Since these developments are not expected to show immediate results, more emphasis should be directed to extending the use of integrated control as a part of a comprehensive integrated pest management program, thus introducing biocontrol into mainstream agriculture.

Biological control may lack the fast biocidal effect exhibited by most pesticides, and it may also result in somewhat lower yields than chemical control. However, the use of this technique will result in yields that are pesticide free and of higher quality. Pesticide-free produce has already been introduced into the markets in the United States and Europe. The market share of this type of environmentally friendly fresh produce is constantly increasing, and pesticide-free fruits and vegetables are commanding higher prices than conventional produce. Thus, the decrease in yield potential that may occur in some crops resulting from a shift to biocontrol is likely to be compensated by increased value of the produce.

REFERENCES

1. Ritter WF: Pesticide contamination of ground water—A review. J Environ Sci Health, Part B, Pesticides Food 25:1–29, 1990.

2. Gough M: Estimating cancer mortality. Environ Sci Technol 23:925–930, 1989.

3. Cook RJ, Baker KF: The Nature and Practice of Biological Control of Plant Pathogens. St. Paul, MN: The American Phytopathological Society, 1983, 539 pp.

4. Costa AS, Muller GW: *Tristeza* control by cross protection: A U.S.-Brazil cooperative success. Plant Dis 64:538–541, 1980.

5. Fletcher JT, Row JM: Observations and experiments on the use of an avirulent mutant strain of tobacco mosaic virus as a means of controlling tomato mosaic. Ann Appl Biol 81:171–179, 1975.

6. Caruso FL, Kuc J: Protection of watermelon and muskmelon against *Colletotrichum lagenarium* by *Colletotrichum lagenarium.* Phytopathology 67:1290–1292, 1977.

7. Sequeira L, Hill LM: Induced resistance in tobacco leaves: The growth of *Pseudomonas solanacearum* in protected tissues. Physiol Plant Pathol 4:447–455, 1974.

8. DeBach P: Biological Control of Insect Pests and Weeds. New York: Reinhold, 1964, 844 pp.

9. Hartly C: Damping-off in forest nurseries. US Dept Agric Bull 934:1–99, 1989.

10. Baker KF, Cook RJ: Biological Control of Plant Pathogens. St. Paul, MN: The American Phytopathological Society, 1974, 433 pp.

11. Weller DM: Biological control of soilborne plant pathogens in the rhizosphere with bacteria. Annu Rev Phytopathol 26:379–407, 1988.

12. Kerr A: Soil microbiological studies on *Agrobacterium radiobacter* and biological control of crown gall. Soil Sci 22:111–116, 1974.

13. New PB, Kerr A: Biological control of crown gall: Field measurments and glasshouse experiments. J Appl Bacteriol 35:279–287, 1972.

14. Schroth MN, Weinhold AR, McCain AH, Hildebrand DC, Ross N: Biology and control of *Agrobacterium tumefaciens.* Hilgardia 40:537–552, 1971.

15. Merriman PR, Price RD, Kollmorgen F, Piggott T, Ridge EH: Effect of seed inoculation with *Bacillus subtilis* and *Streptomyces griceus* on the growth of cereals and carrots. Aust J Agric Res 25:219–226, 1974.

16. Broadbent P, Baker KF, Franks N, Holland J: Effect of *Bacillus subtilis* on increased growth of seedlings in steamed and in nontreated soil. Phytopathology 67:1027–1034, 1977.

17. Turner JT Jr: Relationships Among Plant Growth, Yield, and Rhizosphere Ecology of Peanuts as Affected by Seed Treatment With *Bacillus subtilis.* PhD Dissertation, Auburn University, 1987, 108 pp.

18. Burr TJ, Schroth MN, Suslow T: Increased potato yields by treatment of seedpieces with specific strains of *Pseudomonas fluorescens* and *P. putida.* Phytopathology 68:1377–1383, 1978.

19. Suslow TV, Schroth MN: Rhizobacteria of sugarbeets: Effects of seed application and root colonization on yield. Phytopathology 72:199–206, 1982.

20. Schroth MN, Hancock JG: Disease-suppressive soil and root colonizing bacteria. Science 216: 1376–1381, 1982.

21. Kloepper JW, Schroth MN: Relationship of *in vitro* antibiosis of plant growth-promoting rhizobacteria to plant growth and the displacement of root microflora. Phytopathology 70:1020–1024, 1981.

22. Weller DM, Cook RJ: Suppression of take-all of wheat by seed treatments with fluorescentic pseudomonads. Phytopathology 73:463–469, 1983.

23. Cook RJ, Weller DM, Bassett EN: Effect of bacterial seed treatments on growth and yield of recroped wheat in western Washington, 1987. Biol Cultural Tests Control Plant Dis 3:53, 1988.

24. Chet I: *Trichoderma*—application, mode of action, and potential as biocontrol agent of soilborne plant pathogenic fungi. In Chet I (ed): Innovative Approaches to Plant Disease Control. New York: J Wiley and Sons, 1987, pp 137–160.

25. Papavizas GC: *Trichoderma* and *Gliocladium:* Biology, ecology and the potential for biocontrol. Annu Rev Phytopathol 23:23–54, 1985.

26. Weindling R: *Trichoderma lignorum* as a parasite of other fungi. Phytopathology 22:837–845, 1932.

27. Wells HD, Bell DK, Jaworski CA: Efficacy of *Trichoderma harzianum* as a biocontrol agent for *Sclerotium rolfsii.* Phytopathology 62:442–447, 1972.

28. Chet I: Biological control of soil-borne plant pathogens with fungal antagonists in combination with soil treatments. In Hornby D (ed): Biological Control of Soil-Borne Plant Pathogens. Wallingford, UK: C.A.B. International, 1990, pp 15–25.

29. Hadar Y, Chet I, Henis Y: Biological control of *Rhizoctonia solani* damping-off with a wheat bran culture of *Trichoderma harzianum*. Phytopathology 69:1167–1172, 1979.

30. Elad Y, Chet I, Katan Y: *Trichoderma harzianum:* A biocontrol agent effective against *Sclerotium rolfsii* and *Rhizoctonia solani*. Phytopathology 70:119–121, 1980.

31. Sivan A, Elad Y, Chet I: Biological control effects of a new isolate of *Trichoderma harzianum* on *Pythium aphanidermatum*. Phytopathology 74:498–501, 1984.

32. Elad Y, Zvieli Y, Chet I: Biological control of *Macrophomina phaseolina* (Tassi) by *Trichoderma harzianum*. Crop Protection 5:288–292, 1986.

33. Sivan A, Chet I: Biological control of *Fusarium* spp. in cotton, wheat and muskmelon by *Trichoderma harzianum*. Phytopathol Z 116:39–47, 1986.

34. Sivan A, Ucko O, Chet I: Biological control of fusarium crown rot of tomato by *Trichoderma harzianum* under field conditions. Plant Dis 71:587–592, 1987.

35. Marois JJ, Mitchell DJ, Sonoda RM: Biological control of *Fusarium* crown rot of tomato under field conditions. Phytopathology 71:1257–1260, 1981.

36. Sivan A, Chet I: *Trichoderma harzianum:* An effective biocontrol agent of *Fusarium* spp. In Jensen V, Kjoller A, Sorensen LH (eds): Microbial Communities in Soil. London: Elsevier Applied Sciences Publishers, 1986, pp 89–96.

37. Baker R: Myloparasitism: Ecology and physiology. Can J Plant Pathol 9:370–379, 1987.

38. Handlesman J, Parke JL: Mechanism in biocontrol of soilborne plant pathogens. In Kosuge T, Nester EW (eds): Plant Microbe Interactions, vol 3. New York: McGraw-Hill, 1989, pp 27–61.

39. Chet I, Harman GE, Baker R: *Trichoderma hamatum:* Its hyphal interactions with *Rhizoctonia solani* and *Pythium* spp. Microbial Ecol 7:29–38, 1981.

40. Elad Y, Barak R, Chet I: Possible role of lectins in mycoparasitism. J Bacteriol 154:1431–1435, 1983.

41. Barak R, Elad Y, Mirelman D, Chet I: Lectins: A possible basis for specific recognition in *Trichoderma-Sclerotium rolfsii* interaction. Phytopathology 75:458–462, 1985.

42. Elad Y, Chet I, Boyle P, Henis Y: Parasitism of *Trichoderma* spp. on *Rhizoctonia solani* and *Sclerotium rolfsii:* Scanning electron microscopy and fluorescence microscopy. Phytopathology 73:85–88, 1983.

43. Elad Y, Chet I, Henis Y: Degradation of plant pathogenic fungi by *Trichoderma harzianum*. Can J Microbiol 28:719–725, 1982.

44. Ridout CJ, Coley-Smith JR, Lynch JM: Fractionation of extracellular enzymes from a mycoparasitic strain of *Trichoderma harzianum*. Enzyme Microbiol Technol 10:180–187, 1988.

45. Sivan A, Chet I: Degradation of fungal cell walls by lytic enzymes of *Trichoderma harzianum*. J Gen Microbiol 135:675–682, 1989.

46. Dennis L, Webster J: Antagonistic properties of species-groups of *Trichoderma*. I. Production of non-volatile antibiotics. Trans Br Mycol Soc 57:25–39, 1971.

47. Wright JM: The production of antibiotics in soil. III. Production of gliotoxin in wheatstraw buried in soil. Ann Appl Biol 44:461–466, 1956.

48. Howell CR, Stipanovic RD: Gliovirin, a new antibiotic from *Gliocladium virens* and its role in the biological control of *Pythium ultimum*. Can J Microbiol 29:321–324, 1983.

49. Claydon N, Allan M, Hanson JR, Avent AG: Antifungal alkyl pyrons of *Trichoderma harzianum*. Trans Br Mycol Soc 88:503–513, 1987.

50. Schippers B, Lugtenberg B, Weisbeek PJ: Plant growth control by fluorescent pseudomonads. In Chet I (ed): Innovative Approaches to Plant Disease Control. New York: J Wiley and Sons, pp 19–39, 1987.

51. Sivan A, Chet I: The possible role between *Trichodema harzianum* and *Fusarium oxysporum* on rhizosphere colonization. Phytopathology 79:198–203, 1989.

52. Ahmad JS, Baker R: Rhizosphere competence of *Trichoderma harzianum*. Phytopathology 77:182–189, 1987.

53. Harman GE, Lumsden RD: Biological disease control. In: Lynch JM (ed): The Rhizosphere. New York: Wiley and Sons, 1990, pp 259–276.

54. Hubbard JP, Harman GE, Hadar Y: Effect of soilborne *Pseudomonas* spp. on the biological control agent, *Trichoderma hamatum,* on pea seeds. Phytopathology 73:655–659, 1983.

55. Chet I, Baker R: Induction of suppressiveness *Rhizoctonia solani* in soil. Phytopathology 70:994–998, 1980.

56. Thomashow LS, Weller DM: Role of a phenazine antibiotic from *Pseudomonas fluorescens* in biological control of *Gaeunmanomyces graminis* var. *tritici.* J Bacteriol 170:3499–3508, 1988.

57. Lisanski SG: Production and commercialization of pathogens. In Hussey NW, Scopes N (eds): Biological Pest Control. Poole, UK: Blandford Press, 1985, pp 210–218.

58. Kelley WD: Evaluation of *Trichoderma harzianum*–impregnated clay granules as a biocontrol for *Phytophthora cinnamomi* causing damping-off of pine seedlings. Phytopathology 66:1023–1027, 1976.

59. Harman GE: Deployment tactics for biocontrol agents in plant pathology. In Baker R, Dunn PE (eds): New Directions in Biological Control—Alternatives for Suppressing Agricultural Pests. New York: Alan R. Liss, Inc., 1989.

60. Nelson EB, Harman GE, Nash GT: Enhancement of *Trichoderma*-induced biological control of *Pythium* seed rot and preemergence damping-off of peas. Soil Biol Biochem 20:145–150, 1988.

61. Taylor AG, Klein ED, Whitlow TH: SMP: Solid matrix priming of seeds. Sci Hort 37:1–11, 1988.

62. Harman GE, Taylor AG, Stasz TE: Combining effective strains of *Trichoderma harzianum* and solid matrix priming to improve biological seed treatments. Plant Dis 73:631–637, 1989.

63. Abd-El Moity TH, Papavizas GC, Shatla MN: Induction of new isolates of *Trichoderma harzianum* tolerant to fungicides and their experimental use for control of white rot of onion. Phytopathology 72:396–400, 1982.

64. Papavizas GC: Genetic manipulation to improve the effectiveness of biocontrol fungi for plant disease control. In Chet I (ed): Innovative Approaches to Plant Disease Control. New York: J Wiley and Sons, 1987, pp 193–212.

65. Shapira R, Ordentlich A, Chet I, Oppenheim AB: Control of plant diseases by chitinase expressed from cloned DNA in *Escherichia coli.* Phytopathology 79:1246–1249, 1989.

66. Stasz TE, Harman GE, Weeden NF: Protoplast preparation and fusion in two biocontrol strains of *Trichoderma harzianum*. Mycologia 80:141–150, 1988.

67. Pe'er S, Chet I: *Trichoderma* protoplast fusion: A tool for improving biocontrol agents. Can J Microbiol 36:6–9, 1989.

68. Sivan A, Harman GE: Improved rhizosphere competence in a protoplast fusion progeny of *Trichoderma harzianum*. J Gen Microbiol 137:23–29, 1991.

69. Stasz TE, Harman GE, Gullino ML: Limited vegetative incompatibility following intra and interstrain protoplast fusion in *Trichoderma* spp. Exp Mycol 13:264–371, 1989.

70. Sivan A, Harman GE, Stasz TE: Transfer of isolated nuclei into protoplasts of *Trichoderma harzianum*. Appl Environ Microbiol 56:2404–2409, 1990.

71. Baker R, Scher FM: Enhancing the activity of biological control agents. In Chet I (ed): Innovative Approaches to Plant Disease Control. New York: J. Wiley and Sons, 1987, pp 1–17.

72. Ordentlich A, Nachmias A, Chet I: Integrated control of *Verticillium dhaliae* in potato by *Trichoderma harzianum* and captan. Crop Protect (in press), 1990.

73. Strashnow Y, Elad Y, Sivan A, Chet I: Integrated control of *Rhizoctonia solani* by methyl bromide and *Trichoderma harzianum*. Plant Pathol 34:146–151, 1985.

74. Katan J, Greenberger A, Alon H, Greenstein A: Solar heating by polyethylene mulching for the control of diseases caused by soilborne pathogens. Phytopathology 66:683–688, 1976.

75. Katan J: Soil solarization. In Chet I (ed): Innovative Approaches to Plant Disease Control. New York: J Wiley and Sons, 1987, pp 77–105.

76. Elad Y, Katan J, Chet I: Physical, biological and chemical control integrated for soilborne diseases in potatoes. Phytopathology 70:418–422, 1980.

14

COMPOSTING IN THE CONTEXT OF MUNICIPAL SOLID WASTE MANAGEMENT

MELVIN S. FINSTEIN

Department of Environmental Sciences, Cook College, Rutgers University, New Brunswick, New Jersey 08903

1. MUNICIPAL SOLID WASTE

1.1. Nature of the Waste Stream

Municipal solid waste (MSW) poses a difficult and complex problem for society. Some of the difficulties arise because the MSW stream is quantitatively large and qualitatively heterogeneous, reflecting the myriad consumer products manufactured in modern industrial society. Inconveniently, MSW is largely generated in densely populated areas where its management is most constrained. Thus the problem cuts across a very wide range of human activities and interests. At the same time, MSW represents a uniquely familiar environmental problem, in that everyone contributes to it palpably in the course of daily living.

A useful definition of MSW was offered by the United States Office of Technology Assessment (OTA), the analytical arm of the U.S. Congress (1): ''MSW is solid waste generated at residences, commercial establishments (*e.g.*, offices, retail shops, restaurants) and institutions (*e.g.* hospitals and schools).'' The amount generated in the United States in 1988 was estimated at 180 million tons, or approximately 4 pounds per person per day. This is thought to be increasing by 1% per year.

Environmental Microbiology, pages 355–374, © 1992 Wiley-Liss, Inc.

A lack of uniform methodologies and categorization schemes makes precise comparison among actual MSW streams problematic. Nevertheless, formal sorting studies indicate, not surprisingly, that the composition has changed markedly in recent decades (2). For example, coal ash has nearly disappeared from domestic waste while plastic products have proliferated. Contemporary waste streams vary in composition, depending on such factors as community socioeconomic level and ethnic preference. Seasonal variation also occurs, as in the appearance of more vegetative food and yard waste in summer.

A hypothetical composition of MSW generated nationally (USA) was devised based on production and consumption data from trade associations and other sources (1). This indicates the following nominal percentages by weight as received: paper and paperboard, 36; yard waste, 20; food waste, 9; metals, 9; glass, 8; plastics, 7; textile and wood, 6; rubber and leather, 3; miscellaneous inorganics, 2. The heterogeneity of MSW is a major determinant of management options.

1.2. Mass Landfilling

A newfound popular enthusiasm for recycling notwithstanding (3), the dominant contemporary practice is to landfill MSW *en masse,* or nearly so, as a collective undifferentiated waste. Yet mass landfilling is now generally recognized as an environmental anachronism. A major specific objection is that landfills must be expected to leak eventually, despite multiple hydrologic barriers of clay and synthetic membrane liners. This poses a threat to groundwater. Even as a matter of expediency it is necessary to decrease reliance on landfilling, as it has become difficult to site new facilities in proximity to metropolitan areas.

The tendency is for smaller landfills to be closed and new ones to be larger, though overall capacity is diminishing (4). In 1988 the number of operative landfills in the United States was approximately 5,500 and was projected to decrease to 1,200 by the year 2008 (5). New Jersey, the most crowded state and a highly industrialized one, is already severely impacted. The number of operative MSW landfills in this state decreased from 140 in 1982 to less than 30 in 1990, while in the same period the average tipping fee increased from approximately $5 per ton to $60. Disposal out of state, necessitated by a lack of local capacity, now costs as much as $140 per ton (transport plus tipping). Whereas only a few years ago landfilling economics uniformly discouraged the development of alternatives, at least in some parts of the country the opposite situation now prevails. It is clear that mass landfilling cannot long be sustained as the norm. This situation is generally true worldwide (4).

Even with advanced MSW management, however, *discriminating* landfilling is necessary to dispose of some irreducible amount of nonrecoverable, nonbiologically active waste. This refers to material that cannot reasonably be recycled or usefully composted. The need is to eliminate nondiscriminating landfilling.

1.3. Mass Incineration

A variant of the mass approach is that of mass incineration, the stated main purpose of which is to spare landfill space. Assuming that oversized, bulky durable goods such as major appliances are diverted from the incinerator, volume reduction may be 85%–95% and weight reduction 70%–85% (1).

The future of mass incineration is uncertain, however, owing to a number of socioeconomic, environmental, and philosophical objections. These include very large costs for facility construction and operation, unresolved issues of harmful air emissions and control device adequacy and reliability, and concern about ash toxicity and safe long-term disposal. Ash that is judged to be toxic must be buried in dedicated, unusually costly, landfills. Moreover, mass incineration conflicts with the alternative of recycling and composting. The alternative represents a very different approach to material resource management, and generates many relatively low-paying jobs often needed in community development. On a range of issues, incineration is opposed by a broad array of aroused citizens groups.

1.4. Materials Management

The OTA advocates an approach to the problem of MSW termed *materials management* (1). It has two major aspects. The first concerns prevention, sometimes termed *waste minimization* or *source reduction* (6). This is pursued by means of legislation and regulation leading to financial incentives, or disincentives, for the design and manufacture of environmentally benign products, the banning of certain products, and educating the consumer to make informed purchases. Redesign of a product takes into account not only its primary role, but also its secondary role as postconsumer waste. Quantitative changes are made to reduce the amount of materials used, and qualitative changes are made to avoid using hazardous or toxic substances by finding benign replacements or discontinuing the product line. Nondestructive reuse of the products themselves, as in refillable bottles and washable cloth diapers, also plays a role in source reduction.

The second major aspect of materials management concerns the postconsumer fate of discards. After products become wastes, their constituent materials should be reused as much as possible according to their particular nature. The philosophy of the OTA with respect to this aspect is summarized as follows (1).

The framework developed by OTA suggests that local decisionmakers consider recycling (and composting) first, followed by incineration and landfilling, recognizing that all of these management methods may be viable and complementary in a given situation.

The OTA thus defines composting as a subset of recycling and prefers these practices over incineration and landfilling.

Recycling refers to the conversion of discarded products made of paper, plastic, metal, and glass into secondary raw materials ready for manufacture into new products. Recycling does not fundamentally change the material, in that, for example, waste paper is converted into new paper products.

The role of composting, a microbiological technology, is to transform otherwise nonusable, biodegradable, organic wastes into a lesser volume and weight of stabilized, sanitized, storable, and transportable process residue usable as compost (soil amendment). The starting material is fundamentally altered, in that, for example, food-tainted paper is transformed unrecognizably into an earthy compost. Depending on its quality, the uses of MSW-derived composts might range from landfill cover material (low-grade product) to a soil amendment in horticulture or agriculture (high grade). Compostable materials include paper products that are intrinsically nonrecyclable (e.g., food-tainted and glue-bound items) or lack markets, food and yard wastes, disposable diapers and other sanitary products, and other items consisting in whole or in part of biodegradable materials. It would seem that, on thermodynamic grounds, paper that can be either recycled or composted should be recycled. In this manner organic wastes that are not usable in original form are composted to render them usable.

The idea of recycling for the purpose of conserving material resources, as opposed to mere disposal of wastes, is hardly novel. Material reuse is not currently the main driving force for recycling, however. Rather, it is a shortage of landfill space as manifested in steeply rising transport and tipping fees. The rising costs perhaps reflect a tentative step in the direction of economics adjusting to ecological reality. Regardless, rapid changes in MSW management are being driven by disposal costs.

1.5. Prevention

That industry is beginning to take seriously the idea of source reduction and materials reuse is suggested in the agenda of a recent conference ''Developing Environmentally Friendly Packaging'' (7). Representatives of major corporations spoke on various aspects of packaging redesign to reduce the amount of materials used and to replace toxic substances with environmentally benign ones. Only a few years ago it would have been difficult to imagine a conference of this type. This development is partly driven by legislation, threatened as well as actual, and consumer environmental awareness.

Full expression of this shift in thinking is many years in the future, yet meaningful changes have already been made. These include the marketing of detergent concentrates to save packaging, new materials-saving molding techniques, replacing metal-based inks and pigments (cadmium, lead, mercury) with soybean oil–based products, and replacing mercury in batteries with carbon/zinc. One benefit of such change is a gradual improvement in the quality of MSW-derived composts, independent of segregation and/or separation of potentially toxic materials from composting feedstocks.

1.6. Segregation of Materials at Home

To be reused, waste materials must be reasonably uniform and free of extraneous matter. It is difficult to meet this requirement starting with a mixed, undifferentiated, mass waste stream. It is less difficult with preliminary segregation in households and institutions into a few fractions prior to collection, followed by postcollection separation at special waste management facilities. Segregation and separation in tandem are becoming commonplace where solid waste management is in crisis. This is manifested in various state, county, and local mandatory recycling laws and, in at least some communities, high levels of public participation (8).

Segregation in the household is primarily a matter of the number of "bins" employed (9,10). In a popular two-bin approach, one is for commingled recyclables (*e.g.*, clean paper products, glass, metals, plastics) and the other for the remaining materials operationally defined by this approach as *trash*. A common variant is to bundle the paper separately. This two-bin approach permits subsequent clean separation of the commingled fraction into secondary raw materials. In being simple and relatively nondemanding of the public, it represents an implementable early step away from mass disposal.

A limitation of this two-bin approach is that it defines the trash fraction to include both compostable and nonrecyclable materials. From this mixture a clean compostable fraction is not easily won. The presence of extraneous material does not interfere with composting *per se* in the sense of biological action, but it makes for poor quality end-product compost. Since the emphasis in materials management until now has been on recycling rather than composting, this limitation has not been highly relevant. It becomes relevant as landfilling or incineration of the compostables in admixture with the nonrecyclables (second bin) becomes unacceptable.

Compostables can be segregated while keeping the number of bins at two, by changing the categories. In this scheme a "green bin" is reserved for compostables and a "grey bin" for materials that are recyclable plus those that are neither recyclable nor compostable. In the near-absence of compostables, which tend to be wet and sticky, the recyclables can later be more easily sorted from nonusable trash.

Of course a three-bin system involving commingled recyclable, compostable, and fractious trash is preferable, though more demanding of public participation. Legal mandates aside, a willingness to participate may become the cultural norm as a more environmentally aware generation comes to maturity. Various multibin systems are already being tried. These matters are evolving rapidly through experience and changing circumstance (3,8,9).

1.7. Separation of Materials at a Facility

With respect to postcollection separation at facilities, the most extensive experience is in conjunction with separate commingled recyclables and paper streams at "materials recovery facilities" (3). The biodegradable organics plus nonrecyclable materials are typically landfilled. The recyclable and paper streams may be received in different sections of the facility. In one, the paper is negatively hand-sorted to remove noncompliance items, such as sheet plastic and food-tainted paper. Composites, such as plastic-coated paper, represent a problem. In the other section the commingled stream is subjected to magnetic separation for ferrous metals and eddy-current separation for aluminum, followed by positive hand-sorting from a moving table. Materials such as paper, glass by color, and plastics by major type (*e.g.,* polyethylene according to whether low or high density, polyethylene terephthalate) are removed from the table and dropped through chutes into different bins. Final market preparation steps might include bailing, crushing, or shredding.

Thus the materials management approach has five major postconsumption elements. These are segregation in the household and separation facility, recycling, composting, marketing of products or otherwise utilizing them, and landfilling of an irreducible amount of nonbiodegradable trash.

1.8. Uniqueness of the Composting Element

Composting differs from the other elements in being based on the action of a microbial ecosystem. As such, it intersects with science in a way that the other elements do not. This indicates a need to consider the cause-and-effect relationships that govern the behavior of the composting system and technological approaches to control the process.

2. COMPOSTING

2.1. Composting as a Spontaneous Process

Composting is a solid-phase process based on the phenomenon of microbial self-heating of organic assemblages (11). The material being composted is its own matrix, permitting gas exchange, hence aerobic metabolism. Similarly, the material is its own source of nutrients, water, and inoculum in the form of diverse, indigenous, widespread microbes. Lastly, the material is its own waste sink and thermal insulation. As such, heat generated microbially at the expense of the material is retained. This causes a temperature elevation, which is the hallmark of composting.

Self-heating occurs spontaneously, provided that the conditions noted above are met at least marginally. The underlying events may be portrayed in reference to the simplest of operations, that of informal backyard composting of leaves.

In the ordinary course of events, fallen leaves, like other organic materials, become colonized by diverse bacteria and fungi. As long as the leaves remain scattered about, lacking thermal insulation, any heat generated metabolically is dissipated, and there is no discernible elevation in temperature. This situation is fundamentally altered if the leaves are assembled into a pile large enough to be thermally self-insulating. This sets in motion a train of events directed by an interaction between heat generation and temperature (12).

At first, the temperature elevation stimulates mesophilic growth and decomposition with its attendant generation of heat (positive feedback loop established between microbial heat generation and temperature). At the same time the elevation dynamically selects for microbes with ever higher temperature optima. The rate of heat generation peaks at approximately 35°C (feedback switches to neutral). Elevation of the temperature continues, however, to the point of becoming self-limiting (feedback negative). In this manner the mesophilic community self-destructs at 45°–50°C. But this initiates the growth of thermophilic organisms, leading to a repetition at higher temperature of the pattern represented by the positive–neutral–negative feedback sequence. Ultimately the system may bring itself to a peak of 80°C, at which point the thermophilic community is severely self-debilitated.

Self-heating thus involves a sequential narrowing of the mesophilic then thermophilic population structures. The starting mesophilic population is extremely diverse, though not well characterized (11,13). At moderate thermophilic temperatures (*e.g.*, 55°C) the actinomycete *Streptomyces* and the fungus *Aspergillus fumigatus* may be recovered, while the dominant eubacterial cultivable forms are members of the genus *Bacillus* (14,15). At 65°C, the highest temperature tested, only *B. stearothermophilus* was recovered.

Alternatively, if gas exchange is severely restricted through heavy compaction and loss of matrix structure, or the plugging of pores by extreme wetness, fermentations rather than aerobic metabolism tend to dominate the system. This results in a delayed and/or truncated temperature elevation.

2.2. Importance of High Decomposition Rate

In composting as portrayed above, decomposition progresses relatively slowly owing to inhibitively high temperature, deoxygenation, or both conditions. Both can coexist in different sections of a large uncontrolled pile. Slow decomposition is of little practical consequence in small-scale backyard composting and may be tolerable in large-scale operations dedicated to leaf-composting (16). For composting to play a broader role in MSW management, however, facilities must be designed for high, near-maximal rates of aerobic decomposition.

Rapid decomposition serves to minimize the following interrelated factors: cost of facility construction; cost of routine operation; materials throughput time; inventory and space requirement; needs for structural and mechanical appurtenances; materials handling. It is a truism for many types of industrial activities, including large-scale composting, that time and space are functionally equivalent and, in combination, affect costs.

Moreover, the faster the aerobic destruction of putrescible substances, the less the potential for malodor generation. This aspect of public acceptability alone warrants high rates of decomposition.

2.3. Composting Process Design and Control: Five Cases

The dynamics of composting system behavior and implications for process design and control have been addressed in detail elsewhere (17–26). These matters are summarized below in the form of five cases, all of which are based on real processes. It is assumed that the material has been prepared for composting. This might include some combination of segregation/separation, size reduction/homogenization, and water addition. The fresh material might also be inoculated with a small amount (a few percent of the total) of recycled compost to speed colonization with adapted organisms, to accelerate the temperature increase.

With respect to composting *per se,* the basic manipulable factors are pile size/configuration, mechanical agitation, and forced ventilation by blower. Case 1 involves pile size/configuration as the major process design and control element. With respect to the other cases, an appropriate pile size/configuration is assumed. Case 2 involves mechanical agitation of the material as the means of process control. Cases 3, 4, and 5 involve different approaches to forced ventilation. Though mechanical agitation and forced ventilation can be employed jointly, for simplicity they are discussed separately.

2.3.1. Case 1: Pile Size/Configuration

Informal, small-scale composting of leaves in a backyard setting was described above. Similarly, informal, large-scale composting of leaves (and brush) is feasible at dedicated, technologically rudimentary, municipal facilities at which the major process design element is pile size/configuration (16). In New Jersey alone such facilities number over 200.

Several characteristics of leaves, but not other components of MSW, make them amenable to large-scale, composting at low rates of decomposition. First, leaves appear as a waste in the temperate zone only over approximately 2 months of the year. Consequently, microbial decomposition need only be fast enough for a composting cycle of approximately 9 months. The idea is to remove the previous year's composted residue, without odor nuisance, in time to make room for the succeeding year's leaves. Second, leaves are not highly putrescible. This makes it possible to avoid odor nuisance even with rudimentary composting technology, despite the inevitable fermentative generation of odorous compounds in anoxic, interior regions of the pile during part of the processing cycle. The idea is to avoid the venting of such compounds through ill-timed midcourse or terminal mechanical disturbance (see the next case). In this manner lengthy processing is substituted for high rates of aerobic decomposition realizable through effective biological process control. The third factor that makes leaves a special case is that they are conventionally segregated at the source, facilitating separate management.

In this situation a pile size/configuration is sought to reconcile certain conflicting requirements. These are as follows: In the autumn avoid both excessively high temperatures and profound anoxia (indicates small pile); in the winter avoid extensive freezing (large pile); by termination ensure that an oxygenated front has penetrated, via passive diffusion, into the innermost core (small pile). Penetration of an oxygenated front leaves in its wake biologically stabilized material.

In the climate of New Jersey, these needs can be reconciled reasonably well with the aid of minimal mechanical agitation. The agitation events are timed to avoid the venting of nuisance odors. One protocol is as follows: In the autumn, leaves are formed into windrows, after moistening if necessary, having dimensions of 4 × 2 m; in late December or early January, the material comprising two windrows is agitated and combined into one; in early spring the material is agitated again. By the subsequent autumn the material should be sufficiently well stabilized for use as a mulch or to be formed into large space-saving piles without the potential for nuisance. Other protocols are usable.

Technologically rudimentary large-scale leaf composting thus plays an important, though intrinsically limited, role in MSW management. Extending the role of such facilities has been proposed, particularly for the composting of grass clippings in admixture with leaves. This is problematic, however, in that grass is highly putrescible, appears as a waste over a 6 month period, and the timing of its appearance relative to leaves for combining the two is operationally inconvenient. Given large distances to residences, co-composting grass and leaves at rudimentary facilities might be acceptable. Otherwise, inclusion of grass could jeopardize acceptable operations restricted to leaves.

Rather, materials that are prone to generate malodors and/or pose an inventory problem because of daily input should be composted at high rates of decomposition. One or both of these characteristics pertain to grass, food waste, food-tainted paper, otherwise nonrecyclable paper, and most other biodegradable materials in MSW. This indicates strict biological process control, as described in cases 4 and 5.

2.3.2. Case 2: Mechanical Agitation

Case 2 differs from the first one in that process management basically consists of mechanical agitation at frequent intervals. Special "composting" or "turning" machines are commonly employed. These straddle a windrow or work to one side of it, lifting the material and displacing it backward. This mixes and redistributes the material, breaks up any stratification induced by biological action, and gives a momentary spur to gas exchange and heat removal.

The oxygenation and cooling effects resulting from mechanical agitation are less than might be thought, as demonstrated in the following set of observations (27). A metabolically active windrow was agitated daily. The site of observations was 0.6 m in from both the side and top edges of the windrow. The critical values relative to the fifth agitation were approximately as follows: soon after, 45°C, 19% O_2; 10 hours later, 55°C, 2.5% O_2; 35 hours later, 79°C, 10% O_2. The recovery of O_2 at 35 hours in the absence of further agitation was correctly interpreted as stemming from self-induced inhibitively high temperature suppressive of O_2 consumption. Inferentially, decomposition was at a low overall rate.

Mechanical agitation serves useful purposes, such as size reduction and mixing, but, as demonstrated above, effective control of temperature and O_2 is not one of them. Consequently, pursuit of maximal decomposition rates is not realistic through agitation. Rather, this requires forced ventilation as described in cases 4 and 5.

2.4. Process Control Through Forced Ventilation

The remainder of the section on composting concerns forced ventilation with air delivered by a blower. Ventilation serves to supply O_2 and to remove CO_2,

TABLE 14.1. **Approaches to Composting System Ventilation**

	Case 3	Case 4	Case 5
Approach	Beltsville process	Rutgers strategy	Mushroom tunnel
Air path	Vacuum induced, air-once-through	Forced pressure air-once-through	Forced pressure, process gas recirculated
Blower control	Timer, ultimately in reference to O_2	Baseline timer, temperature feedback	Integrated temperature/ O_2 feedback
Status of key factors			
O_2	System oxygenated	System oxygenated	System oxygenated
Temperature	Inhibitively high	Biologically favorable	Biologically favorable
Water removal	Slight	Substantial	Substantial
Decomposition rate	Low	High	High

heat, and water vapor. The removal of heat and vapor are linked through the mechanism of the latent heat of vaporization (evaporative cooling).

The particular approach taken to ventiliation profoundly affects system behavior. This is illustrated in comparisons among the Beltsville process (28), the Rutgers strategy (29), and the generic "Dutch tunnel" technology used in the mushroom growing industry (25,30) (Table 14.1). The causal relationships as they pertain to cases 3 and 4, which employ air-once-through, are developed below in abbreviated form. Case 5, in which process gas is recirculated, requires separate consideration.

Necessary background information is as follows. Numerical values are approximations based on nominal conditions.

1. The heat is mostly generated through O_2-based microbial respiration.

2. The interaction between heat generation and temperature is operative, as described earlier (in Section 2.1.).

3. Heat losses through the mechanisms of thermal radiation and conduction can be neglected for present purposes.

4. Latent heat of vaporization accounts for over 80% of the heat removal, and sensible heating of the air (temperature increase) accounts for the remainder.

5. Approximately nine times more air is needed to remove heat than to supply O_2 ("air function ratio" ~ 9/1).

6. Owing to the sensible heating, a temperature gradient is established in the direction of airflow.

7. Generation of water metabolically makes up only 10% of loss through vaporization.

8. Biologically generated heat drives 95% of the vaporization, and unsaturation of the inlet air the remainder.

In isolation from biological considerations, heat removal can be represented by the following equation:

$$Q_{vent} = m(h_{out} - h_{in}),$$

where Q_{vent} is the rate of heat removal (energy/time), m the dry air mass flow (mass dry air/time), h_{out} the enthalpy of the outlet process gas (energy/mass dry air), and h_{in} the enthalpy of the inlet air (energy/mass dry air). The enthalpy of air and process gas is primarily a function of relative humidity (RH) and temperature. The RH of the outlet process gas is assumed to be 100%. The factor m corresponds in principle to the extent of ventilation and in practice to the operation of a blower.

Taken literally, the equation would seem to permit any value of Q_{vent} as the product of various combinations of values of m and h_{out}-h_{in}. But, realistically, biological constraints, particularly that of the interaction between microbial heat generation and temperature, permit only a narrow range of outcomes. Biological constraints are taken into account in the discussion.

2.4.1. Case 3: Ventilation Focused on Supplying O₂—Air Once Through

Case 3 is based on the idea of maintaining the system in an oxygenated state, as judged in terms of O_2 levels in the interstitial atmosphere. Owing to the air function ratio and the tendency of the system to self-limit via inhibitively high temperature as described above, high O_2 levels are easily maintained with limited ventilation. A blower with a delivery capacity of tens of cubic meters per dry ton per hour, actuated by a timer on a fixed schedule for only 20% of the time, typically suffices to maintain levels of 10%–15% O_2 (volumetric basis). The underlying chain of causation is as follows: O_2 consumption generates heat; heat is retained, elevating the temperature; the blower delivers sufficient air to maintain an oxygenated condition, while allowing progressive heat retention; the temperature becomes elevated to levels inhibitive of O_2 consumption; a low rate of consumption is approximately balanced by a commensurate rate of ventilative resupply, poising O_2 at a high level. Case 3 thus recognizes the O_2-supplying function of ventilation, but not the heat removal/temperature control function.

In terms of the equation, the temperature component of h_{out} is maximized, at approximately 80°C, because the value of m is small. Once that condition is reached, increasing the value would remove heat faster than its replacement by the temperature-debilitated community. This would result in declines of temperature, the differential h_{out}-h_{in}, and Q_{vent}. Owing to the near self-sterilized condition, recolonization and recovery of heat-generating capacity at the newly moderated temperature would be slow.

The approach represented by case 3 thus sets in motion a chain of events leading to ''microbial suicide.'' The outcome is slow decomposition—the opposite of waste management needs.

2.4.2. Case 4: Ventilation Focused on Removing Heat—Air Once Through

Cases 3 and 4 represent fundamentally different approaches to composting process control and as such have been exhaustively analyzed comparatively (*e.g.*, 16,18–21,28). The fundamental difference between them can be put as follows. Case 3 aims for no particular operational objective other than maintaining an oxygenated condition in the interstitial atmosphere. This condition is implicitly considered as both necessary and sufficient. Case 4 aims to maximize the rate of aerobic decomposition. Maintenance of an oxygenated condition is considered to be necessary but not sufficient. Rather, maintaining both an oxygenated condition *and* a biologically favorable temperature ceiling (*i.e.*, less than 60°C) is necessary and sufficient. Importantly, both functions are served by a single approach to process control, that of ventilative heat removal in reference to temperature (see below).

Mechanically, case 3 involves a blower of prescribed delivery capacity actuated by a timer. With respect to case 4, a timer is used only briefly. This is at the onset during temperature increase and terminally during decrease, to schedule baseline ventilation and to prevent anoxia. When the temperature becomes elevated to a predetermined set point, control of the blower is automatically transferred to a temperature feedback control system. The blower must have an adequate air delivery capacity to meet the *peak* demand for heat removal (hundreds of cubic meters per dry ton per hour). In this manner composting mass and blower interact time-variably as needed to poise the temperature ceiling at a level favoring microbial decomposition (equivalent to heat generation). Eventually microbial activity subsides because of the depletion of substrate and/or available water, and blower control reverts to the timer.

To envision case 4 events, consider that ambient air is forced into the bottom of a biologically active matrix and, after a passage of perhaps 2.5 m, exits from the top. With passage, energy is transferred from the matrix to the flowing airstream in the forms of latent and sensible heat. The sensible component establishes a vertical temperature gradient along the axis of airflow. With material rich in readily available substrates, during periods of peak activity gradients as steep as 20°C/m are observed. There is thus a progressive increase in the moisture content of the air (latent heat) and the temperature gradient (sensible heat), as well as the capacity of the air to hold moisture (increasing temperature). It can be said that the air "chases after" saturation (100% RH), but that its absolute attainment is barred by the temperature gradient.

At the same time temperature feedback control also serves to ensure thorough oxygenation. This stems from the following chain of causation: O_2 consumption generates heat; heat is retained, elevating the temperature; the elevation exerts a demand for ventilation; owing to the wide air function ratio ($\sim$9/1), consumption and resupply are balanced to poise O_2 at high levels. A typical range is 17%–20% O_2 in the interstitial atmosphere. Thus the control parameter, temperature, is linked to O_2 consumption via the generation and retention of heat. Because of this linkage, an unusual feature of temperature feedback control as applied to the composting system is that it also serves as an indirect form of O_2 feedback control.

In terms of the equation, the need is to vary the factor m continuously such as to poise the temperature component of h_{out} at 60°C or less. The purpose is to maximize Q_{vent} on a sustained basis.

Because the dominant heat removal mechanism is the vaporization of water, another consequence of deliberate ventilative heat removal is that the tendency of the composting material to dry is positively related to heat generation. The chain of causation is as follows: decomposition generates heat; most of the heat is removed through the vaporization of water; the loss of vapor dries the material.

This relationship between drying and heat generation is not obviated by two other operative factors. First, metabolic production of water makes up for only a small part of that removed through vaporization. Second, unsaturation of the ambient air contributes only slightly to the drying tendency. These two factors tend to cancel out each other.

Moreover, because the generation of heat is at the expense of organic matter, the tendency of the material to dry is indicative of the decomposition rate. As such, determination of moisture content provides a measure of waste treatment progress. Whether *dryness* comes to limit microbial activity is a different, though related, matter.

2.4.3. Case 5: Ventilation Focused on Removing Heat—Process Gas Recirculated

The final case concerns a generic technology employed in the field of mushroom growing. It has been known as Dutch tunnel technology since its inception in the 1970s (31). Currently over 400 such tunnels are in daily use worldwide, and the number is increasing rapidly. These serve in transforming various combinations of animal and plant wastes into a substrate for growing mushrooms. Commonly used starting materials include horse and poultry manure, chopped corn cobs, wheat straw, and cotton seed meal. The present discussion does not concern mushroom growing. It concerns solely the relevance of this distinct composting technology to the problem of MSW management.

Approximately one-half of the tunnels currently in service were built within the last few years and may be characterized as "modern," in the sense of having computer-assisted biological process control via sophisticated management of ventilation. A key feature is the continuous recirculation of process gas ("used air") to a heat exchanger/condenser external to the tunnel, in conjunction with shunting as needed. The condenser is typically cooled with ambient air or water from a surface or underground source. A generic version of a modern Dutch tunnel and its operation is described below, and is derived from various reports (*e.g.*, 25,30).

The basic structural unit is essentially an elongated box with typical inside dimensions of $33 \times 3 \times 3$ meters. The floor, permeable to air (false floor), is supported above an open space (plenum). The tunnel is filled from one end with material prepared for composting to a height of approximately 2.5 m, leaving a starting head space of 0.5 m. The ends are sealed during processing.

Process gas is recirculated via a route leading from the plenum into the composting matrix, through the head space into the external condenser or by-pass, and then back into the plenum. A ductwork arrangement permits the continuously variable introduction of ambient air to supply O_2. This mixes with recirculated process gas just upstream of the plenum. An equivalent volume of spent gas is displaced, exiting the tunnel from the headspace. Temperature and O_2 are monitored in the headspace and elsewhere, with the signals reporting to a monitor/control computer system. Fresh ambient air is introduced based on the O_2 status, and recirculation gas is shunted to the condenser or bypassed based on the temperature status.

Note that this arrangement, involving continuous recirculation of process gas in conjunction with external heat exchange, partly separates the functions of supplying O_2 and removing heat for temperature control. This is because prior to being returned to the composting matrix the temperature and moisture content of the gas is decreased, restoring its capacity to remove heat and water vapor.

A consequence of recirculating process gas *per se* is that the vertical temperature gradient is less steep, compared with air once through as with case 4. This affords more uniform biological process control throughout the composting material.

At the start of a composting cycle a temperature/O_2 regimen is selected and the computer is instructed to maintain it. This is realized through automatic adjustment of the ambient air inlet gate. Depending on ambient air and composting conditions, the recirculation ratio can be varied over the full range of

100% recirculated–0% ambient to 0% recirculated–100% ambient. In this manner, control of temperature and O_2 level is extremely precise. A report on these and other parameters, as well as cumulative and instantaneous heat output, is available to the operator on demand.

Case 5 offers the additional advantage, compared with case 4, of using less ambient air, hence generating less final exhaust gas requiring management. This stems from two factors. First, the continuous flow of recirculated process gas is itself a driving force for the exchange of O_2 and CO_2. This might be particularly important in constricted, porosity-limited regions of the matrix. Second, the partial separation of the heat removal and O_2-supplying functions permits a more economical use of ambient air while maintaining a condition of thorough oxygenation. In combination, these two factors permit the use of less air. The practical benefits thereof are described below.

3. IMPLEMENTATION

3.1. Packaged Facilities and Unit Processes

A number of companies offer packaged facilities for MSW management that encompass various unit processes. These units are intended 1) to finish the separation of materials started in the household and prepare them for secondary use, 2) to prepare the organic fraction for composting, 3) to effect the composting *per se*. Such packages are often referred to as *composting facilities*, even though this describes only the third unit. This gives the impression that an entire package must be procured, which tends to obscure the actual procurement options. The term *MSW management facility* is preferred for accuracy and because it is more suggestive of the possibility of critically evaluating individual unit processes. The discussion below concerns solely the composting unit process.

3.2. Comparison of Commercially Available Systems in the Waste and Mushroom Industries

A description is developed to represent systems offered in the waste industry that meet two criteria. The first is that they incorporate ventilative biological process control in the form of case 4. None in this industry conform to case 5. The second criterion is that the system be classifiable as enclosed. The significance of enclosure is explained below. Enclosed systems that do not meet the first criterion are excluded because they lack the capacity to approach maximal rates of decomposition. The resultant composite description representing the best from the waste industry is compared to that of a generic tunnel as offered in the mushroom industry (Table 14.2).

TABLE 14.2. Comparison of Commercially Available Systems in the Waste and Mushroom Industries

	Composite waste industry system	Generic mushroom industry system
Biological process control	Assume case 4	Case 5 is based on such systems
Enclosure	Large building encloses entire operation, including work-place; spent process gas mixes with building air; condensation sometimes causes rain or snow in building	Only the composting material enclosed
Materials handling	Various: overhead crane; agitator/translocator; conveyor	Tunnel filled by tele-scoping conveyer with swivel delivery head; emptied via travelling or moving floor
Exhaust gas/air scrubbing	Relatively large volume spent process gas + building air	Relatively small volume spent process gas only
Monitoring of work accomplished	Indirect measurement	Direct instantaneous readout of heat output

Yard waste excepted, enclosure has become a shibboleth in governmental guidance and regulation with respect to composting. This in the sense that only enclosed systems are likely to pass regulatory scrutiny, which, at the same time, typically ignores the crucial issue of biological process control and resultant process performance. Consequently, facilities that are inadequate with respect to process control, or even grossly misconceived in this regard, are approved, provided that they display the trappings of enclosure. This problem is rooted in a gap between the regulatory and scientific cultures and in a notion that odor prevention and public acceptability may somehow be equated with enclosure *per se* (32).

In fact, enclosure can be incidental to process control, as in the waste industry systems, or integral to process control, as in the mushroom industry tunnels (Table 14.2). The problem with case 4 in this context is that the proprietary systems enclose the operation in such a manner as to make a major problem out of air management. Not only is more spent process gas generated in the air-once-through mode, but it mixes with building air to increase greatly the overall volume of exhaust needing management. In contrast, case 5 exploits enclosure by integrating process gas recirculation with biological process control. Significant practical advantages are thereby realized, including the prevention of odors through extremely precise process control and a much reduced volume of spent process gas without admixture with building air. This greatly simplifies any final scrubbing of the exhaust.

These two types of systems represent the outcomes of very different developmental paths. In the waste industry, partial or near-complete enclosure was initially seen as a means of protecting machinery from the elements, as a visual screen, or merely as a given for an industrial operation. Movement in the direction of process control as described in case 4 came after the fact. Meanwhile, regulatory pressure mounted to scrub the exhaust air, which greatly reinforced the trend toward enclosure.

In the mushroom industry, tunnel enclosure was conceived *de novo* as a means to strict biological process control. Only the composting material, not the entire operation, was enclosed. The outcome was an integration of enclosure and process control, as described in case 5.

Generic tunnel technology available in the mushroom industry is scientifically and technically coherent and has an impressive track record under demanding circumstances. It represents the best available control technology and as such has much to offer the field of solid waste management. A document was prepared to facilitate the crossover of this technology to the waste field (33).

ACKNOWLEDGMENTS

The author thanks Dr. Frederick C. Miller for introducing him to the world of "mushroom composting." This is Journal Series Paper No. F-07497-1-91, as supported by state funds.

REFERENCES

1. United States Congress, Office of Technology Assessment: Facing America's Trash: What Next for Municipal Solid Waste? OTA-O-424, Washington, DC, 1989.

2. Senior E: Introduction. In: Senior E (ed): Microbiology of Landfill Sites. Boca Raton, FL: CRC Press, 1990, pp 1–15.

3. The BioCycle Guide to Collecting, Processing and Marketing Recyclables. Emmaus, PA: The JG Press, Inc., 1990, 229 pp.

4. Carra JS, Raffaello C (eds): International Perspectives on Municipal Solid Wastes and Sanitary Landfilling. New York: Academic Press, 1990.

5. United States Environmental Protection Agency: Report to Congress: Solid Waste Disposal in the U.S., vol II. EPA/530-sw-011B, Washington, DC, 1989.

6. Grogan P, Schwartz B: Strategic Shift to Source Reduction. BioCycle 32(1):28–30, 1991.

7. Developing Environmentally Friendly Packaging. Milltown, NJ: The Packaging Group, 1990.

8. Platt B, Doherty C, Broughton AC, Morris D: Beyond 40 Percent: Record-Setting Recycling and Composting Programs. Washington, DC: Institute for Local Self-Reliance, 1990, 252 pp.

9. Bidlingmaier W, Kranert M: Economics from Household to Compost Through a Complete System in the Federal Republic of Germany. In Bidlingmaier W, L'Hermite P (eds): Compost Process in Waste Management. Brussels: Commission of the European Communities, Environment and Waste Recycling Programmes, 1988, pp 1–46.

10. Krogmann U: Separate Collection of Biowaste, Session 6C. Proc 6th Int Conf Solid Waste Management and Secondary Materials. The J Resource Management Technology.

11. Finstein MS, Morris ML: Microbiology of municipal solid waste composting. Adv Appl Microbiol 19:113–151, 1975.

12. MacGregor ST, Miller FC, Psarianos KM, Finstein MS: Composting process control based on interaction between microbial heat output and temperature. Appl Environ Microbiol 41:1321–1330, 1981.

13. de Bertoldi M, Vallini G, Pera A: The biology of composting: A review. Waste Management Res 1:157–176, 1983.

14. Strom PF: Effect of temperature on bacterial species diversity in thermophilic solid waste composting. Appl Environ Microbiol 50:899–905, 1985.

15. Strom PF: Identification of thermophilic bacteria in solid waste composting. Appl Environ Microbiol 50:906–913, 1985.

16. Strom PF, Finstein MS: Leaf Composting Manual for New Jersey Municipalities. New Brunswick: Rutgers University, and N.J. Department of Environmental Protection, Trenton, 1985.

17. Finstein MS, Miller FC, Strom PF, MacGregor ST, Psarianos KM: Composting ecosystem management for waste treatment. Bio/Technology 1:346–353, 1983.

18. Finstein MS, Miller FC, Strom PF: Waste Treatment Composting as a Controlled System. In Rehm H-J, Reed G (eds): Biotechnology, vol 8. Weinheim, FRG: VCH Verlagsgesellschaft (German Chemical Society), 1986, pp 363–398.

19. Finstein MS, Miller FC, Hogan JA, Strom PF: Analysis of EPA guidance on composting sludge. Part I, biological heat generation and temperature. BioCycle 28(1):20–26, 1987.

20. Finstein MS, Miller FC, Hogan JA, Strom PF: Analysis of EPA guidance on composting sludge. Part II, biological process control. BioCycle 28(2):42–47, 1987.

21. Finstein MS, Miller FC, Hogan JA, Strom PF: Analysis of EPA guidance on composting sludge. Part III, oxygen, moisture, odor, pathogens. BioCycle 28(3):38–44, 1987.

22. Finstein MS, Miller FC, Hogan JA, Strom PF: Analysis of EPA guidance on composting sludge. Part IV, facility design and operation. BioCycle 28(4):56–61, 1987.

23. Finstein MS: Report: Activities on composting as a waste treatment technology at the Department of Environmental Science, Rutgers University. Waste Management Res 7:291–294, 1989.

24. Finstein MS, Miller FC, Strom PF: Monitoring and evaluating composting process performance. J Water Pollut Control Fed 58:272–278, 1986.

25. Miller FC, Harper ER, Macauley BJ, Gulliver A: Composting based on moderately thermophilic and aerobic conditions for the production of commercial mushroom growing compost. Aust J Exp Agric 30:287–296, 1990.

26. Miller FC, Finstein MS: Materials balance in the composting of wastewater sludge as affected by process control strategy. J Water Pollut Control Fed 57:122–127, 1985.

27. Randle P, Flegg PB: Oxygen measurements in a mushroom compost stack. Sci Hort 8:315–323, 1978.

28. Willson GB, Parr JF, Epstein E, Marsh PB, Chaney RL, Colacicco D, Burge WD, Sikora LJ, Tester CF, Hornick S: Manual for Composting Sludge by the Beltsville Aerated-Pile Method. EPA-600/8-80-022, Washington, DC, 1980, 65 pp.

29. Finstein MS, Miller FC, MacGregor ST, Psarianos KM: The Rutgers Strategy for Composting: Process Design and Control. EPA-600/S2-85/059. Washington, DC, 1985, 275 pp.

30. Arkenbout J: Air Treatment in Mushroom Growing. In Griensven LJLD (ed): The Cultivation of Mushrooms. The Netherlands, The Mushroom Experimental Station, Horst, 1988, pp 179–211.

31. Gerrits JPG: Compost Treatment in Bulk for Mushroom Growing. Mushroom J 182:471–474, 1988.

32. Finstein MS: Composting Solid Waste: Costly Mismanagement of a Microbial Ecosystem. Am Soc Microbiol News 55:599–602, 1989.

33. Finstein MS, Strom PF, Miller FC, Hogan JA: Element of a Request for Proposal (RFP) for Sludge and Municipal Solid Waste Composting Facilities: Scientific and Technical Aspects. New Brunswick, NJ: Rutgers Cooperative Extension Publ. No. E136, 1990, 14 pp (addendum published October 1990, 5 pp).

MICROBIAL
DESULFURIZATION OF COAL

PIETER BOS
FRED C. BOOGERD
J. GIJS KUENEN

Department of Microbiology and Enzymology, Kluyver Laboratory of Biotechnology, Delft University of Technology, 2628 BC Delft, The Netherlands

1. INTRODUCTION

At regular intervals, we are reminded that the stability of the world's future energy supply of oil is full of uncertainties. During the 1970s this was demonstrated by the two energy crises and in the 1990s by the Gulf crisis. At first sight these events have a strong political background, but their basis lies in the limited reserves of oil. Reserves of other fossil fuels are also limited, but coal seems to be most abundant (1). Therefore the use of coal and coal technology have regained considerable interest in the last decades. Mining and the use of coal are associated with various safety and environmental risks. Apart from the greenhouse effect, which is connected with the use of fossil fuels in general, the contribution of coal usage to acid rain is the most important. Acid rain is caused by the release of nitrogen oxides (NO_x) and sulfur dioxide (SO_2), e.g., in the stack gas of coal-firing installations. Most nitrogen oxides produced are formed from the molecular nitrogen present in the air. The extent of the reaction is determined by the furnace temperature. By using fluidized bed combustion at temperatures around 800°C, the formation of NO_x can be restricted to acceptable concentrations. However, the extent of the SO_2 emmision is determined by the sulfur content of coal. Before discussing the sulfur speciation in coal, it should

Environmental Microbiology, pages 375–403, © *1992 Wiley-Liss, Inc.*

be realized that "the term *coal* encompasses a vast variety of mineral deposits, each of which is not only unique, but also totally lacking in homogeneity. We can only guess at the physical and chemical structure of coals. And the analytical procedures and their significance are often misapplied and misunderstood" (2).

Sulfur in coal can be organically bound, forming part of complex macromolecular structures, or in inorganic forms. The most important inorganic sulfur compound in coal is pyrite, but often other metal sulfides are also present (3). Weathered coals sometimes contain considerable amounts of sulfate. Coals containing elemental sulfur are quite exceptional (4). If elemental sulfur is present its percentage is low. Sometimes the sulfidic minerals, and especially pyrite, are present as coarse particles, but they can also be present as minute crystals (1–5 μm) in the coal matrix. The ratio between organically and inorganically bound sulfur varies greatly among coals. The same holds for the total amount of sulfur present. As far as organically bound sulfur is concerned, many uncertainties exist. It is frequently suggested that most organically bound sulfur is present in thiophenic structures. Recently it was reported by Hippo *et al.* (5) that in seven coal samples tested, disulfide sulfur in structures such as $X–X–S–CH_3$ seems to be most abundant. Obviously much work is needed to clarify the speciation of organically bound sulfur in coal, requiring new analytical methods.

A variety of techniques is available to reduce the environmental risks of coal use, including coal gasification, coal liquefaction, and fluidized combustion in the presence of carbonates (6). It must be realized that the latter technique will give rise to a waste stream rich in gypsum, coal ash, unreacted calcium carbonate, and heavy metals. For electric power–producing installations, stack gas cleaning seems to be the most appropriate process to reduce sulfur dioxide emissions. Again, this results in the production of gypsum. Regenerative processes are being developed in which elemental sulfur rather than a waste product such as gypsum is produced. Another approach is the removal of sulfur compounds before burning. If the coal contains the sulfidic minerals as coarse particles, simple density separation techniques are advisable (7). For the removal of finely distributed sulfidic minerals, the use of physical separation methods such as electrostatic separation (8), high gradient magnetic separation (8) and froth flotation (9) have been suggested. These techniques are based on differences in the physical characteristics of minerals and coal. A prerequisite for effective separation with low carbon losses is often an extremely fine milling to free the minerals enclosed in the coal matrix. In most cases, particle sizes as low as 10 μm are required.

2. BIOPROCESSING OF COAL

In recent years, studies have been initiated to explore the possibilities of the use of biological systems in coal technology. For a complete overview, the reader is referred to Gouch (10), Klein *et al.* (11) and Bos and Kuenen (12). Most investigations in this field are aimed at the development of biodesulfurization processes. Attention is paid to the removal of both inorganically and organically bound sulfur. The present review will be illustrated, especially, by data from the Dutch feasibility study, which was started in 1982. This research was first funded by the Dutch Government within the framework of the Dutch National Coal Programme (NOK) and is now funded by the Commission of the European Communities (CEC). The Dutch studies are characterized by a multidisciplinary approach in which microbial (eco)physiologists are cooperating with bioprocess engineers.

At this point, the reader should note that, apart from biodesulfurization, research is also underway to develop microbial coal liquefaction and gasification processes. However, in this respect it should be realized that any (micro)biological technique aimed at changes in the recalcitrant macromolecular structure of the carbon skeleton of coal will suffer from the slow kinetics of the processes. This will especially be the case if the carbonization process of the coal involved has proceeded most of its way, such as in bituminous coals and anthracites. One should expect the "younger" coals such as lignites to be most susceptable to microbial attack. Studies have revealed that either heavily weathered lignites can be solubilized (13) or that a pretreatment of the coal with nitric acid is required (14).

2.1. Principle Behind the Biodesulfurization of Coal

The principle underlying microbial coal desulfurization, with special emphasis on inorganic sulfur removal, is based on a complex combination of spontaneous (non-biologically) and microbiologically catalyzed oxidations of inorganic sulfidic minerals present in coal. This combination of reactions leads to solution of the sulfidic minerals present in the coal. By separating the coal from the process fluid, a fuel is obtained with a lower sulfur content.

In the global sulfur cycle this combination of spontaneous and microbiologically assisted oxidation reactions plays an important role. A substantial part of the global sulfur is fixed as metal-containing sulfidic minerals mostly in anaerobic environments (15). If these minerals are exposed to moisture and aerobic conditions, they can easily be oxidized spontaneously (in the absence of biocatalysts) at neutral pH. This oxidation is illustrated by reaction 1, which shows the oxidation of the most common sulfidic mineral, pyrite.

$$4\,FeS_2 + 15\,O_2 + 2\,H_2O \rightarrow 2Fe_2(SO_4)_3 + 2\,H_2SO_4. \tag{1}$$

The oxidation of pyrite results in the formation of sulfuric acid. For that reason, the pH of the water phase drops during the oxidation process. At the same time, the rate of spontaneous pyrite oxidation decreases. At pH 3, the rate of spontaneous pyrite oxidation is negligible. As the pH falls, acidophilic sulfidic mineral-oxidizing bacteria become involved at approximately pH 4. In contrast with the spontaneous chemical oxidation, the biologically catalyzed reaction accelerates with decreasing pH until 2–1.5, when the maximal oxidation rate is reached. The biological oxidation of pyrite can also be represented by reaction 1. The ferric iron produced by this reaction can serve as an oxidizing agent for other sulfidic minerals, as is shown in reaction 2.

$$CuFeS_2 + 16Fe^{3+} + 8H_2O \rightarrow Cu^{2+} + 17Fe^{2+} + 2SO_4^2H + 16H^+. \quad (2)$$

The ferrous iron produced in reaction 2 can serve, in turn, as an oxidizable substrate for the bacteria (reaction 3).

$$4Fe^{2+} + O_2 + 4H^+ \rightarrow 4Fe^{3+} + 2H_2O. \quad (3)$$

Bacteria able to oxidize these minerals may develop whenever sulfidic minerals become exposed to moisture and air. Examples of their natural habitats are pyrite-containing, acidic, and potentially acidic soils (*e.g.,* cat clays), as well as places where sulfidic minerals are rising to the surface because of geological activity or erosion. Manmade habitats for these organisms can be found in places related to coal and mineral mining (*e.g.,* exploration sites, storage of coal and minerals, dumps of coal, and metal mining reject materials).

On the one hand, the microbial oxidation of sulfidic minerals has led to environmental problems. In coal and sulfidic mineral mining, the activities of these bacteria lead to acidification of mine effluents (mine drainage water), which poses a severe stress on the quality of the environment in mining regions (16). These effluents also contain considerable amounts of dissolved heavy metals that were originally tightly bound in the mineral. On the other hand, this bacterial activity is exploited on an industrial scale in microbial metal leaching processes (17). Metals present as sulfidic minerals are dissolved by the bacterial activity. Microbial metal leaching processes are currently aimed at low-grade copper ores, and almost 20% of current U.S. copper production is achieved by means of microbial activity. There is also a growing interest in the mechanisms underlying the microbial oxidation of sulfides in gold production (18). Some pyrite concentrates contain low concentrations of precious metals, such as silver, gold, and platinum. Chemical means to expose the pyrite crystal structure in order to make the precious metals attainable for complexing reactions with cyanide are too expensive. It appears that exposure by means of microbial sulfide oxidation is an economically attractive option.

2.2. Bacterial Species Involved in Sulfidic Mineral Oxidation

Originally only mesophilic sulfidic mineral-oxidizing bacteria were isolated from systems in which the microbial solution of sulfidic minerals plays a role

TABLE 15.1. Characteristics of Iron– and/or
Sulfur–Oxidizing Acidophilic Bacteria[a]

| Organism | GC% | Opt. temp. | Growth via oxidation of | | | | | Carbon source | |
			Fe	S^0	FeS_2	$CuFeS_2$	S_4O_6[b]	CO_2	Yeast extract[c]
T. ferrooxidans	58	30	+	+	+	+	+	+	−
T. thiooxidans	52	30	−	+	−/+	−/+	+	+	−
L. ferrooxidans	51–56	35	+	−	+	−/+	−	+	−
Strain BC 13	61	45	−	+	−/+	−/+	+	+	−
Strain TH1/BC1[d]	50	50	+	+	+	+	−	+	+
S. thermo-sulfidooxidans	45–49	50	+	+	+	+	−	+	+
Strain ALV	57	50	+	+	+/−	+	−	+	+
Strain LM2	60	50	+	+	−	ND	−	+	+
Strain TH3	69	50	+	+	+	+	+	+	+
Sulfolobus BC	ND	70	+	+	+	+	+	+	+
Sulfolobus B6-2	ND	ND	−	+	ND	ND	+	+	+

[a]−/+, Growth in mixed cultures. +/−, limited growth relative to +; ND, not determined. Reproduced from Norris (21), with permission of the publisher.

[b]Bacteria growing on tetrathionate also utilize thiosulfate.

[c]Yeast extract supports heterotrophic growth, but, apart from strain TH3, it is not required for growth on iron and sulfur.

[d]Strains TH1 and BC1 share over 90% DNA:DNA homology and represent the same species.

(19,20). More recently moderate and extreme thermophilic acidophiles have been discovered, *e.g.*, from spontaneously heated coal piles, which can also be exploited in sulfidic mineral-leaching processes. Many of the thermophilic sulfur-oxidizing species isolated from geothermal sulfide-containing sources (hot springs and solfateras) display the relevant sulfidic mineral-oxidizing capacity. A recent review by Norris (21) summarizes the main characteristics of the presently known acidophilic iron- and/or sulfur-oxidizing bacteria (see also Table 15.1) (21). It should be noted that the first isolates discovered to be involved in mineral leaching were typical obligate autotrophs such as *Thiobacillus ferrooxidans* and *Thiobacillus thiooxidans*, but that a number of facultative autotrophs are now known to be active in the process (*e.g.*, *Sulfolobus* sp.). The review by Norris (21) also considers oligotrophic heterotrophs, such as representatives of the genus *Acidiphilium*. It is likely that the latter contribute to the stability of the mixed mineral-oxidizing population by consuming organic excretion products produced by the mineral oxidizers (22).

Once spontaneous sulfide mineral oxidation has started and the pH is decreasing, the sulfide mineral-oxidizing microflora will gradually take over the oxidation. Their optimal activity ranges between pHs 1 and 2.5. In the older literature, a succession of different acidophilic bacterial species found in mesophilic habitats was suggested, in which *Metallogenium* species had the highest optimal pH, approximately 4 (23). In later studies, Walsh and Mitchell (24) concluded that their iron-oxidizing, acid tolerant *Metallogenium* strain could be considered as an unusual form of *Gallionella ferruginea*. With decreasing pH, this organism may be succeeded by bacteria such as *T. ferrooxidans* and *T. thiooxidans*. However, it should be realized that there is strong evidence that there are many more different species and genera involved in mineral oxidation at low pH than have thus far been described, since the dominant bacteria of many active cultures are not available as pure cultures.

One of the main problems encountered with the isolation of organisms from these habitats lies in their oligotrophic nature. Organic compounds, especially low-molecular-weight organic acids, are toxic in extremely low concentrations. This toxicity might be explained by the need of these organisms to maintain their cytoplasmic pH at a neutral value. In low-pH environments, organic acids will be in the undissociated form and will be able to pass through the cell membrane by passive transport mechanisms. In the cytoplasm, they will dissociate immediately because of the higher internal pH. The driving force behind the passive transport into the cells is the concentration gradient of the acid in the undissociated form. For that reason, the cells will become saturated with the acid within a short time. The amount of acid will be far too high to be compensated by the buffering capacity of the cytoplasm, and the cells will be inactivated by the low internal pH.

This mechanism also may explain why acidophiles frequently grow poorly on solid media. Agar often contains considerable amounts of low-molecular-weight organic compounds, especially if it is incorporated in a medium with a low pH (25). Some isolation procedures have circumvented the inhibition of the bacteria by using agarose, which is a fraction of agar that is less sensitive to acid hydrolysis. For the isolation of *T. ferrooxidans*, both agarose and silica gel appeared to be appropriate (22). In recent studies of a moderately thermophilic pyrite-oxidizing population, the floating membrane technique was used successfully to isolate the dominant, moderately thermophilic bacterium that was unable to grow properly on solid media (26).

This observation indicates that by using isolation techniques that avoid solid media new, unknown, mineral-oxidizing strains may be isolated in the future. Indeed, one can conclude that the available knowledge of the ecology of mineral-oxidizing bacteria is far from complete. This can also be illustrated by findings in this laboratory (27). With selective fluorescent dye–labeled antibodies for *T. ferrooxidans*, it was not possible to detect this organism microscopically in pyrite-containing nonsterile coal samples during an experiment in which an inoculum from natural, pyrite-oxidizing communities was used. A simple, selective fluorescent dye (ethidium bromide), which reacts with nucleic acids, revealed the presence of a large variety of bacteria after incubation. Indigenous bacteria other then *T. ferrooxidans* probably developed in the culture, and this community was effective in the desulfurization process. Even if these nonsterile coal samples were inoculated with *T. ferrooxidans,* the number of cells of this organism remained very low. Obviously *T. ferrooxidans* cannot compete effectively with the indigeneous pyrite-oxidizing microflora. When the coal slurry had been sterilized before the leaching experiment and then inoculated with a pure culture of *T. ferrooxidans*, large numbers of cells were visible under the fluorescence microscope. Analysis showed that, although the organism could not successfully compete, it nevertheless was able to desulfurize the coal effectively. Although the current literature suggests that *T. ferrooxidans* is the most important bacterium for microbial coal desulfurization, these findings suggest that even though it might be involved it is far from being the most abundant species present. It is quite easy to enrich and isolate *T. ferrooxidans* from such leaching systems by means of a mineral medium containing comparatively high concentrations of ferrous iron (*e.g.*, the 9 K salts medium described by Silverman and Lundgren [28]). However, it is questionable whether this medium is of any relevance for the situation found during the leaching of pyrite from coal, because in such a situation only minute amounts of ferrous iron are present. Most iron is present in the insoluble mineral form and the rest as ferric iron.

In bacterial sulfidic mineral oxidation, either a single species or a mixed culture can be involved. In any case, a combination of the capacities to oxidize ferrous iron and reduced sulfur compounds appears to be a prerequisite. Some bacterial species, such as *T. ferrooxidans* can carry out both reactions. In other cases, the combination of a ferrous iron oxidizer such as *Leptospirillum ferrooxidans* and *T. thiooxidans,* a reduced sulfur compound oxidizer that cannot oxidize ferrous iron, can dissolve sulfidic compounds (29). In contrast with earlier publications, more recent data indicate that some *Leptospirillum*-like organisms can achieve pyrite oxidation without the use of a sulfur-oxidizing capacity (30).

A factor important for the stability of mineral-oxidizing cultures appears to be the presence of oligotrophic heterotrophs. There have been reports that the activity of *T. ferrooxidans* is favored by the presence of these bacteria, which can remove the otherwise toxic organic compounds produced by autotrophs, such as pyruvic acid (22,31,32). The toxicity of organic acids, as explained in a previous paragraph, has a low threshold. For pyruvic acid this threshold occurs at concentrations of 10^{-4} M (33). Some of the acidophilic heterotrophs that occur as satellite populations of *T. ferrooxidans* have been isolated and characterized. There are two different types: obligate and facultative heterotrophs. Representatives of the genus *Acidophilium* belong to the first group and *Thiobacillus acidophilus* to the second. The latter bacterium can grow heterotrophically on a variety of organic substrates, including pyruvic acid, and can use reduced sulfur compounds as its energy source (34). Harrison (22) has postulated that the presence of oligotrophic heterotrophs will contribute significantly to the stability of pyrite-oxidizing cultures.

It has been suggested that the combination of *T. ferrooxidans* and *T. thiooxidans* is a more powerful mineral oxidizer that *T. ferrooxidans* alone. This idea is based on the assumption that, during sulfidic mineral oxidation by nonbiologically catalyzed reactions, elemental sulfur can be produced. Because *T. thiooxidans* is a far better elemental sulfur oxidizer than *T. ferrooxidans,* its presence in a coculture increases leaching rates (35). For more details on the interactions in mixed microbial cultures of mineral-oxidizing bacteria, the reader is referred to Norris and Kelly (32).

3. KINETICS AND MECHANISMS OF PYRITE OXIDATION

In the case of mesophilic bacteria, pyrite oxidation follows first-order kinetics (36,37). Such kinetics would not be expected if the metal sulfides were solubilized by a combined biological and nonbiological oxidation involving chemical attack by ferric iron followed by a biological oxidation of ferrous iron (often referred as *indirect oxidation*). Studies on the kinetics of nonbiological pyrite oxidation by ferric iron have demonstrated that at 30°C the contribution of this reaction to the total oxidation is insignificant. At increasing temperatures, the nonbiologically catalyzed pyrite oxidation gains in importance. For example, at 45°C it will contribute about 8% (38) of the total oxidation and at 70°C over 20%.

Arkesteyn (39) was the first to postulate that in *T. ferrooxidans* both sulfur- and ferrous iron-oxidizing capacity are essential in the oxidation of pyrite. Pyrite oxidation by *T. ferrooxidans* could be stopped by compounds that inhibit either iron or sulfur oxidation, sodium azide, and N-ethylmaleimide (NEM), respectively. Hazeu et al. (40) were able to substantiate these findings by means of biomass yield data (Table 15.2). It was shown that pyrite approaches the higher yield obtained with reduced sulfur compounds. Apparently a substantial portion of the electrons from pyrite, or sulfur, enter the respiratory chain at a level above that used for accepting electrons from ferrous iron. Thus more energy can be conserved during electron transport from sulfur compounds than from ferrous iron. If an indirect reaction mechanism is operating, a cell yield per 1 mol electron of pyrite should be expected to be equal to that of ferrous iron.

Because the sulfur-oxidizing capacity of the acidophilic bacteria is involved in the dissolution of sulfidic minerals, the mechanisms behind these reactions are of obvious interest. In published work most emphasis has been placed on the iron-oxidizing capacity. For an overview, the reader is referred to Ingledew (41). The present knowledge about the routes of sulfur oxidation by acidophiles has

TABLE 15.2. Yields of *T. ferrooxidans* Grown in the Chemostat Under Various Conditions[a]

Limiting substrate(s)	Concentration (mM)	Dilution rate (h^{-1})	pH	Yield[b]
Fe^{2+}	160	0.034	1.6	0.23
$S_4O_6^{2-}$	10	0.030	3.0	0.92
$Fe^{2+}/S_4O_6^{2-}$	15/10	0.030	1.6	0.72
$Fe^{2+}/S_4O_6^{2-}$	70/15	0.032	1.6	0.55
FeS_2	(1/14)	Batch	1.8	0.35–0.5

[a]Reproduced from Hazeu *et al.* (40), with permission of the publisher.
[b]Yield = gram dry weight/mole electron.

been summarized by Pronk et al. (42), who discuss the metabolism of soluble reduced sulfur compounds (Fig. 15.1 is a scheme for the oxidation of reduced sulfur compounds by the acidophilic thiobacilli). Hazeu et al. (43) demonstrated that elemental sulfur can be formed as an intermediate during the oxidation of soluble reduced sulfur compounds by *T. ferrooxidans*. Figure 15.2 shows the proposed structure of hydrophilic sulfur globules formed as intermediates during the oxidation of tetrathionate by *T. ferrooxidans* (44). Some authors have reported the formation of elemental sulfur during the oxidation of sulfidic minerals. In one case (45), it was presented as a possibility to improve leaching processes. The advantage would be that leaching can be performed without excessive formation of sulfuric acid and less oxygen would be required. The authors were able to stimulate sulfur formation by adding low concentrations of silver salts. Beyer *et al.* (46) observed the formation of elemental sulfur during bioleaching of coal and explained the phenomenon by referring to a nonbiologically catalyzed reaction between ferrous iron and pyrite.

$$FeS_2 + 2\ Fe^{3+} + 4\ H_2O \rightarrow 3\ Fe^{2+} + SO_4^{2-} + S + 8H^+. \qquad (4)$$

However, as mentioned above, Boogerd et al. (38) demonstrated that under conditions relevant for microbial desulfurization (pH, ionic strength, and temperature) this reaction does not occur. Under these conditions elemental sulfur is very stable, and its oxidation by ferric iron, although thermodynamically feasible, does not take place at appreciable rates, even at 70°C (47). For these reasons, elemental sulfur as an intermediate in the chemical oxidation of pyrite by ferric iron is not very likely (48,49). These data seem to conflict with published reports (50–52), but the general concept that the oxidation of pyrite by ferric iron will lead to the formation of elemental sulfur as an intermediate is derived from the observations with other metal sulfides that contain the S^{2-} ion. However, sulfur is present as the S_2^{2-} ion in pyrite. Based on molecular orbital

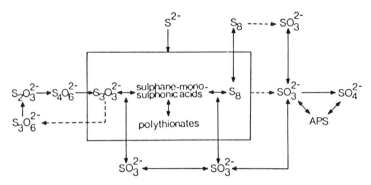

Fig. 15.1. Scheme for the oxidation of reduced sulfur compounds by the acidophilic thiobacilli. The boxed set of compounds and reactions represents a pool of intermediary sulfur compounds. The nature of the intermediary sulfur compounds from sulfide and elemental sulfur are unknown. Also the mechanisms of sulfite formation from elemental sulfur remains to be elucidated. S_8-sulfur may combine with polythionates to form hydrophilic sulfur complexes, as shown in Figure 15.2. (Reproduced from Pronk *et al.* [42], with permission of the publisher.)

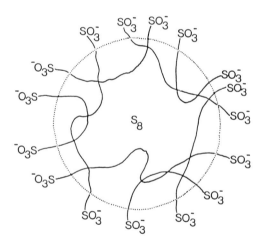

Fig. 15.2. Proposed structure of hydrophilic intermediary sulfur globules formed during the oxidation of tetrathionate by *T. ferrooxidans*. (Reproduced from Steudel et al. [44], with permission of the publisher.)

theory considerations, Luther (53) argued that thiosulfate rather than elemental sulfur should be expected as an intermediate if ferric iron oxidizes S_2^{2-}.

It is not known how acidophilic sulfur-oxidizing bacteria can attack the sulfur moiety in the solid minerals. The nonparallel production of ferric iron and sulfate observed by Andrews (54) during the leaching of pyrite does not necessarily lead to the theory put forward by this author. From the observed preferential oxidation of sulfur, Andrews concluded that there is a migration of sulfur atoms to the pyritic surface, presumably by molecular diffusion. However, these data can also be explained by assuming the formation of jarosite-like precipitates in which the ferric iron/sulfate ratio is higher than in pyrite. A minerals salts medium with a high ionic strength was used for these experiments, and this would favor jarosite precipitation (see also ref. 55). Boogerd et al. (38) showed that if a mineral salts medium with lower concentrations of ammonium and alkali ions was used the release of ferric iron and sulfate from pyrite appeared to be stoichiometric, provided that iron-containing precipitates were absent (38).

In conclusion, the studies of Arkesteyn (39), Hazeu et al. (40), and Boogerd et al. (38) clearly demonstrate that, in the oxidation of pyrite, biological sulfur oxidation must be involved. Otherwise, the observed rates, kinetics, and yields cannot be explained.

4. TECHNICAL FEASIBILITY OF THE BIODESULFURIZATION OF COAL

The idea of exploiting acidophilic pyrite-oxidizing bacteria as a means of cleaning coal is quite old (56). Surveys of the various microbial desulfurization investigations are given by Bos et al. (57), Klein et al. (11), and Bos and Kuenen (12). One might conclude that many groups from various coal-producing nations have dealt with biodesulfurization. However, most studies in the 1970s and the beginning of the 1980s were short-term and monodisciplinary. In many of them, only a few coal samples were tested. The results of those studies have clearly established the concept that inorganic sulfur can be removed from coal by means of biological activity. However, to judge the technical feasibility of biodesulfurization, more extensive and multidisciplinary research was required. Such investigations were started in The Netherlands and in Germany, initially under separate funding by national governments. In 1986, the projects were broadened within the framework of an EC program. Cooperation started between the German and Dutch groups, together with British (Warren Spring Laboratory) and Italian (University of Cagliari, University of Rome) partners. In 1986, in the United States a DOE-research program in biodesulfurization was also initiated (58).

Figure 15.3 shows an outline of the microbial desulfurization process of coal proposed in the Dutch feasibility study. First the coal should be milled to a particle size that will allow the bacteria to reach the pyrite crystals. The coal should then be mixed with the water phase containing mineral salts as nutrients. Because coals contain varying amounts of minerals that will be dissolved during the leaching process, most nutrient requirements are met, and only ammonium and sometimes phosphate should be added (59). The need for these minerals is

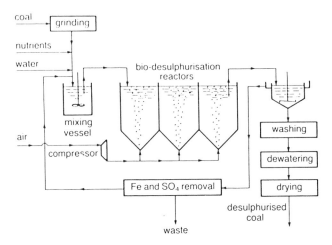

Fig. 15.3. Schematic of the biodesulfurization process. (Reproduced from Huber et al. [91], with permission of the publisher.)

minute, since (mesophilic) autotrophs and extremely low concentrations of bacterial biomass are involved. Only 0.5 mM ammonium and 0.1 mM phosphate is required for the desulfurization of a coal containing 1% (w/w) pyrite leached in a 20% (w/v) coal water slurry (59). The maximal pulp density should be in the range of 20%–25% (w/v). It has been shown that first-order kinetics can be maintained up to pulp densities of 20% (w/v). If higher pulp densities are used, deviating kinetics occur, probably because of gas mass transfer limitations (60). According to the Dutch report (61), preference should be given to gas-agitated reactor systems, since the low-pH process fluid in combination with a high pulp density is highly corrosive and abrasive, and therefore mechanical mixing devices would not be advisable. Regime analysis studies aimed at understanding rate-determining mechanisms revealed that, to create optimal conditions in the reactor systems, the air stream necessary for pyrite oxidation is sufficient to keep the system properly mixed and to prevent sedimentation (61). After a residence time of 8–10 days in the cascade of slurry reactors, the coal should be separated from the process water. For this purpose settlers, hydrocyclones, or filter band conveyers can be used. A washing procedure should be included in order to free the pulverized coal from the adhering process water. The waste water needs further treatment. It should be neutralized. A solid waste containing gypsum, ferric hydroxy sulfates (jarosites), and other metal-containing compounds will be produced. Preference should be given to a waste water treatment in which precipitation is fractionated, resulting in separate products such as pure gypsum and heavy metal-containing concentrates. To achieve this goal, further studies are required.

This process scheme is similar to others produced by other groups (*e.g.*, refs. 62–65). Differences can be found in the type of reactor systems. Andrews *et al.* (65) included a ditch-like air-agitated slurry reactor. Some proposed mechanically mixed reactors (62). Others suggested that the pipelines for coal slurry transport be used as the reactor system (66). Shallow lagoons have also been proposed (67). The main common characteristics of all of these reactor systems is that they should be inexpensive and easy to operate. In view of the fact that huge reactors will be required, the operational costs of microbial desulfurization will mainly be determined by investment costs (63). For that reason, the use of heap leaching for biodesulfurization has also been proposed (68,69). However, it should be stressed that heap leaching will remove a smaller fraction of the inorganic sulfur because the coarse coal lumps involved will prevent the bacteria reaching an important part of the sulfidic minerals. Hyman et al. (69) indicated that about 50% of the inorganic pyrite in their coal could be removed. However, it is likely that because of the suboptimal kinetics, jarosites had formed, as indicated by the nonparallel production of ferric iron and sulfate.

A variation of the biodesulfurization process has been suggested, involving microbial-assisted physical cleaning of coal (70–72). The principle is based on the supposed ability of acidophiles, in short-term contact, to change the surface characteristics of sulfidic minerals, and especially pyrite. Such changes might be due to the oxidation of the pyrite or simply to a selective adsorbance of bacterial biomass to the sulfidic mineral. In this way, the hydrophobicity of the mineral surface might become sufficiently different from that of the coal particles to permit separation techniques based on differences in surface characteristics such as oil agglomeration and froth flotation. Laboratory tests indicated that the principle works, but for separation with low carbon losses an extremely fine milling is required. Moreover, because of the strong adsorbance of the biomass to the solid particles and the impossibility of recycling biomass, the production of active biomass would require huge bioreactors and, again, long residence times (70). More recent studies have demonstrated that the use of acidophiles is not necessary in order to make the changes in surface characteristics. Even baker's yeast can be used for this purpose (73).

We next review the advantages and disadvantages of biodesulfurization of coal in order to be able to evaluate its position among the different techniques designed to reduce the environmental risks of coal use.

5. ADVANTAGES

5.1. Selectivity

It should be stressed that biodesulfurization is a very selective process. The technique concentrates on the attack on inorganic minerals present in the coal.

TABLE 15.3. Differences[a] in Coal Characteristics
Before and After Leaching[b]

Coal	Ash content	Volatile matter	Caloric value	Pyritic sulfur
501GB28	−16.3	−9	+3.7	−82
511US43	−25	−6	−5	−90
513DE38	−29	+4	0	−84
514DE31	−15	+5	−2	−91
515US38	−12	−4	−2	−89

[a]Given as percentages of change.

[b]Coal samples were supplied by the European Centre for Coal Specimens SBN, Eygelshoven, The Netherlands. Reproduced from Bos *et al.* (57), with permission of the publisher.

The acidophilic, pyrite-oxidizing bacteria do not attack the carbon skeleton of the coal. For that reason, biodesulfurization does not result in significant carbon losses or in drastic changes in the caloric value of the coal. This is demonstrated in the data presented in Table 15.3 (57) and contrasts with chemical oxidation processes in which not only the sulfidic minerals are oxidized, but also a part of the carboneous material.

For the almost complete removal of pyrite, a particle size of less than 100 μm is needed (57,74). The particle size required for physical separation techniques such as oil agglomeration, high gradient magnetic separation, electrostatic separation, and froth flotation is, for most coals, much smaller. In contrast with biodesulfurization, many of these methods expect that mineral particles be completely free from the coal carboneous matrix, which requires extremely fine milling. This is true for high gradient magnetic separation, electrostatic separation, and froth flotation, Moreover, it should be stressed that, for effective separation, the difference in hydrophobicity between the fractions to be separated must be large. In most cases, the difference in hydrophobicity between pyrite particles and coal particles is far too small to obtain sufficient cleaning of the coal.

5.2. Heavy Metal Removal

The organisms are oxidizing, either in combination with spontaneous chemical reactions or not, metal sulfides, resulting in a dissolution of the heavy metals present in the coal. This might be viewed as an advantage when the problems related to heavy metals present in bottom and fly ash after combustion of untreated coal are considered. Data on heavy metal removal are presented in Table

TABLE 15.4. Heavy Metal Content in Coals Before and After Leaching[a]

	Cu		Fe		Zn		As	
coal	Before	After	Before	After	Before	After	Before	After
505DE24	5.19	1.28	34,650	1,625	88.48	24.38	7.98	1.24
511US43	9.06	2.24	13,110	6,647	40.05	14.31	1.73	1.35
513DE38	20.42	10.77	7,841	2,628	214.45	94.18	23.86	9.24
514DE31	25.29	20.42	4,581	1,348	29.86	11.5	1.81	1.16

[a]Concentrations of the heavy metals are expressed as ppm. Analysis by neutron activation. Reproduced from Bos et al. (57), with permission of the publisher.

TABLE 15.5. Analysis of the Minerals Present in Coal Sample 513DE38 Before and After Leaching[a]

	Weight distribution expressed as % of the mineral content		
Mineral	Before leaching	After leaching	Group
Pyrite	19.5	<1	
Jarosite	<1	2.4	Soluble minerals
Siderite	11.3	<1	(except for jarosite)
Ca, Mg carbonates	20.7	<1	
Quartz	7.3	9.3	
Silicates	29.3	59.5	Insoluble minerals
Others	12.0	28.8	

[a]Reproduced from Bos et al. (75), with permission of the publisher.

15.4 (57). It can be seen that heavy metal removal, and hence potential beneficial effects, are dependent on the type of coal.

5.3. Ash Reduction

Because of the low pH of the process, a number of common minerals present in the coal are dissolved, thus reducing the ash content. Data describing the reduction of ash forming constituents are given in Table 15.5 (75).

5.4. Almost All Coals Are Suitable

One might conclude from the various reports describing biodesulfurization that a large variety of coals is suitable for the process. In the Dutch research program, more than 25 different coal samples were tested. Without exception, they all appeared to be suitable for desulfurization (61). However, a number of requirements should be fulfilled. The leaching should be carried out in a pH-controlled system. If the coal contains large amounts of carbonates, the pH might have a tendency to rise above values that are suitable for the desired microbial activity. Sometimes the concentrations of carbonates present in the coal exceed those that are required for the titration of the acid formed from sulfidic minerals. A good example is Sulcis coal from Sardinia. Some reports (*e.g.*, ref. 36) suggest that the presence of special metal ions such as silver and tungsten might interfere with microbial activity. Most acidophilic sulfidic mineral-oxidizing bacteria can withstand high concentrations of heavy metals. Their tolerance for heavy metals might differ by a factor 3, or even more, from that of neutrophilic bacteria.

6. DISADVANTAGES

6.1. Slow Kinetics

A disadvantage of microbial desulfurization is the slow reaction rate. If mesophiles are used, residence times of at least 8 days are required to obtain satisfactory inorganic sulfur removal. A prerequisite is that the pyrite should be attainable for microbial attack. Image analysis studies by Vleeskens *et al.* (74) revealed that, if the coal is pulverized to sizes less than 100 μm, over 90% of the pyrite can be removed. Pyrite crystals completely enclosed within the coal matrix are not available to microbial attack. Applying these kinetic data to an installation treating 100,000 tons of coal per year, huge reactors are necessary (57). If a suspension of 20% pulverized coal in water is used, the reactor volume should be at least 10,000 m^3. Even if low-cost, gas-stirred concrete tanks are used, high investment costs are required.

For systems using mesophilic bacteria, the kinetics of pyrite oxidation are first order in the amount of pyrite available for microbial attack (37). To exploit these kinetics most efficiently, a plug flow reactor is advisable. However, the mesophilic acidophiles have a strong tendency to adsorb to the solid particles. If one assumes a continuous process, one should expect a wash-out of active biomass if a plug flow reactor is used. For effective inoculation of the fresh coal, intensive contact, and thus back-mixing in the reactor, is required. This means that to prevent biomass limitation in the reactor system a mixed reaction is essential. To meet these two conflicting requirements, a configuration consisting of a mixed flow reactor followed by a plug flow reactor has been suggested. The minimal residence time in the mixed flow reactor is determined by the growth rate of the biomass in coal water slurries. The residence time in the plug flow reactor can be

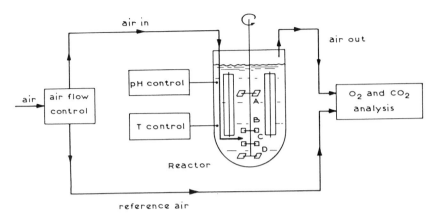

Fig. 15.4. Schematic of a 10 liter batch reactor system used for growth kinetics studies. (Reproduced from Huber et al. [37], with permission of the publisher.)

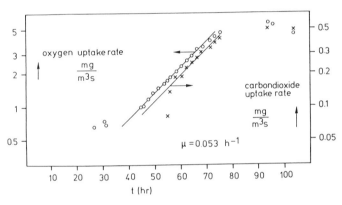

Fig. 15.5. Logarithmic plot of oxygen and carbon dioxide uptake rates as observed in a 10 liter batch leaching experiment. (Reproduced from Huber et al. [37], with permission of the publisher.)

calculated from the kinetics of pyrite oxidation, which can be followed quite easily by determining the iron and/or sulfate produced in the process water.

Determination of the growth rate of pyrite oxidizers in coal water slurries is more cumbersome. One deals here with autotrophic organisms with comparatively low cell yields. Only tens of miligrams of biomass are present in an environment with 20% w/v carboneous material with strong adsorbing capacities. Common ways to follow growth by, for example, protein determination cannot be used. Huber *et al.* (37) found a way to determine growth rates of pyrite-oxidizing bacteria in leaching systems by following the carbon dioxide uptake rate in a batch leaching test of 10 liters. Figures 15.4 and 15.5 show the experimental set-up and data on the carbon dioxide and oxygen uptake rate measurements. The maximal specific growth rate was $0.052 \ h^{-1}$. This means that a residence time of at least 20 hours is required in the first part of the reactor system in which a mixed flow reactor is present. Although earlier Dutch design used a ditch-like plug flow reactor, in the final process design this system was achieved by a cascade of mixed flow reactors (57).

6.2. No Organic Sulfur Removal?

Another important disadvantage is that acidophilic pyrite-oxidizing bacteria are unable to remove organically bound sulfur. The ratio between inorganically and organically bound sulfur varies greatly among different coals. Any process employing acidophilic sulfide oxidizers therefore only solves a part of the problem. The goal of a microbial system that can dissolve and remove organically bound sulfur has been a strong incentive for research. Preference has been given to studies using model substrates containing organically bound sulfur. The most commonly used substrate is dibenzothiophene (DBT), which is supposed to have the basic structure representing organically bound sulfur in coals. Neutrophilic bacteria that possess the ability to break down DBT have been isolated (for an overview, see ref. 11), but cultures that can selectively remove sulfur from DBT without attacking the carbon skeleton are of most interest. Studies are yielding insights into the metabolic pathways leading to sulfate and phenolic compounds (76). Figure 15.6 shows details of some possible pathways (77). Use of recombinant DNA techniques to transfer the genes encoding for this ability to acidophilic organisms has been suggested. The objective is to develop an organism that encompasses the ability to oxidize both organically and inorganically bound sulfur, offering the possibility of a one-step, total sulfur removing process (78).

Studies demonstrating organic sulfur removal by DBT-degrading bacteria have been published (79–81). These reports must be considered with caution. The reported reduction of organic sulfur often falls within the range of analytical error. Determination of organic sulfur in coal is indirect, being the difference between total sulfur and the sum of pyritic and inorganic sulfate sulfur. According to the ASTM standards, "organic sulfur" includes all inorganic sulfur present in sulfides except pyrite, and also elemental sulfur and polysulfides. Critical reviews of organic sulfur in coals can be found in Friedman (2) and Bos and Kuenen (12). Recently, guidelines have been developed in the United States in order to facilitate the interpretation and intercomparison of data from studies on the microbial removal of organic sulfur from coal. They are summarized below (Dr. Greg Olsen and Dr. David Boron, personal communication).

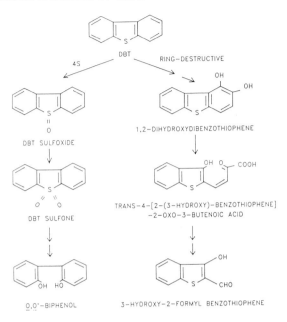

DBT

RING—DESTRUCTIVE

4S

1,2—DIHYDROXYDIBENZOTHIOPHENE

DBT SULFOXIDE

DBT SULFONE

TRANS—4—[2—(3—HYDROXY)—BENZOTHIOPHENE]
—2—OXO—3—BUTENOIC ACID

O,O'—BIPHENOL

3—HYDROXY—2—FORMYL BENZOTHIOPHENE

Fig. 15.6. Proposed pathways for microbial breakdown of dibenzothiophene. (Reproduced from Faison et al. [77], with permission of the publisher.)

1. Experiments should be performed with either depyritized coal samples or with coals with a low natural pyritic sulfur content.

2. Obtain ultimate (C, H, O, N. S, ash, moisture) and ASTM forms of sulfur analysis on the starting coal. The moisture and ash-free organic sulfur value should be used to compare values for organic sulfur before and after microbial treatment.

3. Include a sterile control to account for abiotic changes that occur during exposure of the coal sample to air and water. Conventional autoclaving at 121°C should be avoided. Heating the coal in a vacuum oven at 105°C for 1–2 hours minimizes oxidation and is preferred.

4. At the conclusion of the experiment, separate cells from coal and determine the percent recovery of the coal sample. If more sample is recovered than was added, the complete separation of cells or medium precipitates from coal may not be achieved, and/or the coal may have gained weight due to oxidation. In either case, the sulfur content of the material has probably been diluted. If recovery of coal is low, the coal may have been partly solubilized.

5. Obtain ultimate and ASTM forms of sulfur analysis on the product coal. Again, use the moisture and ash-free value for organic sulfur. Determine if the

coal has been significantly oxidized by the change in oxygen content. A significant increase in oxygen would "dilute" the sulfur content of the coal. If recovery has been a problem, the ultimate analysis of the product will also give clues as to whether significant biomass is associated with the coal. For example, the nitrogen content of cells is much higher than that of coals, and nitrogen values are among the most reliable in coal analyses. In addition, carbon/hydrogen ratios are generally higher in coal than in cells.

6. Since sulfur removal is the focus of the research efforts, every attempt should be made to conduct a sulfur balance. This is a critical requirement, particularly if the information collected is to be used for publication. As a part of this effort, sulfur balances should include the solid coal product, aqueous streams, gas effluents, and biomass.

If organic sulfur removal is a realistic option, one should expect that a great deal of research is needed for the development of a practical process. One must expect that the kinetics of the oxidation of organic sulfur, as part of the complex macromolecular structure of coals, will be extremely slow. Moreover, as was stated earlier in this chapter, very little is known about the exact nature of the organic sulfur present in coals. DBT might not be the most relevant model substrate. Furthermore, the option to build genes from neutrophilic organic sulfur oxidizers into acidophilic bacteria is far from a routine job (78).

6.3. Waste Production

A third disadvantage associated with the microbial desulfurization of coal is the formation of an acid waste water, which needs further treatment. It contains unacceptably high concentrations of ferric iron and sulfate. Because of the ferric iron, which can act as an oxidizing agent for metal sulfides, the waste water will also contain comparatively high concentrations of different heavy metals. The extent and variety are, of course, determined by the mineral composition of the coal involved. This disadvantage is related to one of the advantages of coal desulfurization already mentioned. An important portion of the heavy metals in the coal can be removed by acidophilic sulfur oxidizers, which results in fewer problems related to heavy metal-containing fly and bottom ash from coal firing installations.

Associated with the problem of the waste water is the difficulty of effectively removing the process water from the coal after treatment. Because of the small particle size, dewatering by sedimentation and filtration is not easy. An important part of the sulfur, in the form of sulfate, remains adhering to the coal particles.

**TABLE 15.6. Leaching Results of Samples From the Maamba Coal Mine
Expressed in Percentages[a]**

Sample	Pyritic sulfur		Total sulfur		Iron removal	Pyrite removal
	Before	After	Before	After		
Maamba 1	8.2	0.24	10.1	3.73	86	97
Maamba 2	7.9	0.24	10.2	4.90	85	97

[a]Details of leaching procedure are given by Kos *et al.* (59). Duration, 28 days; slurry percentage, 5%; temperature, 30°C. Reproduced from Bos *et al.* (61), with permission of the publisher.

6.4. Jarosite Precipitation

A fourth disadvantage of the microbial desulfurization process might be the formation of jarosite-like precipitates. Jarosites are basic ferric sulfates. They are extremely insoluble products that decompose at temperatures above 700°C. If pyrite is removed, but jarosite-like precipitates remain in the coal, all efforts to desulfurize the coal are in vain. At pH values even below 2, equilibrium in the conditions relevant for microbial desulfurization favors jarosite. If microbial leaching is carried out at mesophilic temperatures, the rate of jarosite formation is quite slow, and, if one limits the ionic strength of the process water, jarosite formation will be limited or even prevented (59). If one employs moderate or even extreme thermophiles, the rate of jarosite formation increases in such a way that the biodesulfurization process cannot be a success (82). The problem of jarosite formation is often neglected in coal leaching studies. Desulfurization processes are generally followed by analyzing ferric iron concentration in the process fluid or by analyzing the pyritic content before and after leaching. To detect the presence of jarosites, complete balances for sulfur and iron should be made. This is illustrated in Table 15.6. On the basis of the pyrite data, one might conclude that pyrite removal was perfect. However, making iron and sulfur balances shows that a significant part of the iron and the sulfur was left in the coal, most probably as jarosite. As precipitation also depends on the environmental salt concentration, the process must be optimized. In many leaching studies, the 9 K mineral salts medium (28) has been used. This medium contains concentrations of minerals that exceed the need of the microorganisms involved. In view of the extreme insolubility of jarosite (*e.g.*, potassium jarosite), high concentrations of potassium, ammonium, and sodium should be avoided.

It is interesting to note that problems with jarosite formation are closely related to the frequently observed poor leaching results achieved with heavily weathered coals. The weathering of the coal leads to the oxidation of pyrite, resulting in the covering of the pyrite surfaces with jarosite-like precipitates. As a consequence, the pyrite crystals embedded in the ferric iron precipitates become unattainable for microbial attack, and poor leaching results are obtained.

6.5. Heat Production

In the original Dutch biodesulfurization program, attention was focused on low sulfur coals such as are imported into The Netherlands. Because of Dutch governmental regulations, the sulfur content must be reduced further to meet the standard of less than 0.35% total sulfur. Because coals with a sulfur content that can be reduced to this level by depyritization are scarce, interest in biodesulfurization is minimal (61). This differs from the situation in coal-producing nations, where the ratio between inorganic sulfur and organically bound sulfur may be above 1. For that reason, research was initiated with other partners in Europe (Warren Spring Laboratory, U.K.; Bergbau-Forschung GmbH, Germany; University of Cagliari) to study the feasibility of the biodesulfurization of high pyritic coals. Because the oxidation of pyrite is an exothermic process ($\Delta H_r^o = -1,481$ kJ/mol FeS_2), biodesulfurization of high pyritic coals results in an increase of the temperature in the reactor, such that the growth and activity of mesophilic bacteria would be partially or completely inhibited if the reactor was not cooled. For that reason, the use of moderately and extremely thermophilic processes was studied (83). A mixed bacterial community that was isolated from high pyritic reject materials from the Maamba coal mine in Zambia was used. For model studies, it has been possible to isolate the dominant rod-shaped organism that has an optimum temperature of 50°C. It is similar to moderate thermophiles isolated by Norris *et al.* (84) and by Marsh and Norris (85). The behavior of this isolate in coal leaching tests resembles that of mesophilic pyrite-oxidizing cultures. Pyrite oxidation followed first-order kinetics with a rate constant comparable with that of *T. ferrooxidans*. When the organism was grown in a medium in which pyrite replaced pulverized coal, it required yeast extract for growth, behaving under these growth conditions as a chemolithoheterotroph. However, when the carbon dioxide concentration in the supplied air was increased, the organism functioned as a chemolithoautotroph. This phenomenon is shown in Figure 15.7 and is in agreement with earlier observations by Marsh and Norris (86). Because of the production of carbon dioxide from carbonates present in the coal and from weathering processes, sufficient carbon dioxide is probably present in the initial stages of the leaching process to prevent biomass limitation.

Although leaching experiments with coal samples suggest that the use of moderate thermophiles might prevent the costly need to cool the reactors, their application is problematic for several reasons. It appears that moderate thermophiles are far more fragile than their mesophilic and extreme thermophilic counterparts. If energy supply to the cells stops, because of a lack of either oxygen or an oxidizable substrate, their activity is lost within a few hours. The cell walls

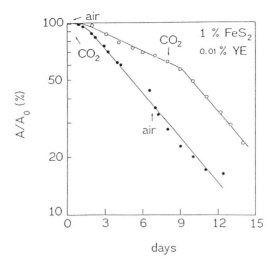

Fig. 15.7. Stimulation of the oxidation of pyrite by a moderately thermophilic acidophilic bacterium by an increase in carbon dioxide concentration in the aeration air. The percentage of pyrite remaining (logarithmic scale) versus time as measured in two batch fermentors is plotted. Both fermentors contained 1% pyrite and 0.01% yeast extract at the start. One fermentor was supplied with air during the first 8 days, after which air enriched with 5% carbon dioxide was supplied. The other fermentor received carbon dioxide–enriched air during the first 7 days. Stimulation of pyrite oxidation can be explained by assuming that biomass limitation can be overcome by induction at higher carbon dioxide concentrations of chemolithoautotrophy. (Reproduced from Boogerd et al. [88], with permission of the publisher.)

of these gram-positive organisms may be less effective in keeping the internal pH poised at the right level in a low pH environment than that of a gram-negative acidophile such as *T. ferrooxidans*. If this is true, the energy requirement for resting cells to maintain the required internal pH would be much higher than for mesophilic, gram-negative acidophiles (87).

Another problem that might be important is that, at elevated temperatures, because of a decreased gas mass transfer, carbon dioxide limitation is likely to occur (88). This could, however, be overcome by supplementing the carbon dioxide supply during initial stages of growth.

7. ECONOMIC FEASIBILITY

Details of the Dutch economic analysis can be found in Bos *et al.* (57). The analysis reveals that biodesulfurization costs are of the same order of magnitude as stack gas cleaning. The Bergbau-Forschung in Essen (recently renamed the

DMT-Gesellschaft) estimated much higher costs (89). The main difference in their cost analysis is the calculation of the energy requirements. Most energy is required for the aeration of the slurry reactors. In the Dutch view, the demand for aeration energy is not directly proportional to reactor volume. The bigger the installation, the smaller the relative need for energy. In the German analysis, energy was considered to be linearly proportional with size. Whether the Dutch or German approach is right can only be decided by large-scale testing. The maximum volume used in coal leaching tests was 240 liters (63,64). A pilot plant with a reactor volume of about 50 m^3 is being built at the ENICHEM Refinery, Porto Torres, Italy.

8. CONCLUSIONS

It must be clearly stressed that, if biodesulfurization of coal is a realistic option in some cases, it is surely not to be generalized. It is suitable for coals in which sulfidic minerals are not present as coarse particles, but as small crystals finely distributed in the coal matrix. These coals are not suitable for treatment with simple, and therefore cheap, sink float techniques. Moreover, to be suitable for biotreatment the coal must have a high ratio between inorganically and organically bound sulfur, combined with low concentrations of acid-neutralizing minerals (carbonates). This conclusion is based on the assumption that truly organically bound sulfur is not accessible to selective microbial attack. Although several authors claim that they can demonstrate organic sulfur removal, the available experimental data are not convincing. For large-scale, coal-using, energy-producing plants, stack gas cleaning should be preferred. This technique removes not only the sulfur dioxide produced from inorganic sulfur, but also that from organic sulfur compounds. Biodesulfurization can be applicable in combination with industrial, small-scale, energy-producing plants. Treatment of the stack gases of these plants is not advisable because of high investment costs. Because biodesulfurization is a wet process, combination with the preparation of coal water slurries may be a viable option. These slurries contain 70% pulverized coal, 30% water, and stabilizers, and they have an oily appearance. They can be used as a replacement for heavy oil (90), can be produced in a central location, and can be distributed easily by pipelines and in tanks. Some European countries (e.g., Sweden and Italy) have begun production of these fuels. Finally, it should be realized that biodesulfurization produces a waste stream that needs further treatment.

ACKNOWLEDGMENTS

For critical reading and correction of the English text, Dr. Lesley Robertson is gratefully acknowledged.

REFERENCES

1. Schilling H-D, Wiegand D: Coal resources. In McLaren DJ, Skinner BJ (eds): Resources and World Development. Chichester: John Wiley & Sons, 1987, pp 129–156.

2. Friedman S: Sulfur analysis in coal—A critical evaluation. In Proc 1990 First Int Symp Biol Processing Coal. Orlando FL, May 1990. Palo Alto, CA: EPRI, 1990, pp 1.3–1.13.

3. Kos CH, Poorter RPE, Bos P, Kuenen JG: Geochemistry of sulfides in coal and microbial leaching experiments. In Proc Int Conf Coal Sci. Dusseldorf, 1981, pp 842–847.

4. Schicho RN, Brown SH, Olson GJ, Parks EJ, Kelly RM: Probing coals for non-pyritic sulfur using sulfur-metabolizing mesophilic and hyperthermophilic bacteria. Fuel 68:1368–1375, 1989.

5. Hippo E, Palmer S, Crelling J, Kruge M: Organic sulfur species distributions in bituminous and low rank coals. In Proc 1990 First Int Symp Biol Processing Coal. Orlando, FL, May 1990. Palo Alto, CA: EPRI,, pp 1.15–1.31.

6. Podolski WF: Fluidized-bed combustion. In Cooper BR, Ellingson WA (eds): The Science and Technology of Coal and Coal Utilization. New York: Plenum Press, 1984, pp 263–305.

7. Palowitch ER, Deurbrouck AW: Wet concentration of coarse coal. Part 1: Dense medium separation. In Leonard JW (ed): Coal Preparation, 4th ed. New York: The American Institute of Mining, Metallurgical and Petroleum Engineers, Inc., 1979, pp 9.1–9.36.

8. Beddow JK: Dry separation techniques. Chem Eng 88:70–84, 1981.

9. Zimmerman RE: Wet concentration of fine coal. Part 3: Froth flotation. In Leonard JW (ed): Coal Preparation, 4th ed. New York: The American Institute of Mining, Metallurgical and Petroleum Engineers, Inc., 1979, pp 10.75–10.104.

10. Gouch GR: Biotechnology and Coal. London: IEA Coal Research, 1987.

11. Klein J, Beyer M, van Afferden M, Hodek W, Pfeifer F, Seewald H, Wolff-Fischer E, Juntgen H: Coal in biotechnology. In Rehn H-J, Reed G (eds): Biotechnology, vol 6b. Weinheim, FRG: VCH Verlagsgesellschaft, 1988, pp 497–567.

12. Bos P, Kuenen JG: Microbial treatment of coal. In Ehrlich HL, Brierley CL (eds): Microbial Metal Recovery. New York: McGraw-Hill, 1990, pp 343–377.

13. Cohen MS, Gabriele PD: Degradation of coal by the fungi *Polysporus versicolor* and *Poria monticola*. Appl Environ Microbiol 44:23–27, 1982.

14. Scott CD, Strandberg GW, Lewis SN: Microbial solubilization of coal. Biotechnol Progr 2: 131–139, 1986.

15. Ivanov MV: Major fluxes of the global biogeochemical cycle of sulfur. In Ivanov MV, Frenay JR (eds): The Global Biochemical Sulfur Cycle. Chichester: John Wiley & Sons, 1983, pp 449–463.

16. Lovell HR: Coal mine drainage in the United States—An overview. Water Sci Technol 15:1–25, 1983.

17. Ehrlich H, Brierley CL: Microbial Mineral Recovery. New York: McGraw-Hill, 1990.

18. Lawrence RW: Biotreatment of gold ores. In Ehrlich HL, Brierley CL (eds): Microbial Mineral Recovery. New York: McGraw-Hill, 1990, pp 127–148.

19. Colmer AR, Hinkle ME: The role of microorganisms in acid mine drainage. Science 106: 253–256, 1947.

20. Colmer AR, Temple KT, Hinkle ME: An iron-oxidizing bacterium from the acid drainage of some bituminous coal mines. J Bacteriol 59:317–328, 1950.

21. Norris PR: Acidophilic bacteria and their activity in mineral sulfide oxidation. In Ehrlich HL, Brierley CL (eds): Microbial Mineral Recovery. New York: McGraw-Hill, 1990, pp 3–27.

22. Harrison AP: The acidophilic thiobacilli and other acidophilic bacteria that share their habitat. Annu Rev Microbiol 38:265–292, 1984.

23. Walsh F, Mitchell R: A pH-dependent succession of iron bacteria. Environm Sci Technol 6:809–812, 1972.

24. Walsh F, Mitchell R: Differentiation between *Gallionella* and *Metallogenium*. Arch Mikrobiol 90:19–25, 1973.

25. Tuovinen OH, Kelly DP: Studies on the growth of *Thiobacillus ferrooxidans*. Arch Mikrobiol 88:285–295, 1973.

26. de Bruyn JC, Boogerd FC, Bos P, Kuenen JG: Floating filters, a novel technique for the isolation and enumeration of fastidious, acidophilic, iron-oxidizing, autotrophic bacteria. Appl Environ Microbiol 56:2891–2894, 1990.

27. Muyzer G, de Bruyn AC, Schmedding DJM, Bos P, Westbroek P, Kuenen JG: A combined immunofluorescence-DNA-staining technique for enumeration of *Thiobacillus ferrooxidans* in a population of acidophilic bacteria. Appl Environ Microbiol 53:660–664, 1987.

28. Silverman MP, Lundgren DG: Studies on the chemoautotrophic iron bacterium *Ferrobacillus ferrooxidans*. I. An improved medium and a harvesting procedure for securing high cell yields. J Bacteriol 77:642–647.

29. Norris PR, Kelly DP: Dissolution of pyrite (FeS_2) by pure and mixed cultures of some acidophilic bacteria. FEMS Microbiol Lett 4:143–146, 1978.

30. Norris PR: Iron and mineral oxidation with *Leptospirillum*-like bacteria. In Rossi G, Torma AE (eds): Progress in Biohydrometallurgy. Iglesias: Associazione Mineraria Sarda, 1983, pp 83–96.

31. Schnaitman C, Lundgren DG: Organic compounds in the spent medium of *Ferrobacillus ferrooxidans*. Can J Microbiol 11:23–27, 1965.

32. Norris PR, Kelly DP: The use of mixed microbial cultures in metal recovery. In Bull AT, Slater JH (eds): Microbial Interactions and Communities, vol 1. London: Academic Press, 1982. pp 443–474.

33. Rao GS, Berger LR: Basis of pyruvate inhibition in *Thiobacillus thiooxidans*. J Bacteriol 102:462–466, 1970.

34. Pronk JT, Meesters PJW, van Dijken JP, Bos P, Kuenen JG: Heterotrophic growth of *Thiobacillus acidophilus* in batch and chemostat cultures. Arch Microbiol 153:392–398, 1990.

35. Dugan PR, Apel WA: Microbial desulfurization of coal. In Murr LE, Torma AE, Brierley JA: Metallurgical Applications of Bacterial Leaching and Related Microbiological Phenomena. New York: Academic Press, 1978, pp 223–250.

36. Hoffman MR, Faust BC, Panda FA, Koo HH, Tsuchiya HM: Kinetics of the removal of iron pyrite from coal by microbial catalysis. Appl Environ Microbiol 42:259–271, 1981.

37. Huber TF, Kossen NWF, Bos P, Kuenen JG: Modelling design and scale up of a reactor for microbial desulfurization of coal. In Rossi G, Torma AE (eds): Progress in Biohydrometallurgy. Iglesias: Associazione Mineraria Sarda, pp 279–289.

38. Boogerd FC, van der Beemd C, Stoelwinder T, Bos P, Kuenen JG: The relative contributions of biological and chemical reactions to the overall rate of pyrite oxidation at temperatures between 30 and 70 degrees C. Biotechnol Bioeng 38:109–115, 1991.

39. Arkesteyn GJMW: Contribution of Microorganisms to the Oxidation of Pyrite. PhD thesis, University of Wageningen, 1980.

40. Hazeu W, Schmedding DJ, Goddijn O, Bos P, Kuenen JG: The importance of the sulfur oxidizing capacity of Thiobacillus ferrooxidans during leaching of pyrite. In Neijssel OM, van der Meer RR, Luyben KCAM (eds): Proc 4th Eur Cong Biotechnol, vol 3. Amsterdam: Elsevier, 1987, pp 497–499.

41. Ingledew W: *Thiobacillus ferrooxidans:* The bioenergetics of an acidophilic chemolithotroph. Biochim Biophys Acta 683:89–117, 1982.

42. Pronk JT, Meulenberg R, Hazeu W, Bos P, Kuenen JG: Oxidation of reduced inorganic sulfur compounds by acidophilic thiobacilli. FEMS Microbiol Rev 75:293–306, 1990.

43. Hazeu W, Batenburg-van der Vegte WH, Bos P, van der Plas RK, Kuenen JG: The production and utilization of intermediary elemental sulfur during the oxidation of reduced sulfur compounds by *Thiobacillus ferrooxidans.* Arch Microbiol 150:574–579, 1988.

44. Steudel R, Hold G, Gobel T, Hazeu W: Chromatographic separation of higher polythionates S_n O_6^{2-} (n = 3, . . . 22) and their detection in cultures of *Thiobacillus ferrooxidans:* Molecular composition of bacterial sulfur secretions. Angew Chem Int Engl 26:151–153, 1987.

45. Bruynesteyn A, Lawrence RW, Viszolyi A, Hackl R: An elemental sulfur producing biohydrometallurgical process for treating sulfide concentrates. In Rossi G, Torma AE (eds): Progress in Biohydrometallurgy. Iglesias: Associazione Mineraria Sarda, 1983, pp 151–168.

46. Beyer M, Ebner HG, Assenmacher H, Frigge J: Elemental sulfur in microbiologically desulfurized coals. Fuel 66:551–555, 1987.

47. Brock TD, Gustafson J: Ferric iron reduction by sulfur- and iron-oxidizing bacteria. Appl Environ Microbiol 32:567–577, 1976.

48. Dutrizac JE, MacDonald RJC: Ferric ion as a leaching medium. Miner Sci Eng 6:59–100, 1974.

49. Biegler T, Swift DA: Anodic behavior of pyrite in acid solutions. Electrochim Acta 24:415–420, 1979.

50. Garrels RM, Thompson ME: Oxidation of pyrite by iron sulfate solutions. Am J Sci 258A:57–67, 1960.

51. King WE Jr, Perlmutter DD: Pyrite oxidation in aqueous ferric chloride. AIChEJ 23:679–685, 1977.

52. Kawakami K, Sato J, Kusunoki K, Kusakabe K, Morooka S: Kinetic study of oxidation of pyrite slurry by ferric chloride. Ind Eng Chem Res 27:571–576, 1988.

53. Luther GW III: Pyrite oxidation and reduction: Molecular orbital theory considerations. Geochim Cosmochim Acta 51:3193–3199, 1987.

54. Andrews GF: The selective adsorption of thiobacilli to dislocation sites on pyrite surfaces. Biotechnol Bioeng 31:378–381, 1988.

55. Andrews GF, Maczuga J: Bacterial removal of pyrite from coal. Fuel 63:297–302, 1984.

56. Zarubina NN, Lyalikova NN, Shmuk EI: Investigation of microbiological oxidation of coal pyrite. Invest Akad Nauk SSr Otedl Tekh Mauk Me Toplivo 1:117–119, 1959.

57. Bos P, Huber TF, Kos CH, Ras C, Kuenen JG: A Dutch feasibility study on microbial coal desulfurization. In Lawrence RW, Brannion RMN, Ebner HG (eds): Fundamental and Applied Biohydrometallurgy. Proc 6th Int Symp Biohydrometallurgy. Amsterdam: Elsevier, 1986, pp 129–150.

58. Olson GJ: Recent progress in coal bioprocessing research in the United States. In Proc Biol Treatment Coals Workshop. July 8–10, Vienna, VA. 1987, pp 24–39.

59. Kos CH, Bijleveld W, Grotenhuis T, Bos P, Poorter RPE, Kuenen JG: Composition of mineral salts medium for microbial desulfurization of coal. In Rossi G, Torma AE (eds): Progress in Biohydrometallurgy. Iglesias: Associazione Mineraria Sarda, 1983, pp 479–490.

60. Andrews G, Darroch M, Hansson T: Bacterial removal of pyrite from concentrated coal slurries. Biotechnol Bioeng 32:813–820, 1988.

61. Bos P, Huber TF, Luyben KCAM, Kuenen JG: Feasibility of a Dutch process for microbial desulfurization of coal. Resources Conservation Recycling 1:279–291, 1988.

62. Detz CM, Barvinchak G: Microbial desulfurization of coal. Miner Congr J 65:75–86, 1979.

63. Beyer M, Höne H-J, Klein J: A multi-stage bioreactor for continuous desulfurization of coal by *Thiobacillus ferrooxidans*. In Norris PR, Kelly DP (eds): Biohydrometallurgy. Proceedings of the International Symposium Warwick, 1987. Kew, UK: STL, 1988, pp 514–516.

64. Uhl W, Höne HJ, Beyer M, Klein J: Continuous microbial desulfurization of coal: Application of a multi-stage slurry reactor and analysis of the interactions of microbial and chemical kinetics. Biotechnol Bioeng 34:1341–1356, 1989.

65. Andrews G, Stevens CJ, Quintana J, Dugan PR: Oxygen mass transfer problems in coal processing bioreactors. In Proc 7th Annu Int Pittsburg Coal Conf, September 10–14, 1990, Pittsburgh, PA, pp 358–372.

66. Rai C: Microbial desulfurization of coals in a slurry pipeline reactor using *Thiobacillus ferrooxidans*. Biotechnol Progr 4:200–204, 1985.

67. Sproull RD, Francis HJ, Krishna CR, Dodge DJ: Enhancement of coal quality by microbial demineralization and desulfurization. In Proc Biol Treatment Coals Workshop. June 23–25, Hendon, VA, 1986, pp 83–94.

68. Beier E: Pyrite decomposition and structural alternations of hard coal due to microbe-assisted pyrite removal. In Proc Biol Treatment Coals Workshop, July 8–10, Vienna, VA, 1987, pp 389–424.

69. Hyman D, Hammack R, Finseth D, Rhee K: Biologically-mediated heap leaching for coal depyritization. In Proc 1990 First Int Symp Biol Processing Coal. Palo Alto, CA:EPRI, 1990, pp 3.101–3.111.

70. Doddema HJ: Partial microbial oxidation of pyrite in coal followed by oil agglomeration. In Rossi G, Torma AE (eds): Progress in Biohydrometallurgy. Iglesias: Associazione Mineraria Sarda, 1983, pp 467–478.

71. Pooley FD, Atkins AS: Desulfurization of coal using bacteria by both dump and process plant techniques. In Rossi G, Torma AE (eds): Progress in Biohydrometallurgy. Iglesias: Associazione mineraria Sarda, Iglesias, 1983, pp 263–305.

72. Butler BJ, Kempton AG, Coleman RD, Capes CE: The Effect of particle size and pH on the removal of pyrite from coal by conditioning with bacteria followed by oil agglomeration. Biohydrometallurgy, 15:325–336, (1986).

73. Townsley CC, Atkins AS: Comparative coal fines desulfurization using the iron-oxidizing bacterium *Thiobacillus ferrooxidans* and the yeast *Saccharomyces cerevisiae* during simulated froth flotation. Process Biochem 21:188–191, 1986.

74. Vleeskens JM, Bos P, Kos CH, Roos M: Pyrite association and coal cleaning: An optical image analysis study. Fuel 64:342–347, 1985.

75. Bos P, Vleeskens JM, Kos CH, Hamburg G: Microbial desulfurization of coal: Characterization of the products. In Moulijn JA, Nater KA, Chermin HAG (eds): International Conference on Coal Science. Amsterdam: Elsevier, 1987, pp 431–434.

76. van Afferden M, Schacht S, Beyer M, Klein J: Microbial desulfurizaton of dibenzothiophene. Am Chem Soc Div Fuel Chem 33:561–572, 1988.

77. Faison BD, Clark TM, Lewis SN, Sharkey DM, Woodward CA, Ma CY: Degradation of organic sulfur compounds by a coal-solubilizing fungus. In Proc 1990 First Int Symp Biol Processing Coal. Orlando, FL, May 1990. Palo Alto, CA: EPRI, 1990, pp 3.43–3.59.

78. Rhee KH, Campbell IM: Overview of the DOE/PETC microbiological coal preparation R&D program: A voyage on the flood tide? In Proc Bioprocessing Coals Workshop—III. August 15–17, Tysons Corner, VA, 1988, pp 150–155.

79. Isbister JD, Kobylinski EA: Microbial desulfurization of coal. In Attia YA (ed): Processing and Utilization of High Sulfur Coals. Coal Science and Technology 9. Amsterdam: Elsevier, 1985, pp 627–641.

80. Kargi F, Robinson JM: Removal of organic sulfur from bituminous coal: Use of the thermophilic organism *Sulfolobus acidocaldarius*. Fuel 65:397–399, 1986.

81. Kilbane JJ: Sulfur-specific microbial metabolism of organic compounds. In Proc Bioprocessing Coals Workshop. August 15–18, 1988. Vienna , VA, 1988, pp 156–166.

82. Bos P: Advantages and disadvantages of microbial coal desulfurization. In Proc 7th Annu Int Pittsburg Coal Conf. September 10–14, Pittsburgh, PA, 1990, pp 350–357.

83. Boogerd FC, van Alphen MMQ, van Anrooij WJ, de Bruyn JC, Bos P, Kuenen JG: The role of growth and maintenance in the oxidation of pyrite in batch culture by moderately thermophilic, facultative chemolithoautotroph. In Salley J, McGready RGL, Wichlacz PL (eds): Proceedings of the 1989 International Symposium on Biohydrometallurgy. Ottawa: Canadian Centre for Mineral and Energy Technology, 1990, pp 735–751.

84. Norris PR, Brierley JA, Kelly DP: Physiological characteristics of two facultatively thermophilic mineral-oxidizing bacteria. FEMS Microbiol Lett 7:119–122, 1980.

85. Marsh RM, Norris PR: The isolation of some thermophilic autotrophic iron- and sulfur-oxidizing bacteria. FEMS Microbiol Lett 17:311–315, 1983.

86. Marsh RM, Norris PR: Mineral sulfide oxidation by moderately thermophilic acidophilic bacteria. Biotechnol Lett 5:585–590, 1983.

87. Bos P, Boogerd FC, Kuenen JG: The use of moderately thermophilic sulfur-oxidizing bacteria in the desulfurization of high sulfur coals. In Proc 1990 First Int Symp Biol Processing Coal. Palo Alto, CA: EPRI, 1990, pp 3.61–3.76.

88. Boogerd FC, Bos P, Kuenen JG, Heijnen JJ, van der Lans RGJM: Oxygen and carbon dioxide mass transfer and the aerobic, autotrophic cultivation of moderate and extreme thermophiles: A case study related to the microbial desulfurization of coal. Biotechnol Bioeng 35:1111–1119, 1990.

89. Klein J, van Afferden M, Beyer M, Pfeifer F, Schacht S: Investigations on microbial degradation of coal and coal-derived substances. In Proc 1990 First Int Symp Biol Processing Coal. May 1–3, Orlando, FL. 1990, pp 3.13–3.12.

90. Armson R: Water makes coal a brighter fuel. New Scientist August 28, pp 44–47, 1986.

91. Huber TF, Ras C, Kossen NWF: Design and scale-up of a reactor for the microbial desulfurization of coal: A kinetic model for bacterial growth and pyrite oxidation. In Proc Third Eur Congr Biotechnol Munchen, September 10–14. Weinheim, FRG: Verlag Chemie, 1984, pp 151–159.

Index